Textbook of

Environmental Studies

for Undergraduate Courses

Third Edition

Erach Bharucha

for

ज्ञान-विज्ञानं विमुक्तये

UNIVERSITY GRANTS COMMISSION

Universities Press

Textbook of Environmental Studies for Undergraduate Courses, Third Edition

Universities Press (India) Private Limited

Registered Office
3-6-747/1/A & 3-6-754/1, Himayatnagar, Hyderabad 500 029, Telangana, India
info@universitiespress.com; www.universitiespress.com

Distributed by
Orient Blackswan Private Limited

Registered Office
3-6-752 Himayatnagar, Hyderabad 500 029, Telangana, India

Other Offices
Bengaluru, Chennai, Guwahati, Hyderabad, Kolkata,
Mumbai, New Delhi, Noida, Patna, Visakhapatnam

First published 2005
Reprinted 2005, 2006, 2007, 2008 (thrice), 2009 (thrice), 2010 (twice), 2011, 2012
Second edition 2013
Reprinted 2013 (twice), 2014, 2015 (thrice), 2016 (thrice), 2017 (twice), 2018, 2019 (thrice), 2020 (twice)
Third edition 2021

ISBN: 978-93-89211-78-8

Typest in Minion Pro 10.5/12.5 by
Bookcraft Publishing Services (I) Pvt Ltd, Chennai

Printed at
B B Press, Noida

Published by
Universities Press (India) Private Limited
3-6-747/1/A & 3-6-754/1, Himayatnagar,
Hyderabad 500 029, Telangana, India

501745

डॉ. अरुण निगवेकर

अध्यक्ष

Dr. Arun Nigavekar
Chairman

विश्वविद्यालय अनुदान आयोग

बहादुर शाह ज़फर मार्ग, नई दिल्ली 110 002

UNIVERSITY GRANTS COMMISSION
BAHADUR SHAH ZAFAR MARG
NEW DELHI 110 002
OFF   :   (011) 23239628
            :   (011) 23221313
FAX   :   (011) 23231797
E-mail   :   narun42@hotmail.com

Foreword (to the First Edition)

Our mother earth is the most precious gift of the Universe. It is the sustenance of 'nature' that is the key to the development of the future of mankind. It is the duty and responsibility of each one of us to protect nature. It is here that the understanding of the 'Environment' comes into the picture. The degradation of our environment is linked with the development process and the ignorance of people about retaining the ecological balance. Indeed, no citizen of the earth can afford to remain aloof from the issues related to the environment. It is, therefore, essential that the study of the environment becomes an integral part of the education process.

The University Grants Commission decided to address the issue of Environmental Studies by introducing a basic course on Environment at the undergraduate level. The decision of the Hon'ble Supreme Court of India made such an approach mandatory. We, therefore, appointed a Committee of Experts to advise us on the curriculum for Environmental Studies at the undergraduate level, and the Committee was also requested to produce a textbook on Environmental Studies for such an undergraduate course. This Committee was headed by Dr Erach Bharucha, a well-known Environmental Scientist of Bharati Vidyapeeth, Pune, and they have come out with a very comprehensive and useful document on this multidisciplinary subject. The material is well presented in an integrated manner and in very simple language, with a large number of examples. This has made this textbook very rich in content and would prove to be a very valuable document both for the students as well as for the teachers. I would like to compliment and also thank the Committee members for doing such a wonderful job. I am confident this book would be of great help to all those who are interested in understanding the importance of the Environment.

(Arun Nigavekar)
Chairman

Preface to the Third Edition

The need for environment management based on public awareness has altered over the last few decades. This Core Module Course is the outcome of a Public Interest Litigation by the Hon. M C Mehta, which prompted the Supreme Court of India to order the Ministry of Human Resource Development and the Ministry of Environment and Forest, Government of India, to respond to India's environmental degradation by developing and implementing a course on environmental studies for all undergraduate students and develop a strategy for public participation. The outcome was a course to sensitise and make the future leaders of this country, belonging to all walks of life, active partners with the Government to provide a better country to live in within the shortest possible time.

Since the First Edition was published in 2005, much has changed in the environment of our diverse and beautiful country. The inherent beauty of her forests, grasslands, wetlands, deserts, rivers and mountains have been severely altered by human activity in the recent past. Much of nature still exists and it must be preserved and protected for its economic, social and environmental assets. This Third Edition aims to enhance one's awareness and stimulate one to act in an environmentally conscious and responsible way towards the earth. Our country is rich in resources but remains essentially poor in managing these resources from an environmentally sensitive point of view. Thus, this book is intended to provide one with the essential reading (knowledge), give an enlightened feeling of one's own surroundings (awareness), leading to a heightened feeling of involvement (concern), which finally paves the way to act responsibly (action). The four steps of *Knowledge* → *Awareness* → *Concern* → *Action* are the cornerstones of environmental studies.

The earlier concept that environmental studies only encompasses the study of nature has now been extended to appreciating and studying not only nature's landscapes (Unit 2), but also landscapes that human civilisations have nurtured as a way to support lives and livelihoods (Unit 7). This mosaic of landuse elements have come to be referred to as our cultural landscapes.

The more recent changes due to our enormous population, poverty, as well as the growing affluent sector that consumes huge quantities of resources and creates waste in insurmountable amounts, is an ever-increasing problem. The Government alone cannot be expected to provide a better life and a cleaner environment for all, without the citizens' contribution. For this, we need to see development through a new set of multi-focal spectacles which combines economic growth, societal equity and environmental long-term management. This concept of development is referred to as *sustainable development*. Thus, in this rapidly growing economy, the three pillars of sustainable development must be engineered into environmental management—an economic, social and environmentally balanced governance system (Unit 1).

We need to preserve our natural resources, not only for ourselves but also for our future citizens. This futuristic management includes the country's environmental assets and the great wealth of her biological diversity. This is an aspect that needs very special attention. Biodiversity is the crux of all living systems, both in natural landscapes – with its wild species of plants, animals, fungi and microscopic creatures – and also in our agricultural crops and livestock as well as urban environments. Our forefathers have created thousands of varieties of food crops – cereals, vegetables, fruits, tubers and medicinal plants – through experiments using traditional knowledge systems (Unit 7). Animal husbandry led to a great diversity of livestock breeds to suit a variety of uses and adapted to local climate and fodder. Not only are many wild species on the brink of extinction, the traditional knowledge of our hunter–gatherer tribal folk, traditional farmers and livestock owners is being lost faster than the loss of wild species.

Modern technology of environmental management has solved many problems resulting from inadequate land management, polluted waterscapes and marine environments, as well as in farming, animal husbandry, fishing, medicine and health issues. Our population is growing older, and more consumerist. Mahatma Gandhi said the earth can support everyone's needs but not everyone's greed. Today we witness what he had foretold decades ago (see Unit 3).

This textbook contains Case Studies which make you aware of a variety of environmental situations, which you need to analyse as having inherent strengths, weaknesses, opportunities and threats. There are activities that encourage you to think creatively, analyse and push yourself towards an environmentally proactive lifestyle. The patterns of socioeconomic stratification from the extremely poverty stricken and deprived, to that of the rich and powerful makes it difficult to suggest changes in all our lifestyles. Encouraging a sustainable lifestyle has to be appreciated by children, youth and adults from all walks of life. You certainly have the ability to act and change a little corner of the world for the better. The evolutionary process has made you one of a unique species (Unit 4). It is every citizen's duty and responsibility, as spelt out in our constitution, to preserve nature and wildlife.

Therefore, this book is important not only for you to read and pass an exam (which is important), but to help live your own life in consonance with nature. Graduates in science, social studies, law, medicine or engineering, all need to know about the environment. The environment is what we live in, and if it is clean and beautiful, it gives us immense happiness. If the environment is messed up, it impoverishes our lives. While environmental management is a combined activity of people in every community, country and the world, it is also a part of the individual's responsibility. This course not only enlightens your mind but aims to touch your heart.

Erach Bharucha
Pune, March 2021

Preface to the First Edition

Perhaps no other country has moved so rapidly from a position of complacency in creating environmental awareness into infusing these newer pro-environmental concepts into formal curricular processes as has happened in India over the last few years. This has undoubtedly been accelerated by the judgment of the Honorable Supreme Court of India that Environmental Education must form a compulsory core issue at every stage in our education processes.

For one who has fought to implement a variety of environment education programs for schools and colleges and for the public at large, this is indeed a welcome change. The author is currently constantly asked to provide inputs to 'environmentalise' textbooks and provide inputs – at the NCERT, SCERT and UGC Levels – to further the cause of formal environment education.

This textbook has been produced as an outcome of a UGC Committee that included the author and was set up to develop a common core module syllabus for environmental studies at the undergraduate level, to be used by every University in the country. The author invites comments from those who wish to contribute towards its improvement in the coming years.

Environment education can never remain static. It must change with the changing times, which inevitably changes our environment.

Each of us creates waves around us in our environment that spread outwards like the ripples generated by dropping a stone in a quiet pond. Every one of us is constantly doing something to our environment and it is frequently the result of an act that we can hardly ever reverse, as after once the stone hits the water one cannot stop the ripple effect from disturbing the pond.

This textbook is written to bring about an awareness of a variety of environmental concerns. It attempts to create a pro-environmental attitude and a behavioral pattern in society that is based on creating sustainable lifestyles. But a textbook can hardly be expected to achieve a total behavioral change in society. Conservation is best brought about through creating a love for nature. If every college student is exposed to the wonders of the Indian wilderness, I believe that a new ethic towards conservation will emerge.

Erach Bharucha
Pune, 2005

Acknowledgements

This edition is completely different from the First Edition published in 2005 at the behest of Dr Arun Nigavekar, former Chairman, University Grants Commission, and the Second Edition in 2013. This book is the outcome of the new Ability Enhancement Compulsory Course (AECC)—Environmental Studies, prescribed by the UGC in 2018 to meet the new environment and development needs of our country. The first and second editions were translated into six languages and were reprinted several times. It has been used by thousands of students so far. In response to their feedback and the paramount importance of sustainable development in India, several current topics have been introduced into this new edition. I thank the many teachers and students who have read the earlier versions and given me ample cause to reflect on what the new textbook for the AECC course should include. It will, I am sure, lead to strengthening knowledge, skill and attitude towards our environmental challenges, which is now an essential life skill for all our youth.

I wish to thank my team at Bharati Vidyapeeth Institute of Environment Education and Research (BVIEER) that has supported me during the writing of this new edition over the last two years. The team includes Dr Shamita Kumar, Dr Kranti Yardi and Shri Anand Shinde. Kranti has gone over the draft several times and suggested many amendments. I am especially grateful for her many inputs that have been included in many drafts over the last year. She has also been instrumental in coordinating with Dr Ms Ernavz Bharucha who has done the initial editing and Ms Vidya Pujari who has patiently done the word processing repeatedly as it was altered several times through computer glitches and version problems.

I wish to thank Ms Anushka Kajbaje for her inputs on the Solid Waste Management section and Ms Archana Kalyani for the 'Chemical pollution' content of *Environmental Pollution* (Unit 5). I would like to thank Ms Gauri Joshi for her inputs on *Environmental Policies and Practices* (Unit 6). My administrative and support services at BVIEER have been extremely helpful—Shri Kate, Shri Bhosale, Shri Vikram Londhe and Shri Appa Chorage. I thank them sincerely.

Any errors in the text are entirely oversights on my part.

The most important inputs of incomparable value have come from my publishers—Universities Press. Mr Madhu Reddy has been a source of inspiration and kindness in all respects. Dr Gita S Dattatri, my esteemed editor, has been highly supportive and extremely patient in carrying through the difficult task of making the book more readable. She has given it a completely new look. I could not have had a more cooperative editor. I profoundly thank Madhu and Gita for their unparalleled inputs for producing and publishing this textbook. I am sure it will be even more successful than our previous editions.

It has been a great team experience. I hope this book will bring about a change in the mindset and actions for the environment among the youth of our country, who are the citizens of tomorrow.

Erach Bharucha
Pune, 2020

Road Map

The tools for understanding this book:
- Read a unit.
- Introspect on how it is related to your own life.
- Observe your own immediate surrounds and visit areas where nature abounds as well as sites where humans have left behind serious environmental problems.
- Work out for yourself on how you can contribute towards a better world.
- Make your own observations on paper. Discuss these ideas in a group to appreciate different viewpoints.
- Work out a course of action.

The tools to use from elsewhere:
- Look for appropriate MOOCs (Massive Open Online Courses).
- Look for preferably Indian audio visuals on Discovery, Animal planet, National geographic, and so on. But also gather a world view from such programs.
- Use the resources provided as additional reading material for topics that interest you and are linked to your own subject from an environmental perspective.
- Become a nature photographer, bird watcher, gardener of indigenous plants, an environmental communicator, or a member of a conservation organisation.

Teaching Methodologies

The syllabus for Environmental Studies for the Ability Enhancement Compulsory Course includes classroom teaching and field work. The syllabus is divided into 8 units, covering 50 lectures. The first 7 units, which cover 45 lectures, are classroom-teaching-based and intended to enhance knowledge skills and attitude towards the environment. Unit 8 is based on field activities, to be covered over five lecture hours, and would provide students with first-hand knowledge on various local environmental aspects. Field experience is one of the most effective learning tools for environmental concerns. This moves education out of the scope of the textbook mode of teaching and into the realm of hands-on learning in the field, where the teacher acts as a catalyst to interpret what the student observes or discovers in his/her own environment. Field studies are as essential as class work and form a unique synergistic tool in the entire learning process.

The course material provided by UGC for classroom teaching and field activities should be effectively utilised.

The Universities/colleges can draw upon the expertise of outside resource persons for teaching purposes.

The Environmental Core Module will be integrated into the teaching programs of all under-graduate courses.

Annual system: The duration of the course will be 50 lectures. The exam will be conducted along with the Annual Examination.

Semester system: The Environment Course of 50 lectures will be conducted in the second semester and the examinations shall be conducted at the end of the second semester.

Credit system: The core course will be awarded 4 credits.

Exam pattern: In case of awarding marks, the question paper should carry 100 marks. The structure of the question paper being:

Part A: Short-answer pattern—25 marks
Part B: Essay-type built-in choice—50 marks
Part C: Field work—25marks

Contents

(according to **Ability Enhancement Compulsory Course Syllabus for Environmental Studies for Undergraduate Courses of all Branches of Higher Education**)

Colour Plates

Introduction to Environmental Studies

My message, especially to young people, is to have the courage to think differently, courage to invent, to travel the unexplored path, courage to discover the impossible, and to conquer the problems and succeed.

— *A P J Abdul Kalam*

Learning Objectives

In this chapter you will learn,

◆ What environmental studies is and why it is multidisciplinary
◆ What the components of the environment are
◆ What the scope of environmental studies is
◆ What sustainability and sustainable development are

Purpose

The interlinkages in nature must be appreciated to unravel the intricacies of the world around us. Only if we internalise the complexity of ecosystems, species and genetic aspects of our environment, and appreciate its rich biological diversity, can we learn to protect and preserve the wealth of environmental assets and services. The earth is like a complex organism made of billions of interconnected parts through multiple functioning pathways. A deeper knowledge of its working will enrich our lives.

Our Role

Each of us has to take responsibility to preserve nature, otherwise, we fail in our duty towards our future generations. We cannot permit them to blame us for despoiling the earth leaving them devoid of life-giving resources.

1.1 INTRODUCTION

Definition: Environmental studies deals with every issue that affects living beings on earth. It is the study of inter-relationships between living creatures and all aspects of their environment. It essentially requires a multidisciplinary approach that brings about an appreciation of our natural world and the impact of humans on its integrity.

It is an applied science that seeks practical answers to the increasingly important question of how to make human civilisation sustainable using the finite resources available on earth. Its components include all natural and social sciences such as biology, geology, chemistry, physics, engineering, sociology, health, anthropology, economics, statistics, computers and philosophy. It deals with human population and the resources on which life depends, and is linked to economics and politics. Governance at global, national, state and local levels is therefore linked to environmental studies. The scope is thus all pervasive and ubiquitous in nature.

1.2 THE MULTIDISCIPLINARY NATURE OF ENVIRONMENTAL STUDIES

Understanding the environment can be appreciated from very different perspectives.

◆ One could look at the environment as a natural scientist—a botanist, zoologist, microbiologist; or as a chemist or physicist who looks at soil, water, air, energy or land, from the view point of the earth's integrity and pollution control (**Table 1.1**).

- Since the environment deeply concerns people, one could see it from the perspective of a social scientist who looks at whether natural resources are being used at sustainable levels or carelessly at unsustainable levels.
- Environmental studies inevitably deals with governance and management of the land, water and air. This is linked to legal issues which need a different perspective. Management, politics, law and ethics play an important part.
- As health and its management are closely associated with the environment and epidemiology, health and safety are also a part of environmental studies.
- Finally, it is the ecologist who sees the links between all these studies – based both on modern science and traditional folk knowledge – and brings home the wide-ranging perspectives into a cogent whole.

Table 1.1 Multiple disciplines that contribute to environmental science

Disciplines	Branches
Natural science	Botany, Zoology, Ecology, Microbiology, Genetics, Forestry, Wildlife biology
Social sciences	Economics, Agriculture, Political science, History, Heritage conservation
Technical education	Information technology, Bioinformatics, Environmental engineering, Computer science, Geoinformatics, Statistics, Architecture and planning
Medical science	Health science, Preventive medicine
Physical science	Chemistry, Physics, Renewable energy
Earth science	Geology, Geography
Environmental Law	Natural resources, Biodiversity laws, Pollution control, Land laws, Energy
Media	Media studies
Environmental economics and sustainability	Economics

1.2.1 Need for Public Awareness

As the earth's natural resources are rapidly dwindling and our environment is being increasingly degraded by human activities, it is evident that something needs to be done. It is not possible for the Government to perform all the necessary clean-up functions. Prevention of environmental degradation must become a part of all our lives. Just as prevention is better than cure for a disease, protecting our environment is economically more viable than cleaning it up once it is degraded or polluted. Individually, we can reduce wastage of natural resources and we can act as watchdogs that inform the Government about sources or polluters that lead to pollution and degradation of the environment; only then will the polluter pay!

Mass media such as newspapers, radio and television strongly influence, create public awareness and reinforce public opinion. Politicians in a democracy always respond positively to a strong public-supported movement. Thus, if you join a Non-Governmental Organisation (NGO) that supports conservation of environmental assets, you might be able to influence politicians to make green policies. 'Spaceship earth' has a limited supply of resources and we need to cooperate with the several Government and Non-Governmental Organisations working towards environmental protection in our country, and become conservation supporters. They have created a growing interest in environmental protection and conservation of nature and natural resources.

Just understanding and making ourselves more aware of our environmental assets and problems is not enough. This should translate to concern about our environment and change the way in which we use every resource. We should shift from wasteful behaviour patterns to environmentally friendly practices, and analyse old methods from new perspectives. Only then will our lifestyles become more sustainable and support our environment (Table 1.2).

Table 1.2 Environmentally significant days

Day and month	Event
02 February	World wetland day
28 February	National science day
21 March	World forestry day
22 March	World water day
18 April	World heritage day
22 April	Earth day
22 May	International biodiversity day
05 June	World environment day
11 July	World population day
16 September	World ozone day
28 September	Green consumer day
03 October	World habitat day
1–7 October	Wildlife week
04 October	Animal welfare day
02 December	Bhopal gas tragedy day

Salim Ali's name is synonymous with ornithology in India and with the BNHS. He also wrote several great books including the famous *Book of Indian Birds*. His autobiography, *Fall of a Sparrow*, should be read by every nature enthusiast. He was our country's leading conservation scientist and influenced environmental policies in our country for over 50 years.

1.2.2 The Environment and its Stakeholders

We are all responsible for our environment and its integrity irrespective of our profession, society or family background. Environmental professionalism includes many different types of people whom we refer to as stakeholders.

Environmental scientists: Professionals who study and implement environmental issues such as environmental assets (air, water, soil and biodiversity), study and limit problems of population growth and poverty, control pollution of air, water and solid waste management, suggest better methods of energy reuse, water distribution, disaster management and environmental health.

Ecologists: Interdisciplinary scientists who study nature (natural history) and study the multitude of inter-linkages between species and with their habitats.

Conservation biologists: Scientists who have been concerned with preserving nature, its species, habitats and genetic diversity as an integrated whole.

Taxonomists: Botanists, zoologists and microbiologists concerned with living organisms and their nomenclature.

Conservationists: People who are conscious of the need to preserve nature in all its glory by setting up nature reserves, understanding the inter-relationships between species and habitat integrity. They use their knowledge and skills to save endangered species and threatened habitats through scientific analysis. A subset includes those who take on eco-restoration of degraded habitats and the translocation and rehabilitation of flora and fauna.

Environmentalists: Scientists or interested intellectuals and people at large who are all interested in protecting the environment around them and fight for a better, livable world. While some are professionals, many are common folk who believe strongly that humanity cannot survive without an undisturbed, productive environment. Anyone from any profession can use at his discretion, pro-environment behaviour that supports nature's life support systems.

Traditional knowledge holders: A select group of people who have expert knowledge of their own environment including locally used resources, medicinal and food resources and the myths and folklore around local resource use.

Socio-environmentalists: This is a growing sector which looks at the inter-relationships between socioeconomic issues and the environment. They consider rehabilitation of displaced people and poverty elevation.

> ### Silent spring
> *The Silent Spring* by Rachel Carson was published on 27 September 1962. This, for the first time, suggested a need for change in society. It showed that democracies and governments operated in ways that damaged peoples' lives. Individuals and groups could question what their governments were permitting environmental offenders to put into the environment. It envisions changes in government policy. Carson believed that the government in the US was part of the problem of environmental degradation. She made the readers and audiences ask 'Who speaks, and why?' and thus brought about a new transformation in society.

Environmental managers

Foresters: The Forest Department administers a large percentage of India's land. In the past, their primary function was to produce timber. Now, they are involved in preserving forests, preventing criminal deforestation, enforcing the Wildlife Protection Act, Forest Conservation Act, Forest Rights Act, and protecting all natural typologies such as wetlands, grasslands, deserts, hills, mountains, lakes, rivers and marine habitats. In this sense, they are the ultimate land use managers.

Agricultural scientists: In the past, they were expected to support the green revolution by increasing production. Now, their role has expanded to preserving Traditional Knowledge Systems (TKS) and indigenous traditionally used crops, maintain seed banks for genetic purposes and train farmers in sustainably managing farmland and their grazing areas.

Animal husbandry: Earlier, their focus was on milk production. This has extended to the need to maintain indigenous breeds of livestock for future genetic management.

Fisheries: Their focus has been on increasing production of fish as part of our food security. However, the introduction of exotic fish into our rivers has led to a steep drop in the abundance of indigenous species. Thus, as resource managers, fisheries development must now look at indigenising fish resources.

Energy managers: This sector was expected to satisfy the energy needs of India's expanding population. The focus was on thermal energy which requires coal or oil, but this causes serious air pollution that affects respiratory health. The sector is thus moving towards cleaner energy sources such as solar and wind power.

Urban planners: They deal with the growing environmental issues of urban habitats that are expanding from erstwhile villages into towns, industrial areas, cities and now into megacities. This is a special sector which includes architects, landscape planners, social scientists, those involved in green housing, water supply, public transport, waste management and associated sectors such as schools, hospitals and markets. An important aspect is planning of green spaces from a biodiversity perspective where the peace and quiet of nature can be enjoyed along with leisure.

Green architecture: Architects are increasingly involved in working out ways to introduce pro-environment concepts into construction by using environment-friendly materials and use of airflow and rain water harvesting. Leed certification is now used to note the environmental quality based on sustainable site development, water saving, energy efficiency and indoor environmental quality.

Policy makers and the judiciary: These are key stakeholders of good environment management. Policy makers respond to green concepts if they realise that it is a part of a large vote bank. To make this possible, the public at large have a great role to play in providing necessary advocacy through public movements.

The judiciary has been extremely proactive in managing our environment and forest, and has supported laws that have made it possible for appropriate executive actions by government officials. The Supreme Court of India has, over the years, supported pro-environmental actions and punished environmental law breakers by using the 'polluters must pay' principle.

CASE STUDY 1.1   The Taj trapezium case

M C Mehta fought for ten long years to protect the Taj Mahal from corrosive pollution. In 1993, after a decade of court battles and threats from factory owners, the Supreme Court ordered 212 small factories surrounding the Taj Mahal to close because they had not installed pollution control devices. Another 300 factories were put on notice to do the same. In 1991, M C Mehta filed a Public Interest Litigation against the Government that led to the inclusion of environmental education in school and college curricula.

1.2.3   Ministry of Environment and Forest and Climate Change (MoEF and CC)

The MoEF and CC is primarily concerned with the planning, promotion and coordination of the implementation of India's environmental and forestry policies and programmes. These relate to the conservation of the country's natural resources including lakes and rivers, its biodiversity, forests and wildlife, ensuring the welfare of its animals and prevention and abatement of pollution. These objectives are well supported by a set of legislative and regulatory measures aimed at the preservation, conservation and protection of the environment. The Ministry also serves as the nodal agency in the country for the United Nations Environment Programme (UNEP), South Asia Cooperative Environment Programme (SACEP), International Centre of Integrated Mountain Development (ICIMOD) and for the follow-up of the United Nations' Conference on Environment and Development (UNCED). The Ministry is entrusted with issues relating to multilateral bodies such as the Commission on Sustainable Development (CSD), Global Environment Facility (GEF) and of regional bodies like the Economic and Social Council for Asia and Pacific (ESCAP) and South Asian Association for Regional Cooperation (SAARC) on matters relating to the environment. As a part of the MoEF's planning of environmental policies, the Environmental Information System (ENVIS) was established in 1982 by the Government of India. The focus of ENVIS is to provide environmental information to decision makers, policy planners, scientists, engineers and research workers all over the country. ENVIS is a decentralised system with a network of distributed subject-oriented centres ensuring integration of national efforts in environmental information collection, collation, storage, retrieval and dissemination to all concerned (Fig. 1.1). A large number of nodes, known as ENVIS centres, have been established in the network to cover the broad subject areas of environment such as pollution control, toxic chemicals, central and offshore ecology, environmentally

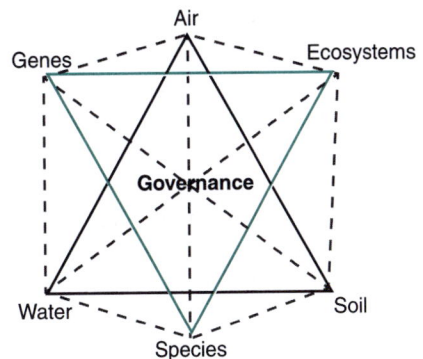

Fig. 1.1   Governance and linkages

sound and appropriate technology, biodegradation of waste and environment management. In May 2014, the Ministry of Environment and Forest (MoEF) was renamed as Ministry of Environment, Forest and Climate Change (MoEF and CC).

1.3 COMPONENTS OF THE ENVIRONMENT

The earth is composed of complex aspects known as spheres that were initiated billions of years ago through sequential changes. It had the ancient sea—the hydrosphere, the land—the lithosphere, and it had air and other gaseous material— the atmosphere (**Fig. 1.2**).

What makes the earth very special is life. Living beings in the biosphere (the living part of nature, biotic) evolved to link all the other three spheres which together form the abiotic (non-living) part of nature. We don't still know if life as we understand it, is present anywhere else

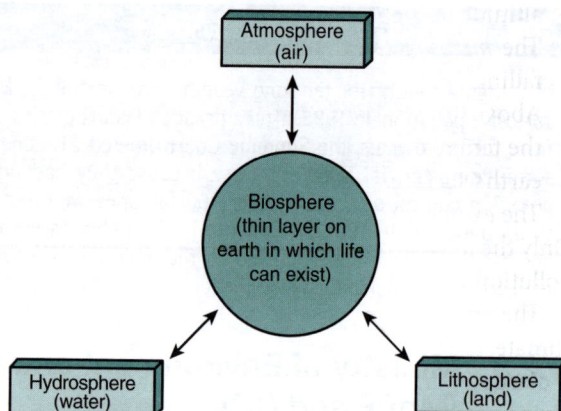

Fig. 1.2   Earth's spheres

in the universe. To appreciate the complexities of life – the biotic part of nature – we need to understand the hydrosphere, lithosphere and atmosphere on which life depends. The vital spheres of our environment are closely interlinked. They function in cycles which drive all the earth's natural processes.

1.3.1   The Atmosphere

The atmosphere provides,

◈ oxygen for human respiration (metabolic requirements),

◈ oxygen for wild fauna in natural ecosystems and domestic animals which are used by humans as food,

◈ oxygen and carbon dioxide, used by plants and animals.

Structure of the atmosphere

The atmosphere forms a protective shell over the earth.

◈ The atmosphere is composed of 79% nitrogen, 20% oxygen and the remaining 1% is a mixture of carbon dioxide, water vapour and trace amounts of several other gases such as neon, helium, methane, krypton, hydrogen and xenon. The atmosphere is divided into several layers. The innermost layer, the *troposphere*, extends up to 6–20 km. It contains about 75% of the mass of the earth's air. The fragility of this layer is obvious by comparing the earth to an apple. This lowest layer would be no thicker than the apple's skin. The troposphere is the only part warm enough for us to survive. Temperature declines with altitude in the troposphere. At the top of the troposphere, temperatures abruptly begin to rise.

◈ This boundary where temperature reversal occurs is called the *tropopause*. The tropopause marks the end of the troposphere and the beginning of the *stratosphere*, which is the second layer of the atmosphere. The stratosphere extends up to 50 km above the earth's surface. While the composition of the stratosphere is similar to that of the troposphere, it has two major differences. The volume of water vapour here is about 1000 times less, while the volume of ozone is about 1000 times greater. The presence of ozone in the stratosphere prevents 99% of the sun's harmful ultraviolet (UV) radiation from reaching the earth's surface, thereby protecting humans from cancer and from damage to the immune system. This layer does not have clouds. Hence, aeroplanes fly in this layer as it creates less turbulence. The temperature rises with altitude in the stratosphere.

- The *mesosphere* extends 50–85 km above the earth. The temperature decreases with altitude, falling up to –110°C.
- Above this is a layer where ionisation of the gases is a major phenomenon, thus causing a rise in the temperature. This layer is called the *thermosphere*. This extends from 85–640 km above the earth's surface.
- The *exosphere* extends beyond 550 km to 1000 km in space.

Only the lower troposphere is routinely involved in our weather and is linked to issues related to air pollution. The other layers are not significant in determining the level of air pollution.

The atmosphere is not uniformly warmed by the sun. This leads to air flows and variations in climate, temperature and rainfall in different parts of the earth. It is a complex dynamic system and if disrupted has serious consequences. Most air pollutants have both global and regional effects. Living creatures cannot survive without air even for a few minutes. To continue to support life, the air must be kept clean and pollution-free.

1.3.2 Hydrosphere

The hydrosphere provides,

- clean water for drinking which is a metabolic requirement for all living processes,
- water for washing and cooking to maintain good hygiene levels,
- water for agriculture which provides us with food,
- food from the sea including fish, crustaceans and seaweed which support millions of people,
- food from freshwater sources including fish, crustaceans and aquatic plants,
- water flowing down from mountain ranges harnessed to generate electricity through hydroelectric projects, and
- water for industries to produce consumer goods.

A major part of the hydrosphere is the marine ecosystem in the ocean. The fresh water in rivers, lakes and glaciers is perpetually being renewed. Some of this fresh water is stored in underground aquifers.

Water pollution threatens the health of communities, as our lives depend on the availability of clean water. This once plentiful resource is now severely depleted and the sad truth is that we are forced to buy this natural resource.

1.3.3 Lithosphere

The lithosphere provides,

- soil with micronutrients which is the basis for agriculture,
- stone, sand and gravel for construction,
- microscopic flora, small soil fauna and fungi present in soil which break down plant litter as well as animal waste to provide nutrients for plants,
- a large number of minerals that can be mined (for industrial use),
- oil, coal and gas from underground sources which provide power for vehicles, agricultural machinery, industry, and are used for cooking, lighting and heating in our homes.

The lithosphere began as a hot ball of matter which formed the earth about 4.6 billion years ago. About 3.2 billion years ago, the earth cooled down considerably and life began on our planet. This resulted from simple organic molecules which began to reproduce. Life probably began in the hydrosphere and evolutionary processes gradually led to diversification of life on land. The crust of the earth is 6–7 km thick and lies under the continents. Of the 92 elements in the lithosphere, only eight are common constituents of crystal rocks. Of these, 47% is oxygen, 28% is silicon, 8% is aluminium and 5% is iron, while sodium, magnesium, potassium and calcium constitute 4% each. Together, these elements form about 200 common mineral compounds. Rocks, when broken down, form soil on which humans are dependent for agriculture. These minerals are also the raw material used in various industries.

1.3.4   Biosphere

The biosphere (the living part of our environment) provides,
◈   food, from crops and domestic animals,
◈   food for all forms of life, which live as interdependent species in a community and form food chains in nature,
◈   biomass such as fuel wood is from forests and plantations, with other forms of organic matter as sources of energy,
◈   timber and other construction materials.

The biosphere is a relatively thin layer on the earth in which life can exist. Within it, the air, water, rocks and soil and the living creatures form the structural and functional ecological units, which together can be considered as one giant global living system—the earth. Within this framework, those characterised by broadly similar geography and climatic conditions, as well as communities of plant and animal life can be divided for convenience into different biogeographical realms. Within these, smaller biogeographical units can be identified on the basis of structural differences and functional aspects into recognisable ecosystems, which give a distinctive character to a landscape or waterscape, such as those of a country, a state, a district or even an individual valley, hill range, river or lake.

The simplest ecosystem to understand is a pond (**Fig. 1.3**). It can be used as a model to understand the nature of any other ecosystem and to appreciate the changes that are seen over time in any ecosystem. The structural features of a pond include its size, depth and the quality of its water. The periphery, the shallow part and the deep part of the pond, each provide specific conditions for different plant and animal communities. Functionally, a variety of cycles like the amount of water within the pond at different times of the year and the quantity of nutrients flowing into the pond from the surrounding terrestrial ecosystem, all affect the 'nature' of the pond.

1.3.5   Natural Cycles Between the Spheres

These four spheres are closely interlinked systems and are dependent on the integrity of each other. Disturbing one of these spheres affects the others. The links between them are mainly in the form of cycles. For instance, the atmosphere, hydrosphere and lithosphere are all connected through the *hydrological* (water) *cycle*. The water evaporating from the hydrosphere (the seas and freshwater ecosystems) forms clouds in the atmosphere. On condensing, this falls as rain, which provides moisture for the lithosphere, on which life depends. The rain also acts on rocks as an agent of erosion and over millions of years has created soil. Atmospheric movements in the form of wind also break down rocks into soil. The most sensitive and complex links are those between the atmosphere, the hydrosphere and the lithosphere on the one hand, and, with the millions of

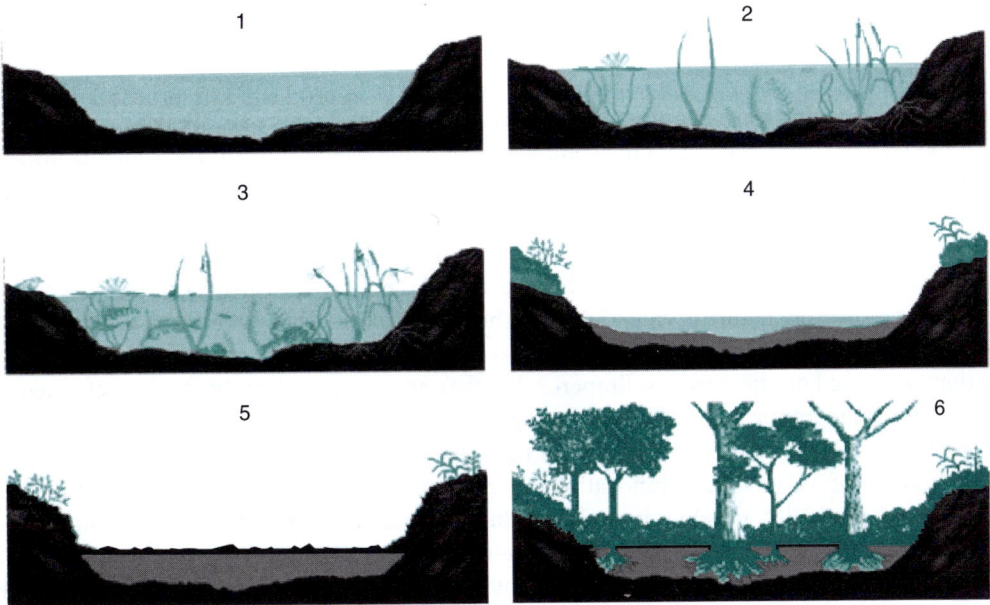

Fig. 1.3 Succession in a pond

living organisms in the biosphere on the other. All living organisms which exist on earth live only in the relatively thin layer of the lithosphere and hydrosphere present on the surface. The biosphere which is formed by living organisms has countless associations with different parts of the three other spheres. The spheres interacting at a regional level thus contribute a variety of ecosystems.

It is, therefore, essential to understand the inter-relationships of the separate entities – soil, water, air and living organisms – and to appreciate the value of preserving intact ecosystems in their entirety.

Activity-based learning

Take a simple object in daily use and track its components back to each of its basic spheres—lithosphere, atmosphere, hydrosphere, biosphere.

Example: This textbook:
◆ Paper from wood—the biosphere
◆ Water for pulping wood—the hydrosphere
◆ Bleach to whiten the paper—a mineral from the lithosphere

1.4   SCOPE AND IMPORTANCE OF ENVIRONMENTAL STUDIES

1.4.1   Scope and Importance

If we study the natural history of the areas in which we live, we would see that our surroundings were originally natural landscapes such as a forest, a river, a mountain, a desert, or a combination of these elements. Most of us live in landscapes that have been profoundly modified by human beings in villages, towns or cities. But even those of us who live in cities must get our food supply from surrounding villages. These in turn, are dependent on natural landscapes such as forests, grasslands, rivers, seashores, for resources such as water for agriculture, fuelwood, fodder and fish. Thus, our daily lives are inextricably linked to our surroundings which inevitably affect the space we live in.

We use water to drink and for other day-to-day activities. We breathe air. We use resources from which food is made and depend on the community of living plants and animals which form a web of life, of which we are also a part. Everything around us forms our environment and our lives depend on keeping its vital systems as intact as possible.

Our dependence on nature is so great that we cannot continue to live without protecting the earth's environmental resources. Thus, traditional societies refer to the earth as Mother Nature, and learned that respecting nature is vital to protect their livelihoods. They have cultural practices that respect nature and all living creatures, protect and preserve their natural resources. Many of India's traditions are based on these values. Emperor Ashoka's edict in the 4th century BC proclaimed that all forms of life are important for our well-being.

Over the past 200 years, modern societies thought that more resources could be generated by the application of technological innovations. Some examples are, growing more food by using fertilisers and pesticides, developing more productive strains of domestic animals and crops by genetic modifications, irrigating farmland through mega-dams and developing industry to provide all sorts of consumer goods. All this has led to rapid economic growth; this pattern of ill-considered development has inevitably led to environmental degradation as well as several other harmful effects on human life and wellbeing.

Industrial development and intensive agriculture that provide the goods for our increasingly consumer-oriented society also use up large amounts of natural resources such as water, minerals, petroleum products, wood and energy. Non-renewable resources, such as minerals and oil, will be exhausted in the near future if we continue to extract these resources without a thought for subsequent generations.

Renewable resources, such as timber and water, can be used at sustainable levels as they can be regenerated by natural processes of nature. However, these resources too will get depleted if we continue to use them faster than nature can replace them. For example, if the removal of timber and firewood from a forest is faster than the re-growth and regeneration of trees, the supply of wood cannot be replenished. A natural forest acts like a sponge which holds water in the rainy season and releases it slowly over the drier periods. Thus, deforestation depletes forest resources, affects our water resources, and leads to flash floods in the monsoon season and leaves rivers dry once the rains are over. We need to understand these multiple effects on the environment resulting from routine human activities.

Our natural resources can be compared with money in a bank. It is our national capital. If we use it rapidly, the capital will be reduced to zero. On the other hand, if we use only the interest, it can sustain us over a longer term. This is the basis of sustainable development. Thus all professionals in their specific fields of work should strive to achieve sustainability through their actions.

Humankind is growing faster than ever before. Growing numbers from around 2.5 billion in the 1950s has reached over 6.3 billion in a world that cannot expand to provide more living space, food, air, water and biological resources. We are using its resources at an ever-growing pace. Land for industry, urbanisation, modern agriculture, pasture lands, energy production from hydropower, thermal, wind and solar power is required for all our needs. We are creating more and more waste products which need large landfills. The environmental issues we need to deal with include:

◆ human population growth (Chapter 7),
◆ environmental goods and services—natural resources (Chapter 3),

- levels of resource consumption and waste (Chapters 3, 5),
- landuse and landscape changes (Chapter 3),
- deforestation and desertification (Chapter 3),
- loss and degradation of wetlands, rivers and lake ecosystems (Chapter 2),
- loss of marine resources (Chapter 2),
- development of an Integrated Protected Area Network (Chapter 4),
- loss of species due to extinction (Chapter 4),
- pollution (Chapter 5),
- reducing our footprint on the environment (Chapter 7),
- our growing demands on energy use (Chapter 3),
- disasters and their management (Chapter 7), and
- risks of climate change and its causes (Chapter 6).

Why do we need to look at individuals for environment management?
- The government cannot address all our environmental issues by itself.
- If we dream of a better world, country, city or village, we can together make it happen.
- Each of us has an environmental responsibility.

Why do we need to look at national concerns?
- Our country has a large population of predominantly poor people.
- Our natural capital has been over used.
- Biodiversity loss and the climate change will impoverish us further.
- We need to act now to fullfil our Sustainable Development Goals (SDGs).
- Our nation is considered a global emerging economy. Thus, we have the problems of the developing and the developed world.
- We still have to address issues related to poverty, education and health.
- We have to address waste and pollution caused by the richer segments of society.
All these concerns are addressed in this book and must be reflected in the way we live our lives.

Natural capital: Just as we have a *financial capital* (economic) in terms of money, we also have a *social capital* which is related to our shared human resources, as well as *natural capital* which includes environmental goods and services.

Learning through critical thinking—The future

What will our descendants think about us when they look back into their past?
Will they see us as greedy, wasteful, unthinking or just ignorant?
Or
Will they see us as a futuristic, thoughtful, enlightened, caring generation?
Create your own dream, nightmare situation of the future...

The world can be looked at through different coloured glasses. Even when each of us tries to delve into the complexities of nature, we see it differently at different points in time. Our perspective constantly changes and even jumps back and forth.

The brain is a trickster. Stare at this transparent cube for several seconds persistently. What happens? The cube is on a two dimensional flat surface. The eyes see it, but the brain interprets it differently all the time. It seems as if it is a real cube seen from different angles. Similarly, people see the world differently.

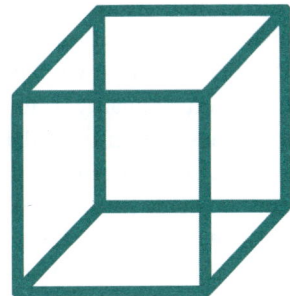

1.4.2 Population Growth and our Resources

One of the most important environmental issues we face today is the recent population boom (**Figs 1.4**). In the past, human numbers grew very slowly as people died by the time they reached their prime. 40–50 years was the average longevity. Subsequently, better health care led to longer lifespans, reduced infant mortality and fall in the death rate. Thus, the growth rate has become exponential leading to the depletion of food resources. Food scarcity was countered by the green revolution through extensive irrigation, more land invested into the agricultural, food processing and transportation sectors.

Little drops of water
Little grains of sand,
Make the mighty ocean
And the pleasant land.

> **Learning by critical thinking**
>
> Have you thought of your own impact on the earth and its resources? This is known as your environmental footprint. It is enormous.

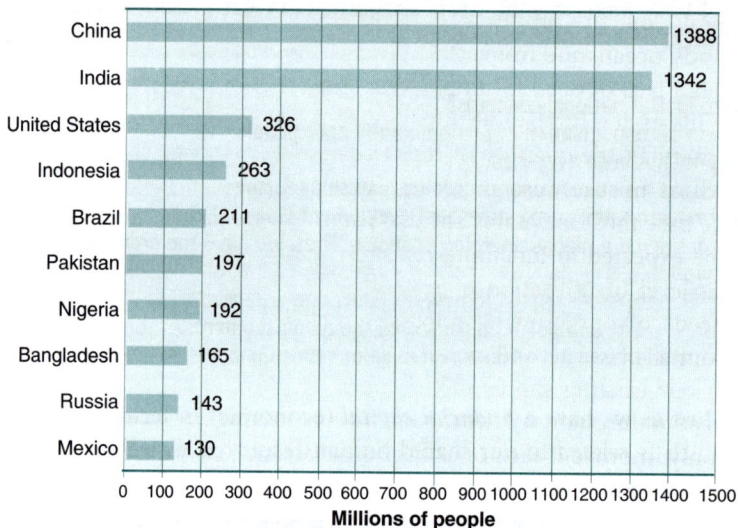

Country	Millions of people
China	1388
India	1342
United States	326
Indonesia	263
Brazil	211
Pakistan	197
Nigeria	192
Bangladesh	165
Russia	143
Mexico	130

Source: Internet World Stats - www.internetworldstats.com/stats8.htm
7,519,028,970 world population estimated for June 30, 2017

Fig. 1.4   Most populated countries in the world

1.4.3 Environmental Footprint

Energy that we use for cooking, heating and cooling homes comes from thermal power and hydropower units through several kilometres of transmission lines and transformers at a great cost. We travel by our own petrol/diesel vehicles or by public transport from one point to another; its impact on air quality is cumulatively enormous because everyone is on the move. We use enormous quantities of non-degradable plastics and waste other resources which need large landfills. The amount of waste generated and to be managed is unimaginable. The large environmental foot print must be made smaller while our small handprint (which are positive actions for the environment) must be made larger for sustainable development to occur (**Fig. 1.5**).

We use paper, wood products and non-timber forest resources to manufacture all sorts of goods that lead to deforestation. We use an enormous quantity of water in everyday life, much of which

Fig. 1.5   Sustainability: Footprint << Handprint

is carelessly wasted and polluted. Can our earth support this high environmental footprint that each one of us is leaving behind every day, along with that of 7.8 billion people on earth (**Table 1.3**)?

The earth has limited natural resources. Water, air, soil, minerals, oil, the products we get from forests, grasslands, oceans and from agriculture and livestock, are all a part of our life-support systems. Without them, life itself is impossible. As we keep increasing in number, the quantity of resources we use (as well as misuse) also increases. The earth's resource base must inevitably shrink. The earth cannot be expected to indefinitely sustain this expanding level of utilisation of resources and energy needs of our growing population. Increased amounts of waste and pollution are contaminating our existing supply of resources and pose a threat to the quality of life for all. This situation will only improve if each of us begins to take actions in our daily lives that help preserve our environmental resources.

Apart from the goods, the environment also provides a number of environmental services. While the goods can be easily assessed for their economic importance, the real economic value of environmental services is extremely complex as this is often hidden. Imagine attempting to put a price on clean air, potable water, and the services provided by forests, wetlands and grasslands in our daily lives. Environmental scientists and economists are struggling to put these complex studies into perspective.

Table 1.3   Ecological footprint (2018 results)

Country	Ecological footprint (global hectares)
China	5,200,000,000
United States of America	2,670,000,000
I India	1,450,000,000

Source: Global footprint network, 2018, National footprint accounts.

Experiential learning

Learning by observing your own surroundings and your own impact. Can you...
◆ reduce your impact?
◆ make positive actions for the environment?
The final outcome will be a result of the relative size of your negative footprint and your positive handprint on our world.

1.4.4   Importance of Management of the Environment

As discussed, a shift to economic growth that was inequitable and environmentally unsound has led to the depletion of the world's rapid natural resource base. Realisation that this inappropriate human behaviour would eventually destroy civilisation, has led to a new wave for proactively managing the environment that could result in a better world for all mankind in the future. MOEF and CC now has the added responsibility of mitigating and adapting to climate change which is the greatest threat to the future of mankind.

1.5 CONCEPT OF SUSTAINABILITY AND SUSTAINABLE DEVELOPMENT

Sustainability is defined as a process or state of the environment that can be maintained indefinitely (Caring for Earth, IUCN, 1991) (**Fig. 1.6**). In 1987, the UN Bruntland Commission defined sustainable development as 'meeting the needs of the present without compromising the ability of future generations to meet their own needs' (Report of the World Commission on Environment and Development).

Until two decades ago, the world looked at economic status alone as a measure of human development. Thus, countries that were economically well developed and where people were relatively richer were called advanced nations, while the countries where poverty was widespread and were

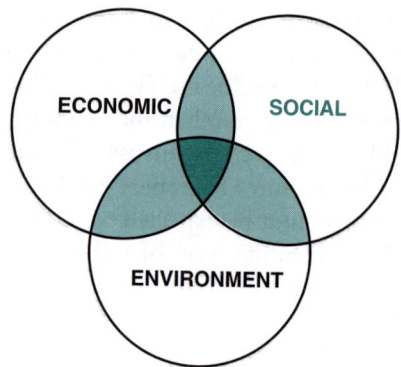

Fig. 1.6 Sustainable development

economically backward were called developing countries. Most countries of North America and Europe, which had become industrialised earlier, are economically more advanced. They not only exploited their own natural resources rapidly, but also used the natural resources of developing countries to grow even larger economies. So as development progressed, the rich countries got richer while the poor nations got poorer. However, even the developed world has begun to realise that their lives were being seriously affected by environmental consequences of development based on economic growth alone. This form of development did not add to the quality of life as the environmental conditions had begun to deteriorate.

By the 1970s, most development specialists began to appreciate the fact that economic growth alone could not bring about a better way of life for people unless environmental conditions were improved. Development strategies in which only economic considerations were used, had begun to suffer from serious environmental problems due to air and water pollution, inadequate waste management, deforestation and a variety of other ill effects that seriously affected peoples' well being and health. There were also equity issues between the 'haves' and the 'have-nots' in society, at both the national and global levels. The disparity in the lifestyles between the rich and the poor was made worse by what has now come to be known as unsustainable development strategies.

> **The sustainable village: Mahatma Gandhi**
>
> Many decades ago, Mahatma Gandhi envisioned the development of a new type of village community based on sound environmental management. He stressed the need for sanitation based on recycling human and animal manure and well-ventilated cottages built of recyclable material. He envisioned clean roads that were free of dust. His main objective was to use village-made goods instead of industrial products. All these principles are now considered part of sound long-term development. He had designed a sustainable lifestyle for himself when these concepts were not a part of general thinking.

The definition of sustainable development expresses the idea of environmental, economic and social equity within the limits of the world's natural resources. It is a process which leads to a better quality of life for all, especially the poor and deprived, while reducing the negative impact on the environment. Its strength is that it acknowledges the interdependence of human needs and environmental requirements.

To ensure sustainable development, any activity that is expected to bring about economic growth must also consider its environmental impact (or environmental costs) so that it is more consistent with long-term growth and development. Many development projects – such as dams, mines, roads, industries and tourism development – have severe environmental consequences in terms of their impacts on natural resource use, land use and biodiversity. All these environmental impacts must be assessed and studied before any development activity is even begun. Thus, for every project, in a strategy that looks at sustainable development, there must be a scientifically and honestly done Environmental Impact Assessment (EIA), without which the project must not be cleared.

To bring about sustainable development, society must create a balance between economic growth, societal equity and long-term environmental stability. Any change in one of these three pillars compromises one or both of the other pillars. This balanced approach to development is reflected in the wellbeing of society. It is a reflection of education, health, housing, access to food, medicine and nutrition, and psychological wellbeing. It is brought about by eliminating poverty, bringing about equality in society to remove stresses, and providing future generations and all other species on earth with an environment conducive to their own growth.

SDGs: To bring about sustainable development at the global level, the nations of the world came up with 17 SDGs (**Table 1.4**). While all the goals are important, there has to be a country-specific focus. For example,

◈ India still has a need to reduce poverty—Goal 1.
◈ India being a mega diversity country rich in flora and fauna, must work on Goals 14 and 15.
◈ Industrialised rich nations must focus on reducing inequality among nations (Goal 10), responsible consumption (Goal 12) and climate action (Goal 13).
◈ Countries with high mortality and morbidity from vector borne diseases must focus on Goal 3 to promote preventive healthcare.

All countries must ensure quality education through communication education and public awareness strategies, with a focus on education for sustainable development (ESD).

1.5.1 New Dimensions for Mainstreaming Sustainable Development into College Curricula

While India surges ahead into a new stage of economic growth, long term benefits of our economy (development) cannot be achieved at the cost of the economically deprived. Development in its true sense must percolate across all sectors of society.

A comprehensive growth strategy is now considered to be of great importance in which all the three pillars of good governance must be considered. This must undoubtedly include economic

Table 1.4 Sustainable Development Goals (SDGs)

S. No.	SDG	Found in Unit
1	No poverty: End poverty in all its forms everywhere—tribal issues (Core) NTFP, BMC-ABS	7
2	Zero hunger: End hunger, achieve food security and improved nutrition and promote sustainable agriculture—sustainable agriculture, farming practices by biofertilisers	7
3	Good health and well being: Ensure healthy lives and promote wellbeing for all at all ages—healthy food, eco-friendly products, pollution control	7
4	Quality education: Ensure inclusive and equitable quality education and promote life-long learning opportunities for all—CEPA	1
5	Gender equality: Achieve gender equality and empower all women and girls—education for girls, CEPA, alternate income generation	1
6	Clean water and sanitation: Ensure availability and sustainable management of water and sanitation for all	3, 5
7	Affordable and clean energy: Ensure access to affordable, reliable, sustainable and modern energy for all	3
8	Decent work and economic growth: Promote sustained, inclusive and sustainable economic growth, full and productive employment and decent work for all	1, 6
9	Industry, innovation and infrastructure: Build resilient infrastructure, promote inclusive and sustainable industrialisation and foster innovation	5, 6
10	Reduced inequalities: Reduce income inequality within and among countries	6
11	Sustainable cities and communities: Make cities and human settlements inclusive, safe, resilient and sustainable	1
12	Responsible consumption and production: Ensure sustainable consumption and production patterns	1, 4
13	Climate action: Take urgent action to combat climate change and its impacts by regulating emissions and promoting developments in renewable energy	6
14	Life below water: Conserve and sustainably use the oceans, seas and marine resources for sustainable development	2, 4
15	Life on land: Protect, restore and promote sustainable use of terrestrial ecosystems, sustainably manage forests, combat desertification, halt and reverse land degradation, and halt biodiversity loss	2, 4
16	Peace, justice and strong institutions: Promote peaceful and inclusive societies for sustainable development, provide access to justice for all and build effective, accountable and inclusive institutions at all levels	6
17	Partnerships for the goals: Strengthen the means of implementation and revitalise the global partnership for sustainable development	1

growth, achieve societal equity and at the same time be environmentally sensitive. This cannot be achieved unless it becomes part and parcel of our educational system. SDG 4 and Aichi Target 1, for biodiversity conservation which are international treaties of which India has been an important signatory, bring forth the need for strong educational and awareness components. The need for an informed society, especially in an emerging economy such as India, is absolutely essential to counter the effects of economically unsustainable growth, inequitable access to resources and education.

While India has had a strong sense of environmental ethics from ancient times through religious and cultural philosophy, this cannot last forever. It is being swamped by newer concepts of philosophy, science and the inevitable homogenising of our cultures due to globalisation. No other country can boast of having set up reserves for animals and tree covered avenues, or promulgated laws which prohibited the killing of a set of species as far back in history as Ashoka's reign in the 3rd century BC. This predates the concept of scheduled species by 2000 years. Sanctuaries for elephants were created several centuries ago. Buddhist and Jain philosophy celebrated and protected all forms of life, long before modern science realised the importance of every species as a part of a well-balanced and healthy ecosystem. Hindu culture is steeped in gods that have taken a variety of animal forms. Mughal rulers studied natural history and supported artists. Ancient tribal communities worshiped the tiger and cobra as major deities. Biodiversity conservation has thus been a part of India's cultural diversity for thousands of years from the earliest drawings of cave dwellers in the Bhimbetka Stone Age hunter–gatherers into our own age in the anthropocene.

However, today, we are in a period when the country is on the threshold of a new era of growth and development in sharp contrast to the ancient historical period. This is a growing void between ancient myth and present reality which needs to be addressed beginning with the current generation.

While the 1950s and 1960s was the age of post independent non-formal nature education, the 1970s and 1980s altered these initiatives into environmental education triggered by degraded ecosystems, increasing levels of pollution of air and water, soil erosion, desertification and deforestation. This led to increasing the pro-environment initiatives through formal school curricula. Students have been increasingly distanced from nature and their own environment and find it difficult to internalise this information. Our new age education must motivate students towards a new way of thinking about the earth's natural resources and take action to find ways of dealing with education for sustainable development.

India has been too slow in giving momentum to alter environmental education by integrating the economic and societal concerns of sustainability into our formal school curricula to include this aspect of environmental education. While some theoretical aspects of sustainability have become known to the teacher community, activating this information into the lives of their students is still vague and unfocused. Education for sustainable development is a dynamic process. It stems from the following steps:

- information,
- awareness of one's own environment,
- developing a concern for the environment, and
- a willingness to act for the environment and its conservation.

Education progressed from nature education to environmental education and finally to sustainable development. As observation of nature is the most effective trigger to create pro-environmental behaviour, the best way to start is to look around a natural landscape and all its beauty to trigger one's environmental consciousness. Observing the pressures on environmental wellbeing will result in concern which will change one's behaviour, thereby reducing our environmental footprint and help create a sustainable world.

Stage I: Nature education (From the 1930s to the '80s)
- Appreciation of flora and fauna through field observations.
- Links of water and food we eat with nature's resources.
- Appreciation of home and school surroundings.

Stage II: Environmental education (From the 1980s to 2010)
- Keeping the environment clean, thus improving our health.
- Appreciating plants, mammals, birds, insects and their role in natural ecosystems.
- Know India's protected areas and biogeographical zones.
- Energy saving and use of alternate energy sources.
- Observe and alter rural/urban environmental disputes.
- Prevention of environment-related diseases.

Stage III: Education for sustainable development (After 2010)
- Sustainable development: economic, social, environmental concerns.
- Importance of people's participation in sustainable development.
- SDGs/Aichi targets: what the nation and its citizens can do.
- Global sensitivity and equity of nations.
- National security and peace.
- Vision for a better world.

Every individual is linked with society and the living environment at different levels. A widening set of influences and influencers encircle our lives. The circles move outwards like the circular ripples when a stone is thrown into still water. Each circle is influenced by every individual (Fig. 1.7). In return each individual is influenced by the circles in the environment.

If a ripple is disturbed, the configurations of all the symmetrically aligned ripples are disturbed. A constant two-way influence is present between us and the economic, social and

Fig. 1.7   Sustainability

environmental dimensions of life. We can prudently behave in a way that can sustain our earth.

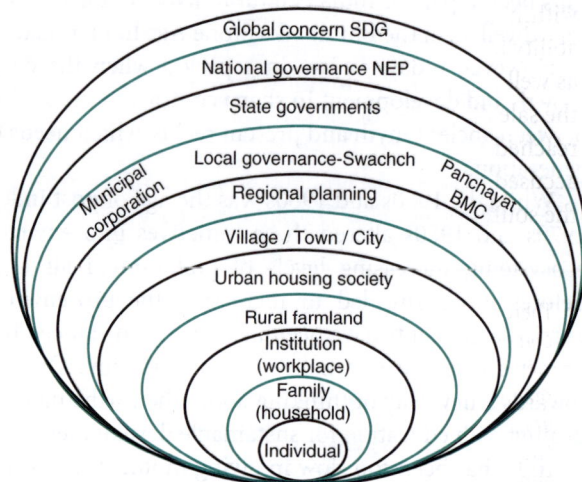

1.5.2   The Need for Sustainable Lifestyles

The quality of human life and the state of ecosystems on earth are indicators of the sustainable use of resources. There are some clear indicators of sustainable lifestyles in human life, such as,
- increased longevity,
- an increase in knowledge, and
- an enhancement of income.

These three together are known to influence the *Human Development Index*. The indicators of the quality of the ecosystems are more difficult to assess. They are:
- a stabilised population, of all living forms or the percentage of species loss,
- species diversity in ecosystems,
- the state of 'naturalness' of ecosystems.

The use of a resource begins with its collection, its processing into a useable product, and transport through a delivery system to the consumer who uses it. It also involves disposal of the waste products produced at each step. Each step in resource use can affect the environment for better or for worse. Control of these steps is known as environmental management. Think of a resource you use and track it through these steps

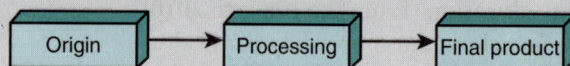

Origin → Processing → Final product

Example: The cotton in the clothes you are wearing. At each step, note:
- What other resources are needed at this step to move the resource you chose to the next?
- What waste products are generated at that step?
- How are they likely to be disposed of?
- What pollutants are generated in the process?

1.5.3   Equitable Use of Resources for a Sustainable Lifestyle

Reducing the unsustainable and unequal use of resources and controlling our population growth are essential for the survival of human society. Our environment provides a variety of goods and services necessary for day-to-day life, but the soil, water, climate and solar energy, which form the abiotic support that we derive from nature, are not distributed evenly throughout the world or within countries. A new economic order at the global and at national levels must be based on the ability to distribute the benefits of natural resources by sharing them more equally among countries as well as among communities within each country. It is at the local level where people subsist by the sale of locally collected resources, that the disparity is greatest. Economic development has not reached those who are dependent substantially on collecting resources. They are often unjustly accused of over exploiting natural resources. They must be adequately compensated for the use of the sources sent to distant regions and thus develop a greater stake in protecting natural resources.

Learning through critical thinking—Perspectives of life on earth

There are several principles that each of us can adopt to bring about a sustainable lifestyle. This primarily comes from caring for the earth in all respects. Love and respect for nature is important for analysing how we use natural resources in a sensitive way. Think of the beauty of the wilderness, a natural forest in all its magnificence, the expanse of a green grassland, the clean water of a lake that supports so much life, the crystal clear water of a hill stream, or the magnificent power of the oceans, and one cannot help but support the conservation of nature's wealth. If we respect this, we cannot commit acts that will deplete our life-supporting systems.

What do you value most? Resource consumption patterns and the need for equitable utilisation

Can individuals justifiably use resources so differently that one individual uses resources many times more lavishly than other individuals who have barely enough to survive? In a just world, there has to be an equitable sharing of resources. There are rich and poor nations, there are rich and poor communities in every country, and there are rich and poor families. In this era of modern economic development, the disparity between the haves and have-nots is widening. Human environments in the urban, rural and wilderness sectors use natural resources that shift from the wilderness (forests, grasslands, wetlands) to the rural sector and from there to the urban sector. Wealth also shifts in the same direction. This unequal distribution of wealth and access to land and its resources is a serious environmental concern.

An equitable sharing of resources forms the basis of sustainable development for urban, rural and wilderness-dwelling communities. As the political power base is in the urban centres, this leads to inequalities and a subsequent loss of sustainability in resource management in the rural and forest sectors (see the charts in the Annexure).

In 1985, Anil Agarwal published the first report on the status of India's environment. It emphasised that India's environmental problems were caused by the excessive consumption patterns of the rich that left the poor poorer. It was appreciated for the first time that tribals, especially women and other marginalised sectors of our society, were being left out of economic development. There are multiple stakeholders in the Indian society who are dependent on different natural resources which cater directly or indirectly to their survival needs. Anil Agarwal brought forth a set of 8 propositions which are of great relevance to the ethical issues that are related to environmental concerns.

◆ Environmental destruction is largely caused by the consumption of the rich.
◆ The worst sufferers of environmental destruction are the poor.

- Even where nature is being 'recreated', as in afforestation, it is being transformed away from the needs of the poor and towards those of the rich.
- Even among the poor, the worst sufferers are the marginalised cultures and occupations and, most of all, women.
- There cannot be proper economic and social development without a holistic understanding of society and nature.
- If we care for the poor, we cannot allow the Gross Natural Product to be destroyed any further. Conserving and recreating nature has to become our highest priority.
- The Gross Natural Product will be enhanced only if we can arrest and reverse the growing alienation between the people and the common property resources. Towards this end, we will have to learn a lot from our traditional cultures.
- It is totally inadequate to talk only of sustainable rural development as the World Conservation Strategy does. We cannot save the rural environment or rural people dependent on it, unless we can bring about sustainable urban development.

Experiential learning

Equitable use of forest resources: We think of forests as being degraded due to fuel wood collection by poor rural communities, but forget that the rich use much greater quantities of timber. Biomass based industries include cotton textiles, paper, plywood, rubber, soap, sugar, tobacco, jute, chocolate, food processing and packaging. These need land, energy, irrigation and forest resources. Do any of us realise this when we utilise, use excessively or waste these resources that we get indirectly from the forests?

Who pays for the cost of environmental degradation? Most sections of society do not feel the direct effects of degradation of the environment till it is too late. Several marginalised sectors of society are most affected by deforestation, or the loss of grassland tracts, or the deterioration of perennial water sources. All these effects can be linked to increasing unsustainable pressures on land and natural resources. Traditional fishermen who are dependent on streams and rivers, and coastal people who fish and catch crustaceans, are seriously affected by the degradation of our aquatic ecosystems. Fuel wood gatherers from different types of forests, and pastoralists who are dependent on common grazing lands suffer when their resources are depleted.

I am often amazed and extremely angry, when people talk about environmental education for the villages. It is the so-called, educated people who need environmental education more than anyone else.
—Anil Agarwal, 'Human-Nature Interactions in a Third World Country'

1.5.4 Environmental Education and Education for Sustainable Development—the Different Perspectives

Urban

Urban dwellers who are far removed from the source of natural resources that sustain their lives, require exposure to a well-designed environmental education programme to appreciate these issues.

Rural

While the rural people have a deep insight regarding the need for sustainable use of natural resources and know traditional methods of conservation, there are several newer environmental concerns that are frequently outside their sphere of life experiences. Their traditional knowledge of environmental concerns cannot be expected to bring about an understanding of issues such as global warming, or problems created by pollution and pesticides. These people thus require a different pattern of environmental education that is related to filling in gaps in their information. With the rapidly changing rural scenario, the development that is thrust on unsuspecting rural communities needs to be addressed through locale-specific environment awareness programmes designed specifically for rural school children and adults. This must include their local traditional knowledge systems as a base on which modern concepts can be built, rather than by fostering concepts that are completely alien to their own knowledge systems.

Common property resources in India once included vast stretches of forests, grazing lands and aquatic ecosystems. When the British found that they were unable to get enough wood for ship building and other uses in England, they converted India's forests into 'reserved forests' for the British Government's own use to grow timber trees. This alienated the local people from having a stake in preserving their own forest resources. This, in turn, led to large scale losses in forest cover and the creation of wasteland.

In the past, in traditional villages that were managed by local panchayats, there were well-defined rules about managing grazing lands, collecting forest resources and protecting sacred groves that supported conservation. There was a more or less equitable distribution of resources controlled by traditional mechanisms to prevent the misuse of common property resources. Any infringement was quickly dealt with by the *panchayat* and the offender was punished. Common property resources were thus locally protected by communities. As land use patterns changed, these mechanisms were lost and unsustainable practices evolved, frequently as a result of an inadequately planned development strategy. Urban citizens, especially the rich, are the consumerist sector of Indian society. They are wasteful in many respects. They waste water, food and electrical and fossil fuel. Urban people have a large environmental footprint. There is a serious gap between the rich and poor in most urban societies in India. The globalised (westernised) aspects are leading to a loss of local cultural values linked to our environment.

1.5.5 The Need for Gender Equity

All over India, especially in the rural sector, women work longer hours than men. The life of a woman is enmeshed in an inextricable cycle of poverty. In attempting to eke out a living from the environment, women constantly have to collect fuel wood both for their homes as well as to sell to nearby urban areas at a low price. They laboriously collect fodder for their cattle. They have to trudge several kilometres to reach a reasonably clean water source. And finally, they must cook meals in a smoky, unhealthy atmosphere on fuel wood, crop and animal waste or use other inefficient sources of energy. All this can take 10 to 12 hours a day of very hard work, every day of the year.

This begets the question of who should control the environmental resources of a rural community. Unfortunately, it is the men who play a decisive role in managing the village commons and their resources, whereas it should really be the local women, whose lives are deeply linked with the utilisation and conservation patterns of natural resources, who should be the decision makers at the local level. Unfortunately, women have not been given an equal opportunity to develop and better their lot. This begins with the lack of attention given to girls whose education is always secondary to that of boys in the family. Unless society begins to see that development cannot be planned from the male perspective alone, we will not be able to create a healthier environment for women and children. To counter this, the government has schemes for educating the girl child. The panchayats are mandated to include women and the leadership is expected to be rotated among both men and women.

Pro-environmental action

- Join a group to study nature, such as World Wildlife Fund for Nature-India (WWF-I), Bombay Natural History Society (BNHS) or another environmental group.
- Begin by reading newspaper articles and periodicals like *Down to Earth*, *WWF-I Newsletter*, *BNHS*, *Hornbill*, *Sanctuary* or other scientific journals which will tell you more about our current environmental issues. There are also several environmental websites.
- Lobby for conserving resources by taking up the cause of environmental issues during discussions with friends and relatives. Practice and promote issues such as saving paper, saving water, reducing the use of plastic, practising the 3Rs principle of *Reduce, Reuse, Recycle*, and proper waste disposal.
- Join local movements that support activities such as saving trees in your area, go on nature treks, recycle waste, buy environmentally-friendly products.
- Practice and promote good civic sense and hygiene such as enforcing no spitting or tobacco chewing, no throwing garbage on the road, no smoking in public places, no urinating or defecating in public places.
- Take part in events organised on World Environment Day, World Biodiversity Day, Wildlife Week and significant other pro-environment days. Visit a national park or sanctuary, or spend time in whatever natural habitat you have near your home.

Can you evolve a strategy for yourself and your immediate circle of friends, neighbours, family and workplace?

SUMMARY

- Environmental studies involves the understanding of human interactions with the environment. It requires an integrated approach to several disciplines of science and social studies.

- We live in a world where various environmental components such as air, land, water, forests, minerals, grasslands and wetlands are limited. We cannot continue to exploit these resources beyond the earth's assimilative capacity. Thus, the sustainable use of resources is of utmost importance.

- It is important for every individual on this earth to bring about sustainable development. Society must create a balance between economic growth, societal equity and long-term environmental stability.

- Everything around us constitutes our environment and our lives depend on keeping the earth's vital systems intact. We need to individually take responsibility towards preserving our environmental resources.

QUESTIONS

1. Explain with an example, the multidisciplinary nature of the environment.
2. Name the three pillars of sustainability. Give any one example of a sustainable project.
3. Explain the need for environment awareness in the society.
4. Explain the role of equitable use of resources in sustainable development.
5. Give five steps to reduce your water footprint and to increase your handprint.

Ecosystems

Learning Objectives

In this chapter you will learn,

◈ What an ecosystem is, its structure and its functions
◈ How energy flows in an ecosystem; food chains and food webs
◈ How various ecosystems function—forest, grassland, desert, aquatic and estuary
◈ About degradation of ecosystems

Purpose

We need to delve into the types of ecosystems and their functioning to learn about our own life. Humans are only a small fragment of the billions of components, both living and non-living, that make up our world. Once we understand how nature's ecosystems work, we will realise how we are damaging and destroying the balance in nature that is essential for human beings to live on earth.

Our Role

Many of our day-to-day activities create some negative impact on the earth's welfare. To reduce this, we need to introspect and act towards minimising wasteful behaviour. Each ecosystem, and all the elements in land and water, are affected by our lifestyle; this is incompatible with nature. We are dependent on urban, industrial, agricultural, pastoral, river, marine and other integrated ecosystems. We need to preserve these for a better future. This is referred to as public–private participation in the management of these ecosystems.

2.1 WHAT IS AN ECOSYSTEM?

An *ecosystem* is a region with a specific and recognisable landscape form, such as a forest, grassland, desert, wetland or coastal area. The nature of the ecosystem depends on its geographical features such as hills, mountains, plains, rivers, lakes, coastal areas or islands and is also controlled by climatic conditions—the amount of sunlight, temperature and rainfall in the region. The geographical, climatic and soil characteristics form its non-living or *abiotic components*. These features create conditions that support a community of plants and animals that evolution has produced, to live in these specific conditions. The living part of the ecosystem is referred to as its *biotic component*.

Ecosystems are divided into *terrestrial* or land-based ecosystems and *aquatic* or water-based ecosystems. These form the two main habitats for the earth's living organisms (see Colour Plates in Unit 4). All the living organisms in an area live in communities of plants and animals. They interact with the abiotic environment and with each other at different points in time for various life-support systems. Life can exist only in a small portion of the earth's land, water and atmosphere. At a global level, the thin skin of the earth on the land, sea and air forms the *biosphere*.

Aspects of an ecological system

Every ecological system has three aspects.

◆ **Ecosystem structure** describes the physical appearance. The structure tells us what its different features are and how it is typical and different from other systems. The differences one observes from the base to canopy of a forest or edge to deep part of a pond or lake, constitute the structure of an ecosystem. For example, evergreen, deciduous or coniferous forests differ in structure as their component trees differ. A natural or culturally/human modified system differs from a natural system.

◆ **Ecosystem composition** describes the community of plants and animals within and is linked to the abundance and variety of species and their genetic differences. For example, a forest with different species of plants (trees, shrubs, climbers, epiphytes). Fauna adapted to the different microhabitats (including macro and micro fauna).

◆ **Ecosystem function** describes how the ecosystem works in nature—its food chains, food webs, food pyramids and biogeochemical cycles that run it (water, nutrient, energy and chemical cycles). The ecological succession or variation is based on changes in climate, temperature, rainfall, energy transfer and biotic (human) influences.

At a sub-global level, this is divided into *biogeographical realms*. For example, Eurasia is called the *Palearctic* realm, South and Southeast Asia (of which India forms a major part) is the *Oriental* realm, North America is the *Neoarctic* realm, South America forms the *Neotropical* realm, Africa the *Ethiopian* realm, and Australia the *Australian* realm.

Observe the components of the structure, species variability of the composition, and the functions. This includes the activities and linkages present during the day or night, seasonal differences such as migration, animal behaviour, and floral assemblages linked to pollinators and seed dispersers. It also includes the effects of tree felling, loping that lead to changes in structure, composition of species and functional aspects.

Try to experience the abiotic and biotic features and compare them to other ecosystems you may have observed.

India is divided into ten major biogeographic zones, each of which has its own distinctive terrestrial and aquatic ecosystems. Thus, the structural, compositional and biogeochemical nature of this multitude of ecosystems is reflected in each of the ten biogeographic zones. This is the reason for India's rich and varied plant and animal life.

Definition: The living community of plants and animals in any area together with the non-living components of the environment – soil, air and water – constitute an ecosystem.

2.2 STRUCTURE AND FUNCTION OF ECOSYSTEMS

2.2.1 Structure of Ecosystems

Natural ecosystems include forests, grasslands, deserts and aquatic ecosystems such as ponds, rivers, lakes and the sea. Human-modified ecosystems include agricultural land and urban or industrial land use patterns. A mosaic of such elements in the environment is now referred to as a *landscape*. Human-modified landscapes are referred to as cultural landscapes. A mix of landscape elements can be natural as well as cultural.

Each ecosystem has a set of common features that can be observed in the field:

◆ **What does the ecosystem look like?**
One should be able to describe specific features of the different ecosystems in one's own surroundings. Field observations must be made in both urban and natural surroundings.

◆ **What is its structure?**
Is it a forest, a grassland, a water body, an agricultural area, a grazing area, an urban area or an industrial area? What you should look for are its different characteristics. A forest has various layers from the ground to the canopy. A pond has different types of vegetation from the periphery to its centre. The vegetation on a mountain changes from its base to its summit.

◆ **What is the composition of its plant and animal species?**
List the well-known plants and animals you can see. Then, document their abundance and numbers in nature—very common, common, uncommon, rare. For example, in a cultural landscape, wild mammals will rarely be seen in large numbers, cattle however would be common. Some birds are common as they adapt to human society; find out which are the most common species. Insect species are very common and most abundant. In fact, there are so many that they cannot be easily counted.

◆ **How does the ecosystem work?**
Can you observe the functioning of the water cycle or the nutrient cycle? Some ecosystem functional activity is perpetually happening around us.

2.2.2 Functions of Ecosystems

The ecosystem functions through several biogeochemical cycles and energy-transfer mechanisms.

The integral components of the ecosystem (biotic and abiotic) interact with each other through several functional aspects. Plants, herbivores and carnivores are part of food chains with three to four links in each chain. All these chains together form a web of life on which humans depend for all their needs. Each of the food chains uses energy that comes from the sun and powers the ecosystem. The sun is the primary source of energy.

> **Observational learning**
>
> Observe and document the components of the ecosystem, which consist of its abiotic features such as air, water, climate and soil, and its biotic components—the various plants, fungi and animals.

At each level of the food pyramid, energy is used up for the living activities of plants and animals. Plants directly use energy from the sun to photosynthesise, they are thus, the producers in an ecosystem. Animals live on plants (herbivores), or on other animals (carnivores). The herbivores and carnivores are together the consumers. Detrivores include soil organisms which break down dead material (detritus) such as leaves and dead wood into smaller fragments; decomposers such as fungi and microbes act on these and convert them into simple inorganic nutrients used by plants.

2.2.3 Interconnectedness in Nature—Human and Ecosystem Resources

Most traditional societies used their environmental resources fairly sustainably. Though inequality in resource utilisation has existed in every society, the number of individuals that used a large proportion of resources in the past was extremely limited. In recent times, the proportion of rich people in affluent societies has grown rapidly. Inequality has thus become a serious challenge. Whereas in the past many resources such as timber and fuel wood from the forest were extracted sustainably, this pattern has drastically changed during the last century. The economically powerful sections began to use greater amounts of forest products, while those people who lived in the forest became increasingly poor. Similarly, the building of large irrigation projects has led to wealth in those areas that had canals, while those who remained were dependent on rain-fed monsoon crops or dependent on a constant supply of water from the perennial river itself, and have found it difficult to survive.

The key to this issue is the need for an equitable distribution of all types of natural resources. A more even sharing of resources within the community can reduce these pressures on the natural ecosystems.

2.2.4 Ecosystem Goods and Services

Ecosystems provide a range of goods and services. These can be broadly categorised as direct and indirect values that we benefit from in every ecosystem inhabited by people.

(i) **Direct values:** These are resources that people depend upon directly and are easy to quantify in economic terms. They can be categorised as follows:
   - Consumptive use value—fruit, fodder and firewood used by people who collect them from their surroundings, for their own personal use and are not for sale.
   - Productive use value—the commercial value of timber, fish, medicinal plants, and other resources such as medicinal plants that people collect for sale.

(ii) **Indirect values:** These are services produced by nature that are not easy to quantify in terms of a clearly definable price.
   - Functional value—ecosystem functions such as cycles in nature linked to air, water, soil nutrients, climate regulation, flood and storm protection and erosion prevention.

Maintaining natural cycles is a service that has enormous economic value but usually remains cryptic and hidden.

◆ Non-consumptive use value—scientific research, bird-watching, ecotourism.

◆ Option value—maintaining options for the future, so that by preserving them one could reap economic benefits in the future.

◆ Existence value—these involve the ethical and emotional aspects of the existence of nature.

The non-living components of an ecosystem are the amount of water, inorganic substances and organic compounds, and climatic conditions, which depend on the geographical conditions and location.

The living organisms in an ecosystem are inseparable from their habitat. Plant life ranges from extremely small bacteria (which live in the air, water and soil), algae (which live in fresh and saltwater), the terrestrial plants (which range from grasses and herbs that grow after the monsoon every year), to the giant long-lived trees of the forest. Some trees may be hundreds of years old. Other plants and animals may be linked to annual seasons of migration across continents.

2.3   ENERGY FLOW IN AN ECOSYSTEM

Every ecosystem has several interrelated mechanisms that affect human life. These are the water cycle, carbon cycle, oxygen cycle, nitrogen cycle and energy flow. While every ecosystem is controlled by these cycles, each ecosystem's abiotic and biotic features are specific to that ecosystem.

All the functions of the ecosystem are in some way related to the growth and regeneration of its plants and multiplication of its animal species. These interlinked processes can be depicted as various cycles. All these processes depend on energy from sunlight. During photosynthesis, carbon dioxide is absorbed by plants and oxygen is released into the atmosphere in the presence of sunlight. Animals depend on this oxygen for their respiration. The water cycle depends on rainfall, which is necessary for plants and animals to live. The sun drives the cycle through transpiration of water given out by plants. The energy cycle recycles nutrients into the soil on which plant life grows. Our own lives are closely linked to the proper functioning of these cycles of life. If human activities go on altering them, no life can survive on earth.

The cycles are based on the flow of energy through the ecosystem.

◆ *Producers*: The energy from sunlight is converted by plants through photosynthesis into growing of leaves, flowers, fruit, branches, trunks and roots. Since plants can grow by converting the sun's energy directly into their tissues, they are known as *producers* in the ecosystem.

◆ *Consumers*: The plants are used by herbivores as food, which gives them energy. A large part of this energy is used up for the metabolic functions of these animals such as breathing, digesting food, supporting the growth of tissues, maintaining blood flow and regulating body temperature. Energy is also used for activities such as searching for food, finding shelter, breeding and rearing young ones. The carnivores, in turn, depend on the herbivores on which they feed. *Omnivores* eat both animals and plants. Herbivores, carnivores and omnivores are together called consumers.

◆ *Decomposers*: When plants and animals die, this material is returned to the soil after being broken down into simpler substances by decomposers, such as insects, worms, bacteria and fungi. Animals excrete waste products after digesting food, which goes back to the soil and plants absorb these nutrients through their roots. This links the energy flow to the nitrogen cycle.

Thus, different plant and animal species are linked to one another through food chains. Each food chain has three or four links. However, as each plant or animal can be linked to several other plants or animals through many different food chains, these interlinked chains can be depicted as a complex food web. This is thus called the '*web of life*' since it shows that there are thousands of inter-relationships in nature.

◆ *Energy flow:* The energy in the ecosystem can be depicted in the form of a food or energy pyramid. The food pyramid has a large base of plants called producers. The pyramid has a narrower middle section that depicts the number or the biomass of herbivorous animals, which are called *first-order consumers*. The apex depicts the small biomass of carnivorous animals called *second-order consumers*. Humans are one of the animals at the apex of the pyramid *(omnivores)*. Thus, to support humans, there must be a large base of herbivorous animals and an even greater quantity of plant material.

The transfer of energy from the source in plants through a series of organisms, by eating and being eaten, constitutes the food chain. At each transfer, a large proportion of energy is lost in the form of heat. These food chains are not isolated sequences, but are interconnected. This interlocking pattern is known as the food web. Each step of the food web is called a *trophic level*. These trophic levels together form the *ecological pyramid* (**Fig. 2.1**).

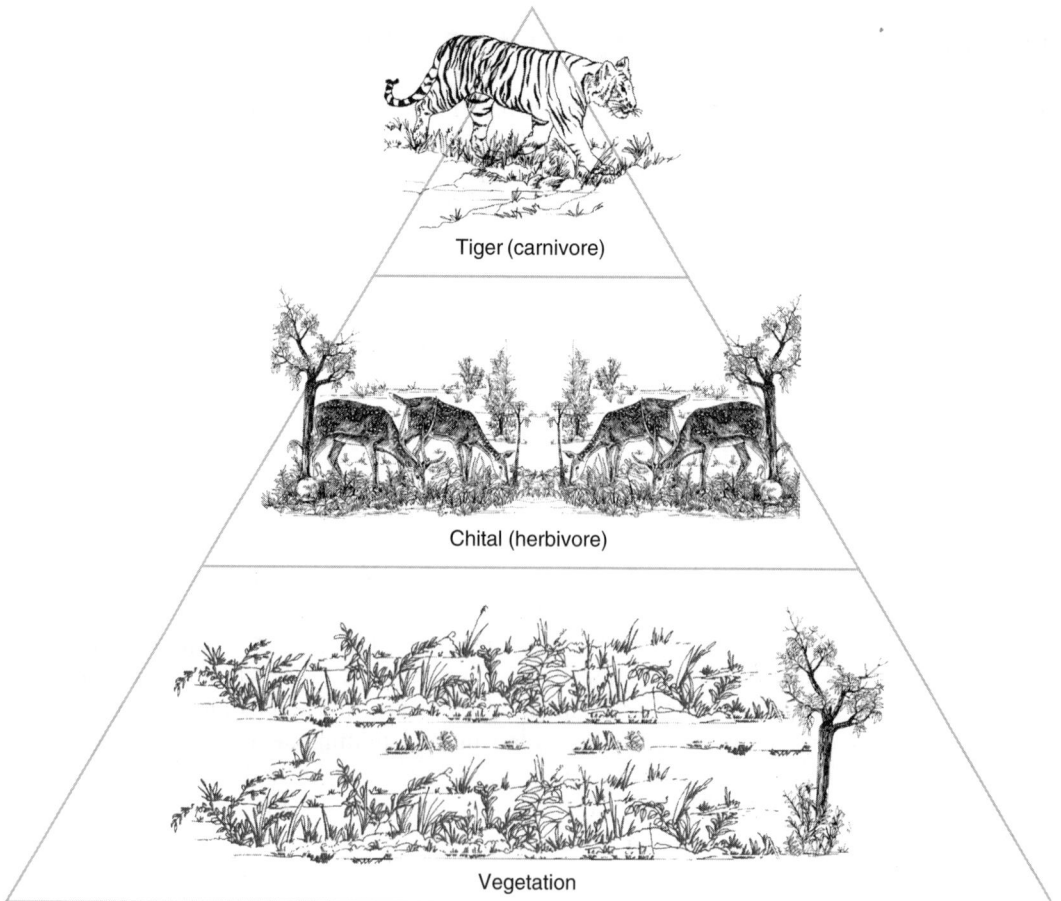

Tiger (carnivore)

Chital (herbivore)

Vegetation

Fig. 2.1   Food/Energy pyramid

2.4   FOOD CHAIN, FOOD WEB AND ECOLOGICAL SUCCESSION

Every living organism is in some way dependent on other organisms and on their habitat. Plants are food for herbivorous animals, which are in turn food for carnivorous animals. Some organisms such as fungi live only on dead material and inorganic matter. Plants are producers in every ecosystem—terrestrial and aquatic. The herbivores are primary consumers as they live on the producers. They include animals such as the deer and elephants in forests), blackbuck (in grasslands), and chinkara or Indian gazelle (in semi-arid areas that use the sparse grass). In the sea, there are small fish that live on algae and other plants. Domestic animals such as cattle, sheep and goats are the herbivores in agricultural systems that feed on shrubs and agricultural residue of crops.

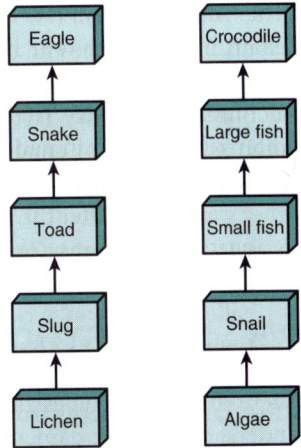

Fig. 2.2   Simple food chains

At a higher level are carnivores or secondary consumers, which live on the herbivores. In our forests, the carnivores include tigers, leopards, jackals, foxes and wild cats. The marine carnivores in the sea range from microscopic forms to giant mammals such as the whale. Decomposers or detritivores include worms, insects, bacteria and fungi which break down dead, organic material into smaller particles and finally into simpler substances that are used by plants as nutrition. Thus, decomposition is a vital function in nature without which all the nutrients would be tied up in dead matter and no new life would be produced.

Food chain: It can be defined as a chain consisting of producers, consumers, decomposers, each at a different trophic level. Each chain has four to five links (**Fig. 2.2**).

Food web: In an ecosystem, there are a very large number of interlinked chains; together, these form a food web. If the links in the chains that make up the web of life are disrupted due to human activities that lead to the loss or extinction of species, the web breaks down.

> **Learning by critical thinking**
>
> ◆ When human activities alter the balance of ecosystems, these perturbations can lead to the disappearance of uncommon species. When this happens to an endemic species that is not widely distributed, it becomes extinct forever.
> ◆ If a species is linked to many other species (key stone species), its disappearance can disrupt the whole ecosystem.
> ◆ Think about which human activities have the potential to disrupt whole ecosystems or lead to extinction.

2.4.1   Ecological Succession

Ecological succession is a process through which ecosystems tend to change over a period of time. Succession can be related to seasonal environmental changes, which create alterations in the community of plants and animals living in the ecosystem. Other successional events may take much longer periods of time, extending to several decades. If a forest is cleared, it is initially colonised by a secondary group of species of plants and animals, which gradually change through an orderly process of community development. One can predict that a cleared or open area will gradually be converted into a grassland, then a shrub land and finally a woodland and a forest, if permitted to do so without further human interference. There is a tendency for succession to produce a more or less stable state at the end of several succession stages.

The most frequent successional changes occur seasonally in a pond ecosystem; this ecosystem fluctuates from a dry terrestrial habitat to the early aquatic stage in the monsoon. This is the colonisation stage by small aquatic species such as algae and zoo plankton. This gradually passes through several stages into a mature aquatic ecosystem, with amphibians, fish, aquatic weeds and other vegetation. After the monsoon, the pond shrinks and reverts to its dry stage in summer when its aquatic life remains dormant.

2.5 CASE STUDIES OF ECOSYSTEMS

There are several types of ecosystems (Table 2.1). For each of these ecosystems, we need to understand four fundamental issues:

◆ What is the nature of the ecosystem? What is its structure, composition of species and ecological function?

◆ Who uses the ecosystem and for what purpose? Is the utilisation sustainable or unsustainable?

◆ How is this ecosystem degraded? What can be done to protect it from deteriorating in the long term?

◆ How can the ecosystem with all its life – flora and fauna – be conserved?

Table 2.1 Types of ecosystems

Terrestrial ecosystems	Aquatic ecosystems
Forest	Pond
Grassland	Lake
Semi-arid areas	Wetland
Deserts	River
Mountains	Delta
Islands	Marine

Observational learning

When you observe an ecosystem, look for its abiotic and biotic elements. What is each component doing? How is it linked to humans by way of goods we use and the services it provides?

Critical thinking

Once you have done this, think of what would happen if a species became extinct due to human activities. What would the effect on humans be if those goods and services from the ecosystem are lost?

2.5.1 Forest Ecosystems

Forests are formed by a community of plants. Structurally a forest is defined by its trees, shrubs, climbers, ground cover, herbs and grasses.

Natural vegetation is varied and looks vastly different from a group of planted trees in orderly rows. The most 'natural' undisturbed forests are located mainly in national parks and wildlife sanctuaries. The landscapes that make up various types of forests look very different from each other. Their distinctive appearance is one of the fascinating aspects of nature. Each forest type forms a habitat for a specific community of animals that are adapted to live in it.

What is a forest ecosystem? The biotic component of a forest includes both large (macrophytes) and microscopic plants and animals. Thus, a forest ecosystem has two parts:

Action learning

What can we do to prevent disturbing these ecosystems of nature?

Observational learning

Look at the different layers in a forest – tree canopy, leaves, branches, trunk and roots – from the top to the ground level. There is also a second storey of smaller trees. There are shrubs and plants at ground level, and there are dead leaves on the ground ready to be recycled as nutrients for fresh plant growth.

- *The non-living or abiotic aspects of the forest*: The type of forest depends on the abiotic (non-living) conditions. Forests on mountains and hills differ from those along rivers and valleys. The vegetation is specific to soil characteristics, amount of rainfall, local temperature, and other factors which vary according to latitude and altitude.
- *The living or biotic aspects of the forest*: Plants and animals form communities that are specific to each forest type. For example, coniferous trees occur in the Himalayas, deciduous forests in the areas of moderate rainfall, evergreen forests in high rainfall tracts, mangrove species in river deltas and thorn trees in arid areas.

Forest fauna includes the following species.

- Insect life such as beetles and ants abound in forests.
- The snow leopard lives in the Himalayas, while the leopard and tiger live in the forests in the rest of India. Wild sheep and goats live high up in the Himalayas and many of the birds of the Himalayan forests are different from those in the rest of India. Insects and amphibians are richest in evergreen forests. Wetlands are favoured habitats of fish and migrant waterfowl.
- The evergreen forests of the Western Ghats and Northeast India have the richest diversity of animal species.
- The semi-arid thorn forest areas have a rich diversity of reptiles as they tolerate extremes of temperature.

Forest flora includes the trees, shrubs, climbers, grasses and herbs in the forest. These are classified into species that flower (angiosperms) and others that are non-flowering (gymnosperms).

- Ferns, bryophytes and fungi do not have flowers and seeds. Cornifers and cycus are examples of gymnosperms.

The plant and animal species are closely dependent on each other and together they form different types of forest communities. Humans are a part of these forest ecosystems and the local people depend directly on the forest for several natural resources that provide their 'life-support systems'. People who do not live in the forest buy forest products such as wood and non-wood products – fruits, roots, nuts, gum, resin and medicinal plants – extracted from the forest by forager communities. Thus, they use forest produce indirectly.

Forest types in India

Forests in India can be broadly divided into coniferous forests and broad-leaved forests. They can also be classified according to the nature of their tree species, for example, evergreen, deciduous, xerophytes or thorn trees and mangroves. They are also described according to the most abundant species of trees, such as sal or teak forests. In many cases, a forest is named after the first three or four most abundant tree species.

Coniferous forest

- *Coniferous forests* grow in the Himalayan mountain region, where the temperatures are low. These forests have tall stately trees with needle-like leaves and downward-sloping branches, so that the snow can slip off the branches. They have cones instead of flowers and seeds, and are called gymnosperms.

- *Broad-leaved forests* are of several types, such as evergreen forests, deciduous forests, thorn forests and mangrove forests. Broad-leaved trees usually have large leaves of various shapes and are found in the middle and lower latitudes.
- *Evergreen forests* grow in the high rainfall areas of the Western Ghats, North-eastern India, and the Andaman and Nicobar Islands. These forests grow in areas where the monsoon lasts for several months. Some places even get two monsoons, as in some parts of South India. Evergreen plants shed a few of their

Broad-leaved forest

Evergreen forest

leaves throughout the year. There is no dry leafless phase as in the case of deciduous forests. An evergreen forest looks green throughout the year. The trees overlap each other to form a continuous canopy. Thus, very little light penetrates down to the forest floor. Only a few

Deciduous forest

shade-loving plants can grow in the ground layer in areas where some light filters down from the closed canopy. The forest is rich in epiphytes such as orchids, and ferns. The barks of the trees are covered in moss. The forest abounds in animal life and is very rich in insect life.

- *Deciduous forests* are found in regions that have a moderate amount of seasonal rainfall that lasts for only a few months. Most of the forests in which teak trees grow are of this type. The deciduous trees shed their leaves during the winter and hot summer months. In March or April, they regain their fresh leaves just before the monsoon, when they grow vigorously in response to the rains. Thus, there are periods of leaf-fall

and canopy regrowth. The forest frequently has thick undergrowth as light can penetrate easily onto the forest floor.

◆ *Thorn forests* are found in the semi-arid regions of India. The trees, which are sparsely distributed, are surrounded by open grassy areas. Thorny plants, called xerophytic species, are able to conserve water. Some of these trees have small leaves, while other species have thick, waxy leaves to reduce water loss during transpiration. Thorn-forest trees have long or fibrous roots to enable them to reach water at great depths. Many of these plants have thorns, which reduce water loss and protect them from herbivores.

Thorn forest

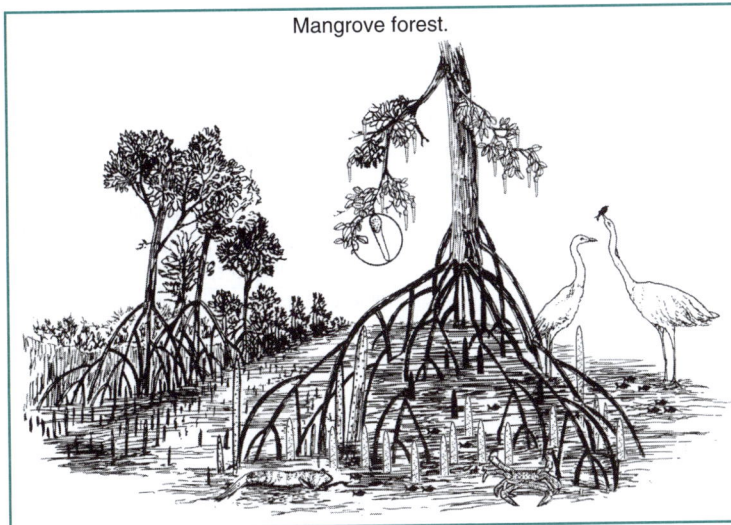

Mangrove forest.

◆ *Mangrove forests* grow along the coast, especially in the river deltas. These plants are uniquely adapted to be able to grow in a mix of saline and fresh water. They grow luxuriantly in muddy areas covered with silt that the rivers have brought down. They have breathing roots that emerge from the mud banks.

Total forest and tree cover, according to the 2017 state of forest report, is spread across 802,088 sq. km, which is 24.39% of the geographical area of the country. However, in the north-east region (NER), the current assessment shows an actual decrease of forest cover to the extent of 630 sq. km. Between 2015 and 2017, India has added 6,778 sq. km of forest cover and extended 1,243 sq. km of tree cover. The present report underlines an actual increase of 86.89 sq. km of forest cover in all the tribal districts of India.

Forest utilisation

Natural forests provide local people with a variety of products, if the forest is used carefully and sustainably by giving the forest sufficient time to regrow.

◆ The forest products collected by people include food like fruits, roots and herbs as well as medicinal plants (**Table 2.2**).
◆ People depend on fuel wood to cook food, collect fodder for domestic animals and use wooden building material for housing and farming.
◆ Medicinal plants have been known for generations and are used to treat several ailments.

◆ Non-Timber Forest Produce (NTFP) such as fibre and gum are used to make household goods. Articles such as brooms, cane furniture and bamboo goods that are made by tribal folk are sent to urban markets.

◆ Timber from different species of trees is put to a variety of uses. For example, soft wood is used for the yoke of a bullock cart while hardwood is used for its axle. Timber is a renewable source of building material for housing.

◆ Traditional types of agriculture use forest material like branches and leaves, which are burnt to form wood-ash which acts as a fertiliser for cereal crops such as rice.

These forest products are of great economic value as they are collected, sold and marketed. The communities which are heavily dependent on forest resources are known as *foragers*. They have many skills that farming, pastoral or urban communities do not know about. Most urban people know very little about many of these articles which they use—where they come from or who collects them from far away wilderness areas.

Table 2.2 Forest communities

Forest type	Plants (examples)	Common animals (examples)	Rare animals (examples)
Alpine	Ground flora	Snow leopard and wolf	Kiyang
Himalayan coniferous	Pine, deodar	Wild goats and sheep, Himalayan black bear	*Hangul*, Himalayan brown bear, musk deer, Himalayan wolf
Himalayan broad-leaved	Maple, oak, bamboo	*Hangul*, Dachigam tigers	Red panda
Evergreen North-east	Bamboo, alder, hollock tree, sal	Tiger, leopard, *sambar*	Pigmy hog, rhinocerous
Western Ghats	Jamun, ficus	Malabar whistling thrush	Lion-tailed macaque
Andaman and Nicobar	Dipterocarpus, orchids	Hornbill	Marine turtles, coral, invertebrates
Deciduous-dry	Teak, *ain*, terminalia, bamboo	Tiger, *chital*, barking deer, babblers, flycatchers, hornbills	Ratel, rusty spotted cat
Deciduous-moist	Sal, bamboo	*Sambar*, *chital*, tiger, leopard	Barasingha (Kanha)
Thorn and scrub, semi-arid forests	*Ber*, babul, neem	Blackbuck, *chinkara*, four-horned antelope, partridge, monitor lizard	Bustard, florican, Asiatic lion (Gir)
Mangrove delta forests	Avicenia	Crocodiles, shorebirds—sandpipers and plovers, fish, crustaceans	Water monitor lizard

Forest services include the control of the flow of water in streams and rivers. Forest cover reduces the surface run-off of rain water and allows ground water to be stored in underground aquifers. Forests also prevent the erosion of soil. Once soil is lost by erosion, it can take thousands of years for soil to re-form. Forests regulate the local temperature. It is cooler and more moist under the shade of the trees in the forest. Most importantly, forests absorb carbon dioxide and release oxygen that we require to breathe. All animals are dependent on oxygen. An important function is that forests sequester (collect and store) carbon that burning of fossil fuel releases into the atmosphere. Carbon dioxide released by burning fuel creates smoke in our atmosphere. The release of carbon dioxide causes the *greenhouse effect* that is leading to serious climate changes in the earth's atmosphere. This is mankind's greatest challenge today as it leads to ill effects such as rise of sea level, storms and cyclones, cloud bursts and changes in flora that affects all animals.

The wild relatives of crop plants and fruit trees have special characteristics that are used to develop new crops and newer varieties of fruit. These newer varieties give greater yields, and are more resistant to diseases. New industrial products are being produced from the wild plants of the forest. Many new medicines come from wild plants.

Direct uses of forest products
- Fruits: Mango, jamun, amla
- Roots: Dioscoria
- Medicine: Gloriosa, foxglove
- Fuel wood: Many species of trees and shrubs
- Small timber for building huts and houses
- Wood for farm implements
- Bamboo and cane for baskets
- Grass for grazing and stall-feeding livestock

Indirect uses of forest products
- Building material for construction and furniture for the urban sector
- Medicinal products collected and processed into ayurvedic drugs
- Gums and resins processed into a variety of marketable products
- Raw material for industrial products and chemicals
- Paper from bamboo and softwoods
- All herbal products for medicines, cosmetics that are manufactured and sold as being safe and non-toxic

What are the threats to the forest ecosystem?

We cannot use more resources than forests can produce during a growing season as they grow very slowly. The increasing use of wood for timber, wood-pulp for making paper and the extensive use of fuel wood, results in continual forest loss. If timber is felled beyond a certain limit, the forest cannot regenerate. The varied flora if substituted by monoculture plantations for timber or other products, impoverishes the local people as the economic benefit usually flows to people who use it extensively or market it by manufacturing products.

Developmental activities together with urbanisation, industrialisation and the increasing use of consumer goods, made from forest resources lead to the over-utilisation of the forest ecosystem altering food chains, the web of life and the food pyramid. Forests are rapidly shrinking as the need for agricultural land increases. It is estimated that India's forest cover has decreased from about 33% to 11% in the last century. Large parts of good forests are lost by mining and building dams.

What if forests are destroyed?

When forests are cut, tribal people who depend directly on them for food, fuel and other household uses find it very difficult to survive. Tribal communities, especially, do not get enough fuelwood and small timber for making houses, farm implements and biomass wood ash (shifting) cultivation. Urban people who depend on food from forests and agriculturists who in turn depend on neighbouring forest ecosystems find it difficult to get these resources as well. The forest resource base on which local people had traditionally survived for many generations, is rapidly being destroyed due to overexploitation and changes in landuse and its loss is felt most by the local forest dwelling communities who are deprived of their basic needs and livelihood.

The insects that live and breed in the forests, such as, bees, butterflies and moths, decrease in number once forests are degraded. As their population decreases, they are unable to effectively pollinate agricultural crops and fruit trees. This leads to a decline in agricultural and horticultural yield.

When the forest gets completely degraded, it impacts other forest functions. For example, the loss of forest cover leads to irreversible changes such as soil erosion, large scale run-off of surface water during monsoons which in turn leads to flash-floods. The rain that falls on deforested land flows directly into nearby rivers. Thus, rain water does not reach underground aquifers and there is a shortage of water once the monsoon is over as wells and tanks are dry. The exposed soil is rapidly washed away during the rains once the protective forest cover is removed. This seriously affects agriculture in such areas.

> **Observational learning**
>
> In deforested areas, during rain, the water in streams is brown in colour as the soil is being washed away. In contrast the water in forested streams is usually crystal clear.

Threats due to illegal extraction of wood from many forests results in severe forest degradation and even desertification. As the forest resources are exploited, the forest canopy is opened up, the ecosystem is altered, and its wildlife is seriously threatened. Wild animals lose their habitat, leading to the extinction of endangered species. As the forest is fragmented into small patches, the gaps in the forest change the habitat for animals, and the more sensitive species cannot survive under these changed conditions. Wild plant and animal species become extinct, and these can never be brought back.

How can forest ecosystems be conserved?

We can conserve forests only if we use its resources carefully. This can be done by leading *sustainable lifestyles*. Some examples include reducing, reusing and recycling goods made out of forest products. Reusing paper and packaging, switching to alternative sources of energy instead of fuel wood or thermal power (since this is produced from coal mined in forest areas) should be followed. There is a need to grow more trees to replace those that are cut down from forests every year for timber. Afforestation needs to be done continuously, from which fuel wood and timber can be judiciously used. An important aspect to remember is that a large and unsustainable overuse of Non-Timber Forest Products (NTFP), especially medicinal plants, is damaging forests and leading down the pathway to extinction. The natural forests with all their diverse species must be protected as national parks and wildlife sanctuaries to preserve the full range of plants and animals.

2.5.2 Grassland Ecosystems

Grasslands are included in a wide range of landscapes in which the vegetation is predominantly grass and small annual ground flora specifically adapted to India's different climatic conditions and soil characteristics. A majority of grasslands occur where forest cover has been degraded by fires and cattle grazing. Many healthy grasslands have been converted for other uses, such as tree plantation, urbanisation or industry.

What is a grassland ecosystem?

Grasslands cover areas where rainfall is usually low and/or soil depth and quality is poor. The low rainfall prevents the growth of a large number of trees and shrubs, but is sufficient to support the growth of grass cover during the monsoon. Many grasses and other small herbs become dry and the exposed portion of the grass above the ground dies in summer. In the next monsoon, the grass cover grows back from the root-stock and seeds of the previous year. This change gives grasslands a highly seasonal appearance, with periods of increased growth being followed by a dormant phase. There are tropical (Savannas), temperate (Pampas, Prairies, Veldts, Steppes) and polar grasslands in the world.

A variety of grasses, herbs and several species of insects, birds and mammals have evolved so that they are adapted to these wide-open grass-covered areas. These animals can live in conditions where food is plentiful after the rains, where they store this as fat for use during the dry period during which there is very little nutritious grass left for them to eat. In ancient times humans began to use these grasslands as pastures to graze domesticated livestock and thereby became pastoralists.

Many grassland ecosystems have been overgrazed and converted to wasteland as we have a very large number of cattle, sheep and goats. These animals overgraze and produce severe stress and food shortage for wild herbivorous animals.

Types of grasslands in india

Grasslands form a variety of ecosystems located in different climatic conditions, ranging from near-desert conditions to patches of Shola grasslands that occur on hill-slopes alongside the extremely moist evergreen forests in South India. In the Himalayas, there are the high altitude cold pastures. There are tracts of tall elephant grass in the low-lying terai belt south of the Himalayan foothills. There are extensive tracts of semi-arid grasslands in Western India, parts of Central India and in the Deccan plateau.

Semi-arid scrubland

Grassland structure

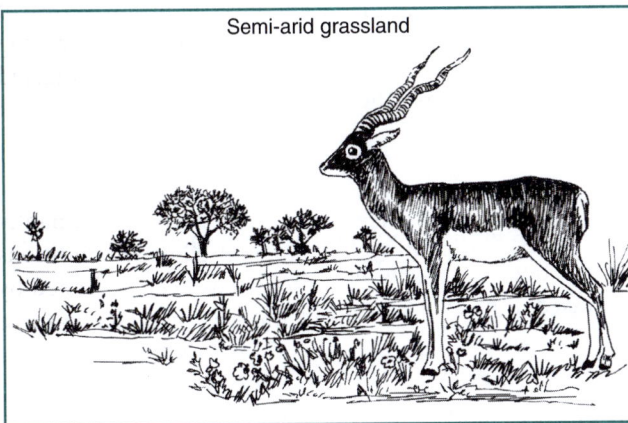
Semi-arid grassland

The physical features include both abiotic and biotic aspects. The Himalayan pasture belt extends up to the snowline; the grasslands at a lower level form patches along with coniferous or broad-leaved forests. Himalayan wildlife requires both forest and grassland ecosystems as vital parts of their habitat. The animals migrate upwards into the high-altitude grasslands in the summer and move down into the forest to browse on trees and shrubs in the winter when the snow covers the grasslands. These Himalayan pastures have a large variety of grasses and herbs. The Himalayan hill-slopes are covered with thousands of colourful flowering plants as well as a large number of medicinal plants, during the few months when there is no snow. The terai consists of patches of tall grasslands interspersed with sal forest ecosystems. The patches of tall elephant grass, which grow to a height of about five metres, are located in the low-lying water logged areas. The sal forest patches cover the elevated regions and the Himalayan foothills. The terai also includes marshes and wetlands in low-lying depressions. This ecosystem extends in a belt south of the Himalayan foothills.

The semi-arid plains of Western India, Central India and the Deccan are covered by grassland tracts with patches of thorn forest. Several mammals such as the wolf, blackbuck, chinkara, and birds such as bustards and floricans are adapted to these arid conditions. The scrublands of the Deccan plateau are covered with seasonal grasses and herbs on which its fauna are dependent. It teems with insect life on which the insectivorous birds feed.

Thorn forest

Shola grassland

The Shola grasslands consist of patches on hill-slopes that occur alongside the Shola forests on the Western Ghats, the Nilgiri and Anamalai ranges. These form a patchwork of grasslands on the hill slopes with forest habitats along the streams and low-lying areas.

Grasslands are not restricted to low-rainfall areas. Certain types of grasslands are formed when clearings are made in different forest types. Some grassy areas are located on the higher, steep hill-slopes with patches of forest that occur along the streams and in depressions. The grasslands that are subjected to repeated fires do not permit the forest to regrow.

Grassland composition

Grasses are the main producers of biomass in each of these regions. Each grassland ecosystem has a wide variety of species of grasses and herbs. Some grass and herb species are more sensitive to excessive grazing and their growth is suppressed if the area is over-grazed.

Grassland functions

Biogeochemical factors control the quality of grasses and fauna, grasslands sequester carbon above the ground in the grazing season and retain it in the roots during the non-grazing season. This is a vital system that addresses the effects of climate change.

Others are destroyed by repeated fires and cannot regenerate. These over-used or frequently burnt grasslands are degraded and are poor in plant species diversity. They are incorrectly labelled as wasteland as most of them are used by local people as their pastureland. Such areas should be used sustainably by rotating grazing cycles so that they have time to regenerate. They should not be planted with trees as this would destroy an important ecosystem.

Uses of grasslands

Grasslands are the grazing areas of many rural communities. Farmers keep cattle or goats for milk or dung. Shepherds have herds of sheep that migrate across grasslands; they are highly dependent on grasslands to supply food for their livestock. Fodder is collected and stored to feed cattle when there is no grass left for them to graze in summer. Grass is also used to thatch houses and farm-sheds. The thorny bushes and branches of the few trees that are seen in grasslands are used as the main source of fuelwood. Grasslands maintain unique biodiversity of specialised flora and fauna. They serve as a storehouse for carbon. Grassland Protected Areas provide recreational use and wildlife viewing.

Threats to grassland ecosystems

In many areas, grasslands have been used for centuries by pastoral communities. Over-utilisation and changes in the use of 'common grazing lands' of rural communities have led to their degradation. Grasslands have a limited ability to support domestic animals and wildlife. When animals over-graze, the grasses are converted into flat stubs with very little green matter. Degraded grasslands have fewer grass species as the nutritious species are entirely used up by the large number of domestic animals and are thus unable to regenerate. The large number of domestic animals reduces the 'naturalness' of the grassland ecosystem, leading to its deterioration. The grassland cover in the country, in terms of permanent pastures, is only 3.7% of the total land (Food and Agriculture Organisation, 2002).

In the Deccan, grasslands have been converted to irrigated farms and are now mainly used to grow sugarcane, which is a water-intensive crop. After continuous irrigation, such land becomes saline and useless in a few years, open to evaporation that brings salts to the surface.

More recently, many of these residual grassland tracts have been converted into industrial areas. This provides short-term economic gains but results in long-term economic and ecological loss. Other human activities such as fires also affect grasslands adversely. When fires are lit in the grasslands, the burnt grass gets a fresh flush of small green shoots which the domestic animals graze on. If this is done too frequently, the grasslands begin to deteriorate. Finally, the grasslands become bare; the soil gets compacted by trampling cattle, or is washed away during the monsoon by rain and whipped into dust storms during the hot dry summer. The land is degraded, as there is no grass to hold the soil in place. It becomes a wasteland and desertified.

Learning by critical thinking

Reflection on grassland ecology: degradation of grassland ecosystem is due to
- overgrazing,
- repeated fires,
- conversion to other type of land use,
- tree plantation, and
- thinking of them as wastelands.

Why is our grassland fauna vanishing?

Most people think that only our forests and their wildlife must be protected. However, other natural ecosystems, such as grasslands, are disappearing even more rapidly.

Many of the grassland species which were found 50–60 years ago have disappeared from several parts of India. The cheetah is extinct, the wolf is highly threatened, and the blackbuck and chinkara are being extensively poached for meat, birds such as the beautiful Great Indian Bustard and florican are on the brink of extinction. Unless grasslands and their species are protected, this wonderful productive ecosystem will disappear. Natural and undisturbed grasslands are left in very few locations. However, many of the grasslands have been formed as a result of human activities. Left to themselves, they would get back to scrubland or even dryland forests.

What if our grasslands disappear?

If our grasslands are destroyed, we will lose a highly specialised ecosystem to which plants and animals have adapted themselves over millions of years. In addition, the local people will not be able to support their livestock herds.

The extinction of a wild species of flora or fauna is a great loss to human kind. The genes of wild grasses are extremely useful for developing new crop varieties. New medicines could well be discovered from wild grassland plants. It is possible that genes from wild herbivores like wild sheep, goat and antelope may be used for developing new strains of domestic animals. All these possibilities will be lost along with the grasslands.

How can grassland ecosystems be conserved?

Grasslands should not be over-grazed. By rotation, certain areas should be closed for grazing after a few years. It is better to collect grass for stall-feeding cattle, than allowing free grazing of cattle. Repeated fires must be prevented and rapidly controlled. In hilly areas, soil and water management in each micro-catchment will help the grassland to return to a natural, highly-productive ecosystem.

To protect the most natural undisturbed grassland ecosystems, sanctuaries and national parks must be created. Their management should focus on preserving all their unique species of plants and animals. Grasslands should not be converted into plantations of trees. The open grassland is the special habitat of its locally adapted fauna. Planting trees in these areas reduces the natural features of this ecosystem, resulting in the destruction of this unique habitat for wildlife.

What should we do?

◆ There is a pressing need to preserve the few natural grasslands that still survive, by creating national parks and wildlife sanctuaries in the different types of grasslands. Public awareness and political will through your advocacy can bring this about.

◆ Animals such as the wolf, blackbuck and chinkara as well as birds such as the Great Indian Bustard and florican have now become rare. They must be carefully protected in the few national parks and wildlife sanctuaries that have natural grassland habitats.

◆ Grasslands for grazing cattle and sheep must be managed appropriately help bring about a better understanding about pastoral people.

◆ We need to create awareness among people that grasslands are of great value. If we are all concerned about our disappearing grasslands and their wonderful wildlife, the Government will be motivated to protect them.

◆ Keeping grasslands alive should be made a national priority.

2.5.3 Desert Ecosystems

Deserts and semi-arid lands are extremely specialised and sensitive ecosystems in arid areas that are easily destroyed by human activities. These are areas where the precipitation is far less than the evaporation rate. The plants and animals of these areas can live only in this harsh ecosystem as they have adapted over several generations. There are tropical, temperate and cold deserts in the world.

What is a desert or semi-arid ecosystem?

Deserts and semi-arid areas are mainly located in Western India and the Deccan plateau. The climate in these vast tracts is extremely dry. Cold deserts occur in Ladakh, and are located in the high plateaus of the Himalayas.

Desert

Desert structure

The most typical desert landscape in India is the Thar desert in Rajasthan. This has patches of sand dunes and semi-arid scrubland. It also has areas covered with sparse grasses and a few shrubs, which grow if and when it rains. In most areas of the Thar, rainfall is scanty and sporadic. In some areas, it may rain only once every few years.

Composition of desert ecosystems

This ecosystem has grasses tolerant to long periods of aridity. It contains vegetation of thorny shrub often referred to as scrubland. Its trees are xerophytic and sparsely distributed. In the adjoining semi-arid tract, the vegetation consists of specialised shrubs and thorny trees such as kher and babul. This forms the major food for wild herbivores and domestic animals.

The Great and Little Rann of Kutch are highly specialised arid ecosystems. In summer, they are similar to a desert landscape. However, as these are low-lying areas near the sea, they are converted to salt marshes during the monsoons. During this period, they attract an enormous number of aquatic birds such as ducks, geese, cranes and storks. The Great Rann is the only known breeding colony of the greater and lesser flamingos in our country. The Little Rann of Kutch is the only home of the wild ass in India.

Desert and semi-arid regions have a number of highly-specialised insects and reptiles. The rare animals include the Indian wolf, desert cat, desert fox and birds such as the Great Indian Bustard and florican. Some of the more common birds include the partridge, quail and sand-grouse.

Functions of desert ecosystems

The biogeochemical cycles are heavily influenced by the arid and seasonal nature of this ecosystem. Its flora and fauna are adapted to these functions and do not live in other ecosystems.

How are desert and semi-arid ecosystems used?

Areas of scanty vegetation with semi-arid scrubland have been used for camel, cattle and goat grazing in Rajasthan and Gujarat, and for sheep grazing in the Deccan plateau of Maharashtra and Karnataka. This is the land of traditional livestock owners who have adapted their lifestyles to this remarkable ecosystem.

Areas that have a little moisture, for example, along the water courses, have been used for growing crops such as jowar (sorghum) and bajra (millet). The natural grasses and local varieties of crops are adapted by evolution to grow at very low moisture levels. These can be used for genetic engineering and developing semi-arid land crops in the future.

What are the threats to desert ecosystems?

Several types of development strategies as well as human population growth have begun to affect the natural ecosystem of the desert and semi-arid lands. The conversion of these lands through extensive irrigation systems has changed several of the natural characteristics of this region. Canal water evaporates rapidly, bringing the salts to the surface. The region becomes highly unproductive as it has been turned into a saline tract. The over-extraction of ground water from tube wells lowers the water table, creating an even drier environment. Thus, human activities are destroying the unique features of this important ecosystem. Many of the special wild floral and faunal species that evolved here over millions of years may soon become extinct.

How can desert ecosystems be conserved?

Desert ecosystems are extremely sensitive. The ecological balance that forms a habitat for their endemic plants and animals is easily disturbed. Desert people have traditionally protected their meagre water resources. The Bishnoi community in Rajasthan is known to have protected their khejdi trees and the blackbuck for several generations based on the tenets of their guru. 200 years ago, when the ruler of this region ordered his army to cut down trees for his own use, several Bishnoi women were mercilessly killed while trying to protect their trees.

There is a pressing need to protect residual patches of this ecosystem within national parks and wildlife sanctuaries in desert and semi-arid areas. Canal systems in Rajasthan and Gujarat are destroying this important natural arid ecosystem by using the region for intensive agriculture. In Kutch, areas of the Little Rann, which is the only home of the wild ass, will be destroyed by the spread of industrial tracts.

Development projects alter the desert and arid landscape. There is a sharp reduction in the habitat available for its specialised species, bringing them to the verge of extinction. We need a sustainable form of development that takes the special needs of the desert into account.

2.5.4 Aquatic Ecosystems

The aquatic ecosystems comprise marine environments of the sea and freshwater systems in lakes, rivers, ponds and wetlands.

What is an aquatic ecosystem?

In aquatic ecosystems, plants and animals live mainly in water. Many plant and animal species are adapted to live in different types of aquatic habitats.

Fresh water aquatic ecosystem

Structure of an aquatic ecosystem

The special abiotic features are its physical aspects such as the quality of the water, which includes its clarity, salinity, oxygen content and rate of flow. Aquatic ecosystems may be classified as being still-water ecosystems (lentic) or flowing-water (lotic) ecosystems. The mud, gravel or rocks that form the bed of the aquatic ecosystem alter its characteristics and influence its plant and animal species composition.

Aquatic ecosystems are classified into freshwater, brackish and marine ecosystems, which are based on the salinity levels. The freshwater ecosystems that have running water are streams and rivers. Ponds, tanks and lakes are ecosystems where water does not flow. Wetlands are special ecosystems in which the water level fluctuates dramatically in different seasons. They have expanses of shallow water with aquatic vegetation, which forms an ideal habitat for fish, crustaceans and water birds. Marine ecosystems are highly saline, while brackish areas have lower salinity levels such as in river deltas and coastal areas. Coral reefs are very rich in species and are found in only a few shallow tropical seas. The richest coral reefs in India are around the Andaman and Nicobar Islands and in the Gulf of Kutch. Brackish water ecosystems in river deltas are covered by mangrove forests and are among the world's most productive ecosystems in terms of biomass production. The largest mangrove swamps are in the Sundarbans in the delta of the Ganges river.

Composition of an aquatic ecosystem

The vegetation of this ecosystem is related to the level of sunlight, the depth of water, its quality and seasonality. Thus, each type has its own flora and fauna.

Function of an aquatic ecosystem

Water is where life first evolved into complex food chains and pyramids through multiple chemical reactions. Each aquatic ecosystem has varied functional aspects in its biogeochemical nature.

(i) Pond Ecosystems

The pond is the simplest aquatic ecosystem to observe cyclic changes. There are differences between a temporary pond that has water only in the monsoon season, and a larger tank or lake that is an aquatic ecosystem throughout the year. Most small ponds become dry after the rains are over and are covered by terrestrial plants for the rest of the year.

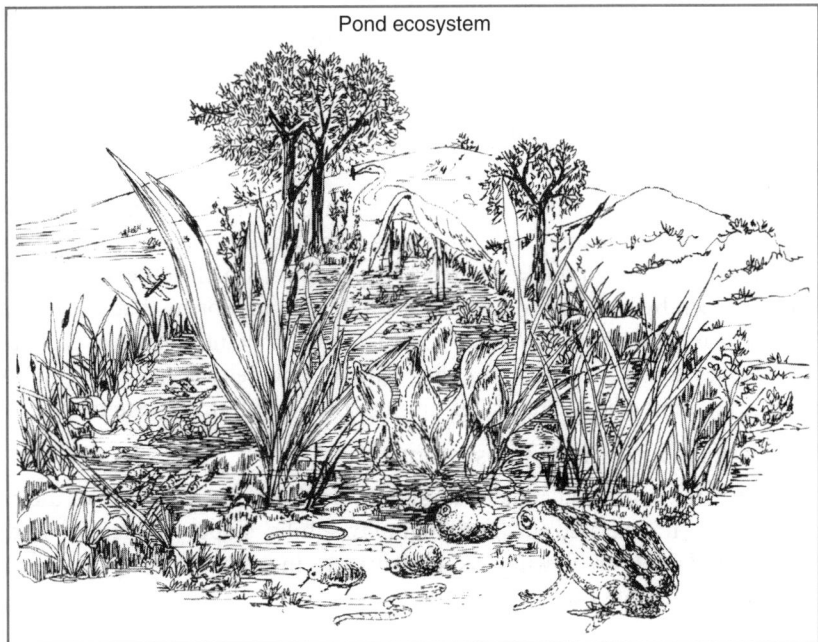

Pond ecosystem

When a pond begins to fill during the rains, its life forms, such as algae and microscopic animals, aquatic insects, snails and worms, emerge from the floor of the pond where they have remained dormant during the dry phase. Gradually, more complex animals such as crabs, frogs and fish return to the pond. The vegetation consists of floating weeds in the deeper part. Rooted vegetation spreads on the periphery. These plants are rooted in the muddy floor under the water of the expanding pond as the rain progresses. These are referred to as emergent vegetation and consist of reeds and aquatic grasses.

As the pond fills in the monsoon, a large number of food chains are formed. The algae (phytoplankton) are eaten by microscopic animals (zooplankton), which are in turn eaten by small fish, on which the larger carnivorous fish depend. These are in turn eaten by birds such as kingfishers, herons and birds of prey. Aquatic insects, worms and snails feed on the waste material excreted by animals in the water and on the dead or decaying plant and animal matter. They act on the detritus, which is broken down into nutrients which aquatic plants can absorb, thus completing the nutrient cycle in the pond.

Temporary ponds begin to dry up after the rains and the surrounding grasses and terrestrial plants spread into the moist mud that is exposed. Animals like frogs, snails and worms remain dormant in the mud, awaiting the next monsoon. This annual seasonal change from terrestrial to aquatic and back to terrestrial system is known as *succession*.

(ii) Lake Ecosystems

Lakes are freshwater ecosystems that may be natural or more frequently, artificially created by the construction of dams and tanks. Damming rivers alters a flowing water ecosystem to a still water ecosystem. This is usually developed for irrigation, or for water storage for urban or industrial use and hydroelectric power generation. There are several types of lakes—oligotrophic, dystrophic, eutrophic, endemic, volcanic, meromictic and artificial. Their biodiversity includes algae, which derives energy from the sun. This is transferred to microscopic animals which feed on the algae. Herbivorous fish depend on algae and aquatic weeds. Small animals such as snails are eaten by carnivorous fish, which in turn are preyed upon by larger carnivorous fish. Some specialised fish, such as catfish, feed on the detritus on the muddy bed of the lake; they are called *bottom feeders* or *bottom dwellers*. Fisherfolk depend heavily on this freshwater ecosystem, and farmers use it for water for their fields.

Sunlight penetrates the water surface of shallow parts of a lake ecosystem used by the aquatic plants. From the aquatic plants, energy is transferred to herbivorous animals and carnivores that live in water. These animals excrete waste products, which settle at the bottom of the lake and are broken down by small animals such as molluscs and worms that live in the mud in the lake bed. This acts as the nutrient material used by aquatic plants for their growth. During this process, plants use carbon from CO_2 for their growth and release oxygen. This oxygen is then used by aquatic animals, which filter water through their respiratory system.

(iii) Stream and River Ecosystems

The structure of flowing aquatic ecosystem depends on the nature of the area through which the water flows. It is related to the topography – hills, slopes or plains – and the amount of water. Streams and rivers are flowing water ecosystems in which all living organisms are specially adapted to different rates of flow.

Composition of ecosystem

Some plants and animals, such as snails and other burrowing animals, can withstand the rapid flow of hill-streams. Other species, like water beetles and skaters, can live only in slower moving water.

Some species of fish, like the Mahseer, move upstream from rivers to hill-streams for breeding. They need crystal clear water to be able to breed successfully. As perennial streams dry up due to deforestation and misuse, the Mahseer of the Himalaya and Western Ghats have become highly endangered. Many hill-stream fish are likely to become extinct unless protected.

The community of flora and fauna of streams and rivers depends on the clarity, flow and oxygen content as well as the nature of stream beds. The stream or river can have a sandy, rocky or muddy bed, each type having its own species of plants and animals. As deforestation occurs in the hills, the water in the streams that once flowed throughout the year becomes seasonal. This leads to flash floods during the rains and water shortage when the streams dry up after the monsoon.

Functions of the riverine systems

The biogeochemical cycles of the aquatic system is closely linked to that of the stream or river bank's terrestrial ecosystem. There is a constant two way energy flow as many fauna such as amphibians, molluscs and reptiles tend to use both the aquatic and terrestrial habitats.

(iv) Marine Ecosystems

The Indian Ocean, Arabian Sea and Bay of Bengal constitute the marine ecosystems around peninsular India.

Structure of marine ecosystems

In the coastal areas, the sea is shallow while further away it is deep. The deeper part of the ocean is divided into the euphotic (abundant light), bathyal (dim light) and abyssal (no light) zones. Each is a different ecosystem. The producers in this ecosystem vary from microscopic algae to large seaweeds. There are millions of zooplankton and a large variety of invertebrates. Coral covered shallow areas are exceptionally rich in species.

Composition of marine ecosystems

Fish, crustaceans, turtles and marine mammals live off the coast in our waters. The shallow areas near Kutch and around the Andaman and Nicobar Islands have some of the most incredible coral reefs in the world. Coral reefs are second only to tropical evergreen forests in their richness of species. Fish, crustaceans, starfish, jellyfish and polyps (that deposit the coral) are just a few of the thousands of species that form this incredible world under the shallow seas. The deforestation of the adjacent coastal mangroves leads to silt being carried out to sea where it is deposited on the coral, which is rapidly getting destroyed. Global warming bleaches the coral and leads to the loss of thousands of invertebrates species.

Functions of marine ecosystems

This is a vital and dynamic part of this ecosystem that covers a large part of the earth's surface. It is highly productive. Millions of algae, zooplanktons, fish and so on, live in it and are dependent on its biogeochemical cycles, tides and the weather. There are many types of coastal ecosystems, which are highly dependent on the tide.

The splash zone that is at the shore line is dependent on tidal conditions and changes at high and low tide in species composition. The marine ecosystem is used by coastal fisherfolk for their livelihood. In the past, fishing was done at a sustainable level and the marine ecosystem continued to maintain its abundant supply of resources for many generations. Now, with the growth of intensive fishing using giant nets and mechanised trawler boats, the fish catch in the Indian Ocean has dropped significantly.

(v) Coastal Ecosystem

Structure of the coast Beaches can be sandy, rocky, shell-covered or muddy. These are eco-sensitive environments and are easily destroyed. Thus, the government has put in place coastal zone regulations to preserve the ecosystem of the coast. On each of these different types of ecosystems, several specific species have evolved to occupy their own separate niches. There are different crustaceans, such as crabs, that make holes in the sand. Various shore birds feed on their prey by probing into the sand or mud. Several different species of fish are caught by fishermen that are food for coastal people. The mangroves in deltas have brackish water. They are highly sensitive and productive ecosystems in which fish and crustaceans breed.

2.5.5 Estuary Ecosystems

This is one of the most dynamic and productive of coastal ecosystems found on both our West and East coasts. These are aquatic brackish water ecosystems which act as hotspots of coastal biodiversity of great economic, social and biodiversity value. These are breeding areas for marine fish, crustacea and feeding areas for specialised estuarine avi-fauna. Thousands of birds migrate to our estuaries in the winter from across the Himalayas.

Types of estuarine ecosystems

All estuaries have a mix of fresh and saline sea water. Thus, they are connected to river systems and the sea through a connection at one end.

◆ In coastal plains, river deltas open into a wide aquatic ecosystem that in turn opens through an outlet into the sea. The sea water flows in and out of this outlet with each tide. Thus, it is extremely dynamic. In the monsoon, fresher water enters the estuary while in summer it is more saline. This is related to spawning of fish and crustacea.

◆ Most estuaries have a central deeper channel and smaller channels with shallow edges which are the feeding zones of aquatic wading birds. This is seen at the Thane estuary near Mumbai.

◆ Several of our estuaries are similar to a coastal lake with a connection to the sea. This is seen at the Chilika lake in Odisha.

◆ River deltas are often linked to estuarine systems as the tide flows into the delta from the sea from multiple areas. This is seen in the Sundarbans.

Environmental assets

The estuaries are a great source of income for local fisherfolk. The wealth of natural resources such as fish, crabs, shrimps, seagrass and algae, are all used as subsistence and productive resources. Millions of people depend on these for their food security and protein values. Estuaries have traditionally been used as ports for trading vessels to have a safe haven. Currently thermal power plants use them for sea water for cooling their plants. Many industries after treating their effluents release them in estuaries.

As climate change is altering sea levels, major changes will be seen in the sensitive ecosystem of estuaries. Changes in the flow of rivers due to dams and development projects also damage these fragile ecosystems that need to be conserved as vital protected areas as Sanctuaries, National Parks and Ecologically Sensitive Areas. Any development activity in the vicinity of estuaries must include a carefully carried out ecological assessment to study possible damage to these systems.

Thane creek: A large part of the Thane estuary is owned by the Godrej industries who have nurtured and maintained its ecology. They have planted mangroves in their part of the creek which has led to creating a place for fish and Crustacea to breed. It is a haven for thousands of birds. The mangrove now prevents siltation and acts as barrier against storms and cyclones. It is a major green lung for Mumbai.

How are aquatic ecosystems used?

Humans use aquatic ecosystems for clean fresh water. We need clean water to drink and for other domestic uses. Water is usually impounded by large dams to ensure a constant supply throughout the year. Agriculture and industry are highly dependent on these large quantities of water. Further, dams are built across rivers to generate electricity. A large proportion of this energy is used by urban people, by agriculturists in irrigated farmlands and in enormous quantities for industry.

Fisherfolk use the aquatic ecosystems to earn a livelihood. People catch fish and crabs, and collect a variety of edible aquatic and shoreline plants. These are used locally as food or sold in the market.

Marshes and wetlands are of great economic importance for people whose livelihood depends on their fish, crustaceans, reeds, grasses and other marketable natural resources. Sand and clay are extensively mined from aquatic ecosystems; this is very damaging. Most of the silt is used for urban and industrial development which damages the ecology of the aquatic ecosystem.

What are the threats to aquatic ecosystems?

◈ Water pollution occurs from urban sewage and poorly-managed solid waste. Sewage leads to a process called eutrophication, which destroys life in the water when the oxygen content is severely reduced. Fish and crustaceans cannot breathe and are at times killed in large numbers. A foul odour is produced. The natural flora and fauna of the aquatic ecosystem is damaged and eliminated.
◈ Pollution also occurs from waste water of irrigated agriculture that uses chemical fertilisers and industry that produces toxic chemicals.
◈ In rural areas, the excessive use of fertilisers causes an increase in nutrients, which also leads to eutrophication. Pesticides used in adjacent fields pollute the water and kill its aquatic animals.
◈ Chemical pollution from industry kills a large number of life forms in adjacent aquatic ecosystems. Contamination by heavy metals and other toxic chemicals affects the health of people who live near these areas and depend on this water.
◈ Thermal power plants release hot water into aquatic ecosystems that seriously affects its natural ecosystem.
◈ Over-fishing has led to a serious decline in the quantity of catch and has become a long-term loss of income for the fisherfolk.

Other than water quality, the quantity of water in fresh water ecosystems also poses a significant threat. Dams built across rivers greatly alter the flow of natural river ecosystems, causing rivers downstream to run dry. Changing the nature of an aquatic ecosystem from flowing water to a static one destroys natural biological diversity, causing habitat loss for the species that require running water. In some semi-arid areas that are artificially irrigated, the high levels of evaporation leads to severe salinisation as salts are brought up to the surface layers of the soil. Land is therefore eventually rendered unproductive. The social implications of large dams are a serious conflict between communities. All over the world, thousands of people have been displaced and lost their livelihoods because of the construction of large dams.

How can aquatic ecosystems be conserved?

For the sustainable use of an aquatic ecosystem, the first need is to prevent water pollution, as cleaning up or treating polluted water is an expensive reactive approach. Prevention of pollution through public awareness and generating a new ethic of avoiding activities that damage water quality and depletes its year round quantity is the key to a sustainable approach. Options for meeting water and energy needs must be explored. Proper planning and assessment must be carried out to protect affected people who are displaced by large dams to ensure a more equitable distribution of benefits from dams.

Aquatic ecosystems, especially wetlands, need protection by including them in sanctuaries or national parks in the same way in which we protect our natural forests. These sanctuaries in aquatic ecosystems protect a variety of forms of life as well as rare fish which are now highly endangered, such as the *Mahseer*. Wetland sanctuaries and national parks are of the greatest importance, as they are among the most threatened of our ecosystems.

2.6 DEGRADATION OF ECOSYSTEMS

Ecosystems on land and in water are the basis of life itself! Natural ecosystems in the wilderness provide a variety of products and are regions in which many of the vital ecological processes are present. Without the natural cycles, human civilisation would not be able to exist on earth. The natural ecosystems not only provide a wealth of resources but provide highly valued ecosystem services. They reduce pollution and provide carbon sequestration to counter the ill effects of climate change.

However, ecosystems are frequently disrupted by human actions, leading to the extinction of species of plants and animals that can live only in the different natural ecosystems. Some species, if eliminated, seriously affect the ecosystem (*keystone species*). Extinction occurs due to changes in land use. Forests are destroyed for timber or fuel wood. Wetlands are drained to create agricultural lands. Semi-arid grasslands which are used as pastures are converted to irrigated fields, or tree plantations. Both have disastrous consequences on the natural ecology of the semi-arid area. The waste water and chemicals from industries and the household waste from urban settings leads to pollution and poses a threat to the existence of several species of flora and fauna.

The reason for the depletion of natural resources is two-fold—our rapidly exploding population that needs increasing resources to sustain itself, and the growth of affluent societies that consume and waste a very large proportion of resources and energy. The increasing extraction of resources is at the cost of our natural ecosystems, leading to a derailing of their important functions. We all use a variety of resources in our daily lives. If traced back to their source, we find that the resources were originally obtained from nature and natural ecosystems. Our insensitivity in using resources has produced societies that nature can no longer sustain. If we think before wasting resources such as water, reuse and recycle paper, use less non-biodegradable material, we can cumulatively help conserve our natural resources.

SUMMARY _____

- Ecosystems represent the non-living components of the environment such as climate, soil, air and water, as well as the living community of flora and fauna in any area.

- The structural aspects of ecosystems include producers (plants which manufacture food), consumers (plants, animals and invertebrates that live on producers) and decomposers (worms, insects, bacteria and fungi which break down organic material into smaller particles and usable chemicals on which the producers depend).

■ Every ecosystem has several interrelated mechanisms that cycle energy and nutrients through the biosphere—the water cycle, carbon cycle, nitrogen cycle, phosphorus cycle, sulphur cycle and the energy cycle. Human activities play a significant role in altering these cycles.

■ Carnivores feed on herbivores which in turn feed on plants. At every stage of this food chain, energy is transferred from one living organism to another.

■ Terrestrial ecosystems in their natural state are found in different types of forests, grasslands, semi-arid areas, deserts and seacoasts. Over thousands of years, these ecosystems have been modified by humans. In recent times they have been converted into intensively irrigated agricultural ecosystems and urban industrial centres.

■ Aquatic ecosystems comprise marine environments of the seas and freshwater systems in lakes, rivers, ponds and wetlands.

■ Landuse changes, habitat loss, disruption of environmental cycles and population pressure are some of the driving forces of ecosystem degradation.

■ Ecosystems provide a range of goods and services essential to human life. Therefore, it is crucial that we protect and conserve our natural ecosystems and its resources.

QUESTIONS

1. Describe at least five forest ecosystem goods and services that humans benefit from.
2. What are the current threats to river ecosystems and how can these ecosystems be conserved?
3. Name the types of grasslands in India and two animal species found in these grasslands.
4. What are the main threats to aquatic ecosystems, and how can we protect them?
5. Describe the functions of the wetland ecosystem.
6. What are food chains in nature? Give one example.
7. What is ecological succession?
8. What is nutrient cycling and what is its role in nature?

Natural Resources: Renewable and Non-Renewable Resources

The earth provides enough to satisfy every man's need but not every man's greed. — *Mahatma Gandhi*

Learning Objectives

In this chapter you will learn,

◆ What natural resources are, their types
◆ What deforestation is, its causes and impacts
◆ About water, over-exploitation and its effects, conservation
◆ How the heating of earth occurs, circulation of air; air mass formation and precipitation
◆ About energy resources, alternate energy sources

Purpose

Natural resources provide us with all our needs and the services of nature. The food, the air we breathe, the water we need for our day-to-day activities, are all environmental goods that we use. Natural services clean up the air, recycle and clean water, recycle nutrients in soil and are the foundation of every ecosystem.

As human activities in the modern world (anthropocene) use more resources than nature can support, our environmental footprints on earth keep increasing and our future generations will be deprived of all these life support systems.

Our Role

Each of us, irrespective of what we do, can contribute our bit to the earth's environment by sustainably using all types of resources and preventing their overuse and misuse. As we are a part of all of the earth's systems, we can support all nature's processes that provide us with environmental services.

3.1 NATURAL RESOURCES

The environment provides us with a variety of goods and services necessary for our day-to-day life. These natural resources include air, water, soil and minerals (which are linked to the climate), and solar energy. These resources form the non-living or *abiotic* part of nature. The *biotic* or living parts consist of plants and animals, including fungi and microbes. Plants and animals can only survive as communities of different organisms, all of which are closely linked to each other in their own habitat. They require specific abiotic environmental conditions. Thus, forests, grasslands, deserts, mountains, rivers, lakes and the marine environments, all form habitats for specialised communities of plants and animals to live in.

The interactions between the abiotic aspects of nature and specific living organisms together form ecosystems of various types. Many of these living organisms are used as food resources by other organisms. Still others are linked to food less directly, such as bees, which act as pollinators, and species that are dispersal agents of the seeds of different plants, soil-fauna such as worms, insects and molluscs, cycle nutrients for plant growth. Fungi and termites break up dead plant material so that microorganisms can act on the detritus to replenish soil nutrients.

3.1.1   Renewable and Non-renewable Resources

Natural resources are classified as,

◆ *renewable resources,*
◆ *non-renewable resources*, and
◆ *perpetual resources.*

> ### Natural capital
>
> All resources from nature constitute our *natural capital,* which is not factored into routine economic thinking in business and industry. This is the cost of the river that gives water, the forest that provides resources and services such as cleaning up of the polluted air, also the cost of solar energy that drives all natural cycles. The cost of cleaning up pollution and the cost of waste management is far greater than prevention of pollution and minimising waste. Thus, natural capital is a new way of thinking cyclically and using the resources – air, water, soil, energy and minerals – judiciously. This gives a national price for nature's services—nutrient recycling, climate control, biodiversity, pollution minimisation that nature inherently provides.
>
> Neither the resources nor the services are clearly accounted for in most economic evaluations.

Renewable resources

A renewable natural resource if harvested in a sustainable manner, can be regenerated after use. Nature can replenish these resources (if used at a sustainable level, and) if given enough time to re-form. These include water, air, soil and nutrients in nature. Ecosystems act as resource producers and processors. Solar energy is the main driving force of ecological systems, providing energy for the growth of plants in forests, grasslands and aquatic ecosystems. A forest recycles its material by continuously returning its dead material, leaves, branches and animal waste to the soil. Grasslands recycle material much faster than forests, as the grass dries up after the rains every year. All the aquatic ecosystems also depend on solar energy and have cycles of growth when plant life spreads and aquatic animals breed. The sun drives the water cycle by evaporation and transpiration.

It is important to understand that although water and biologically living resources such as forests, grasslands and wetlands are considered renewable, they are, in fact, renewable only within a certain limit. If over-utilised and/or degraded beyond that limit, they lose their capacity to regenerate.

Air: The air we breathe is a renewable resource as it contains oxygen, nitrogen, carbon dioxide and other gases that nature can replenish through the oxygen cycle.

Water: Fresh water (even after being used) evaporates due to the sun's energy, forms water vapour and is re-formed in clouds, which fall to the earth as rain. The water cycle, through evaporation and precipitation, maintains hydrological systems which form rivers and lakes and supports a variety of aquatic ecosystems.

Forest resources: People who live in or near forests know the value of the forest resources first-hand, because their lives and livelihoods depend directly on these resources. However, the rest of us also derive great benefits from the forests, which we are rarely aware of. The water we use depends on the existence of forests on the watersheds around river valleys. Our homes, furniture and paper are made from wood from the forest. We use many medicines that are based on forest produce and we depend on plants for the oxygen they emit and to remove the carbon dioxide we breathe out, from the air. Resources from other ecosystems such as grasslands, wetlands, coasts, mountains and marine systems are of great value to humankind.

Non-renewable resources

◆ Non-renewables are resources in nature that have taken millions of years to form and are therefore irreplaceable after they are used up.
◆ These include soil, minerals and fossil fuels (such as coal, petroleum and natural gas) which have been formed millions of years ago and will be used up completely in the next few decades.
◆ They are often present in only a fixed amount on earth.
◆ Timber and other construction materials are extracted from the biosphere.
◆ All these are now being consumed at a faster rate than the environment's capacity to regenerate them. Thus, we need to change our energy sources from non-renewable fossil fuels to energy from the sun, wind and biofuel.

Perpetual resources

Perpetual resources are those resources that are everlasting. They are renewable resources that will not run out regardless of how much they are used. They are inexhaustible and do not require any form of renewal. Examples of these are solar, wind, tidal wave power, geothermal energy and energy from flowing water.

3.2 LAND RESOURCES AND LANDUSE CHANGE

Resource use over time

About ten thousand years ago when hunter–gatherers living in forests and grasslands gradually changed into agriculturalists and pastoralists, they began to alter the environment to suit their own requirements. As their ability to grow food and use domestic animals developed, the 'natural' ecosystems they lived in were converted into agricultural and grazing land. Most traditional agriculturists depended extensively on natural sources—soil for nutrients, rain, streams and rivers for water. These were livelihood resources. Later, they began to use wells to tap underground water sources and impounded water and created irrigation canals.

To develop more productive irrigated land by building dams and canals, they had to use newer technology. In modern times, these 'anthropocene' agriculturists began using fertilisers and pesticides to further boost the yield from the same amount of land. However, we now realise that all this has led to several undesirable changes in our environment. Humans have been overusing and depleting natural resources. The over-intensive use of land has exhausted the capability of the ecosystem to support the growing demands of more and more people, all requiring more intensive use of resources. Industrial growth, urbanisation, population growth and the enormous increase in the use of consumer goods have further stressed the environment, as they create great quantities of solid waste. Air, water and soil pollution have begun to seriously affect human health.

On the other hand, human civilisations have also attempted to protect natural resources. Traditional cultures have retained sacred groves and venerated hills, mountains, streams and rivers. In recent times in India, the government has notified as many as seven hundred and seventy protected areas in our national parks, sanctuaries, conservation reserves and community reserves. Many wild species of plants and animals and the genetic variations in traditional cultivars and indigenous livestock have been protected. However, this is still insufficient for long term safeguarding of environmental resources.

Anthropocene: The age in which humans began to dominate the earth and alter her nature and natural resources through the increased use of land and resources. It began in the industrial age and has continued into the present times. The use of nature's resources is beyond the carrying capacity of our earth. The pattern of economic development in the anthropocene is unsustainable.

Natural resources include water, air, minerals, oil and products used from terrestrial ecosystems forests, grasslands, and aquatic ecosystems such as rivers, lakes and the sea. Our resources also come from cultural ecosystems—agriculture, pastures, mining and fishing.

During the last 100 years, a better healthcare delivery system and improved nutrition has led to rapid population growth, especially in developing countries. This phenomenal rise in population has placed great demands on the earth's natural resources. Large stretches of land – forests, grasslands and wetlands – have been put to agricultural, urban and industrial use. These changes have brought about dramatic alterations in landuse patterns and the rapid disappearance of valuable natural ecosystems. The need for more water, more food, more energy and more consumers, is not only the result of a larger population, but also that of an increasingly consumerist society, that uses more products and wastes more resources.

Industrial development is aimed at meeting growing demands for all consumer goods. However, these consumer goods also generate waste in ever larger quantities. As shown in **Fig. 3.1**, the growth of industrial complexes has also led to a shift of people from their traditional, sustainable, rural way of life to urban centres that developed around industries. During the last few decades, several small urban centres have become large cities; some have even become giant megacities. This has increased the disparity between what the surrounding land can produce and what the large numbers of people in these areas of high population density consume. Urban centres cannot exist without resources such as water from rivers and lakes, food from agricultural areas, domestic animals from pasture land and timber, fuelwood, construction material and other resources from forests. Rural agricultural systems are in turn dependent on nature's forests, wetlands, grasslands, rivers and lakes. The resulting movement of natural resources from the wilderness ecosystems to the agricultural sector, and to the urban user, has led to a serious inequality in the distribution of resources among human communities, which is both unfair and unsustainable.

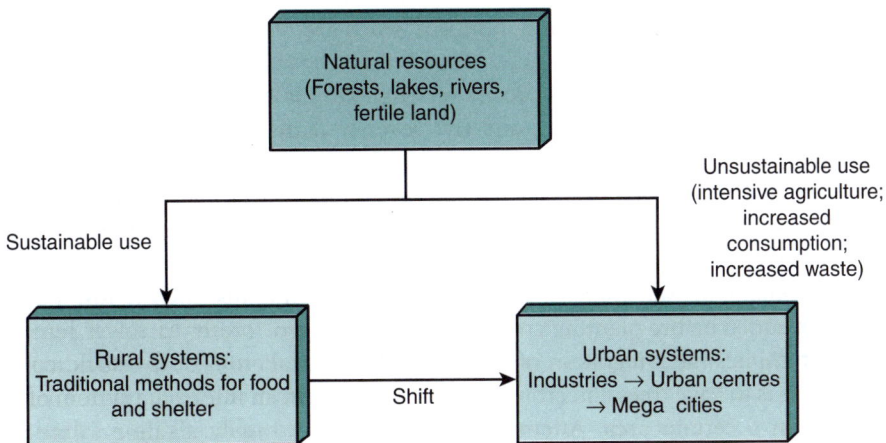

Fig. 3.1   Shift in pressure of natural resources from rural to urban systems

Modern landuse planning

Current methods of planning land use and land cover is done through satellite images and geoinformatics which is used to develop regional plans for urban, rural and wilderness ecosystems.

Planning land use

① Urban, ② Industrial, ③ Agricultural, ④ Waste management, ⑤ Discharge to aquatic ecosystem

Wasteland reclamation is implemented to counter the loss of soil and vegetation cover which leads to the loss of soil through erosion. This ultimately creates wastelands. This is one of the pressing problems of our country as loss of soil has already ruined large amounts of cultivable lands. If it remains unchecked, it will affect the remaining unused lands. Unless we safeguard our good lands, we may eventually face a serious shortage of food grains, vegetables, fruit, fodder and fuelwood. Hence, conservation of soil, protecting the existing cultivable lands and reclaiming the already depleted waste lands figure prominently among the priority tasks of planning for India's future. Unfortunately, some wasteland reclamation programmes have been unsuccessful as reclaimed lands have reverted to their original poor condition, due to mismanaged and unscientific methods of reclamation.

In choosing wasteland reclamation methods, attention must be paid to the cost factor. A proper study of environmental aspects and human impact which are responsible for the development of wastelands must also be made.

Wastelands can be classified into, (i) easily reclaimable, (ii) reclaimable with some difficulty, (iii) reclaimable with extreme difficulty, and (iv) severely damaged eco-sensitive lands that are irreversible for generations.

Easily reclaimable wastelands can be used for agriculture. Those that can be reclaimed with some difficulty can be utilised for agro-forestry. Waste lands that are reclaimed with extreme difficulty can be used for forestry or to recreate natural ecosystems. Wastelands can be reclaimed for agriculture by reducing the salt content in the soil by leaching and flushing. Gypsum, urea, potash and compost are added before planting crops in such areas. Agro-forestry involves putting land to multiple uses. It implies the integration of trees with agricultural crops or livestock management. The main purpose is to have trees and crops interspersed to form an integrated system of biological production within a certain area. Attempts to grow trees in highly alkaline saline soils have been largely unsuccessful. Tree seedlings planted with a mixture of original good soil, gypsum and manure, leads to better growth. It is preferable to use indigenous species of trees so that the programme recreates the local ecosystem with its own natural mix of constituent plant species.

Need for wasteland development

Wasteland development provides a source of income for the rural poor. It ensures a constant supply of fuel, fodder and timber for local use. It makes the soil fertile by preventing soil erosion and conserving moisture. The programme helps to maintain an ecological balance in the area. The regrowth of forest cover maintains the local climatic conditions. The regenerated vegetation helps in attracting birds which feed on pests in the surrounding fields and function as natural pest controllers. The trees help in holding moisture and reducing surface run-off thus controlling soil erosion. Trees and vegetation recover and sequestrate carbon from the atmosphere thus countering the effects of climate change.

The problems of wastelands need to be identified at the micro level. For this, district, village and plot-level surveys of the wasteland are necessary. A profile of the maps indicating the detailed distribution and information on the wasteland is essential. With the help of local government institutions such as the village panchayats, along with block development officers and revenue department functionaries, a plan based on the community's needs must be produced. This must be a participatory exercise, involving all the different stakeholders in the community. A think-tank of administrators, ecologists and local NGOs must also be involved in the process. Next, the factors that are responsible for the formation of wastelands should be identified. Based on these factors, the wasteland is classified into marginally, partially or severely deteriorated lands. The steps to be followed include the following.

- Locale-specific strategies for reclaiming the wasteland must be planned. Government officials along with the local NGOs must assist farmers by demonstrating improved methods of cultivation, arranging for loans for the small, marginal and landless farmers and the people from the weaker sections of society.
- Involving the local women will prove to be of great value.
- Publicity campaigns integrated with training farmers and frontline government and forest department staff on the various aspects of wasteland utilisation should be organised.
- Environmental scientists can help by suggesting necessary changes in cropping patterns particularly for drought-prone areas.
- Selection of appropriate crops for fodder and trees that provide local people with NTFPs according to the nature of the wasteland.
- Soil must be tested in laboratories to provide guidance to the farmers on the proper land management techniques to be employed.
- Newer technological advances in irrigation and other expertise for improving productivity without creating unsustainable patterns of development should be provided to the local people, especially the weaker sections and landless farmers. Guidelines to control waterlogging must be provided.
- Collective efforts have to be made to check soil losses through water and wind erosion, and to prevent the collapse of the irrigation system through siltation.

It is important that plans concerning wasteland reclamation and utilisation prepared at various stages must be properly integrated for a successful long-term outcome. The demands of the increasing human population for environmental goods and services have imposed severe pressures on the available land resources, especially on the forests and green cover. This is closely linked to the wellbeing of the rural population that constitutes a large percentage of the people who depend on local natural resources for their survival. Thus, the development of agroforestry-based agriculture and forestry has become the prime prerequisite for the overall development of economy in the country. The pressure on the land is already very high and the only hope of increasing productivity lies in bringing about appropriate improvement in the various categories of wasteland spread over the country.

What can you do?

◆ Plant more trees of local or indigenous species around your home and workplace and encourage your friends to do so. Plants are vital for our survival on earth.

◆ If your urban garden is too small for trees, plant local shrubs and creepers instead. These support bird and insect life that form a vital component of the food chains in nature. Urban biodiversity conservation is feasible and can support a limited but valuable diversity of life.

◆ If you live in an apartment, grow a terrace or balcony garden using potted plants. Window boxes can be used to grow small flowering plants, which also add to the beauty of your house.

◆ Whenever and wherever possible, prevent trees from being cut. If it is not possible for you to prevent this, report it immediately to the concerned authorities. Old trees are especially important for the ecosystem.

◆ Insist on keeping our hills free of settlements or similar encroachments. The degradation of hill-slopes leads to severe environmental problems.

◆ When shopping, choose products with limited packaging. It will not only help cut down on the amount of waste in landfills, but also helps reduce our need to cut trees for paper and packaging.

◆ Look for ways to reduce the use of paper. Use both sides of every sheet of paper and send your waste paper for recycling.

◆ Buy recycled paper products for your home; for example, sheets of paper and envelopes.

◆ Reuse cartons and gift-wrapping paper. Recycle newspaper and waste paper instead of throwing it away as garbage.

◆ Participate in the events that highlight the need for creating sanctuaries and national parks, nature trails, open spaces and for saving forests.

◆ Support Project Tiger, Project Elephant and other such conservation projects, and join NGOs that deal with environmental protection and nature conservation.

Don't

◆ Try not to present flower bouquets, instead give a potted plant and encourage your friends to do so.

◆ Don't unnecessarily collect unnecessary pamphlets and leaflets just because they are free.

3.2.1 Land Degradation

The extreme pressure on agriculture and pasture land eventually degrades the nutritive value of the soil. The vegetation begins to suffer from a lack of nutrients which seriously impacts human life.

Land as a resource: Landforms such as hills, valleys, plains, river basins and wetlands include different resource-generating areas that the people living there depend on. Many traditional farming societies had ways of preserving areas from which they used resources. For example, in the 'sacred groves' of the Western Ghats, requests to the spirit of the grove for permission to cut a tree or extract a resource were accompanied by simple rituals. The request could not be repeated for a specified period.

If land is utilised carefully, it can be considered a renewable resource. The roots of trees and grasses bind the soil. If forests are depleted or grasslands overgrazed, the land becomes unproductive and wasteland is created. Intensive irrigation leads to waterlogged and salinised soil, on which crops cannot grow. When highly toxic industrial and nuclear waste is dumped without careful management, the damage may be irreversible for many generations. Land thus gets converted into a non-renewable resource.

Land on the earth is as finite as any other non-renewable natural resource. While we humans have learnt to adapt our lifestyle to various ecosystems, we cannot live comfortably for instance, on polar ice caps, under the deep sea, or in space in the foreseeable future. We need land for building homes, cultivating food, maintaining pastures for domestic animals, developing industries to provide goods, and supporting the industries by creating towns and cities. Equally importantly, humans need to protect wilderness areas in forests, grasslands, wetlands, mountains and coasts to protect the economically and culturally valuable biodiversity.

Thus, the rational use of land demands careful planning. One can develop most of these different types of landuse categories almost anywhere, but protected areas (national parks and wildlife sanctuaries) can only be situated where some of the natural ecosystems are still undisturbed. These PAs are important aspects of good landuse planning.

Farmland is under threat due to more and more intense utilisation. Every year, around 5–7 mha of land worldwide is added to the existing degraded farmland. When soil is used more intensively for farming, it gets eroded even more rapidly by wind and rain. Over-irrigating farmland leads to salinisation as the evaporation of water brings the salts to the surface of the soil preventing the growth of crops. Over-irrigation also creates water logging of the topsoil, so that the roots of crops are affected and the yield deteriorates. The use of unnecessarily large quantities of chemical fertilisers poisons the soil and eventually the land becomes unproductive.

As urban centres grow and industrial expansion occurs, the agricultural land and forests shrink. This is a serious loss and will have unfavourable long-term effects on human civilisations.

3.2.2 Soil Erosion

The characteristics of natural ecosystems, such as forests and grasslands, depend on the type of soil. Different types of soil support different varieties of crops. The misuse of an ecosystem leads to the loss of valuable soil through erosion by the monsoon rains and, to a smaller extent, by wind. The roots of the trees in the forest hold the soil. Deforestation thus leads to rapid soil erosion. Soil is washed away into streams, transported into rivers and finally lost to the sea. The process is more evident in areas where deforestation has led to erosion on steep hill slopes as in the Himalayas and in the Western Ghats. These are called *ecologically sensitive areas* or ESAs. To prevent the loss of millions of tons of valuable soil every year, it is essential to preserve what remains of our natural forest cover. It is equally important to afforest denuded areas using local species of plants. The link between the existence of forests and the presence of soil is greater than the forest's physical soil binding function alone. The soil is enriched by the leaf-litter of the forest. This detritus is broken down by soil microorganisms, fungi, worms and insects, which help to recycle nutrients in the system. Further loss of our soil wealth will impoverish the country and reduce its capacity to grow enough food in the future.

Land as a resource is now under serious pressure due to an increasing 'land hunger'—to produce sufficient quantities of food for an exploding human population. It is also affected by degradation due to misuse. Land and water resources are also polluted by industrial waste and rural and urban sewage, apart from being diverted for short-term economic gains to agriculture and industry. Natural wetlands of great value are being drained for agriculture and other purposes and semi-arid land is being irrigated and overused.

The most damaging change in land use is demonstrated by the rapidity with which forests have vanished in recent times, both in India and in the rest of the world. In the long term, the loss of forests and the services they provide is far greater than the short-term gains obtained by converting forested lands to other uses.

3.2.3 Desertification

Severe soil erosion leads to desertification especially in semi-arid areas. The chemical fertilisers and pesticides that have been used, along with irrigation to boost agricultural productivity, finally result in a completely unproductive ecosystem. Excessive irrigation leads to salt accumulation in the top soil which prevents crops in the desertified land.

3.3 DEFORESTATION

Forests once extended over large tracts of our country. With the spread of agriculture, only patches of forests were left which were controlled mostly by tribal people. They hunted animals and gathered plants and lived entirely on forest resources. Deforestation became a major concern during the British rule when a large amount of timber was extracted for building ships. This led the British to develop scientific forestry in India. They, however, alienated local people by creating *reserved* and *protected forests*, which curtailed access to the resources. This led to a loss of stake in the conservation of the forests and their gradual degradation and

Table 3.1 Forest and tree cover of India

Class	Area (km²)	Geographical area (%)
Forest cover		
Very dense forest	99,278	3.02
Moderately dense forest	3,08,472	9.39
Open forest	3,04,499	9.26
Total forest cover (includes 4662 km² under mangroves)	**7,12,249**	**21.67**
Scrub	46,297	1.41
Non-forest	25,28,923	76.92
Total geographical area	**32,87,469**	**100.0**

Source: India state of forest report 2019

fragmentation across the country. The data in **Table 3.1** is represented by a pie chart for easier comprehension (**Fig. 3.2**).

Another period of over-utilisation and forest degradation occurred soon after India gained independence, as state sponsored agricultural expansion became a priority. The following years saw India's residual forest wealth dwindle sharply.

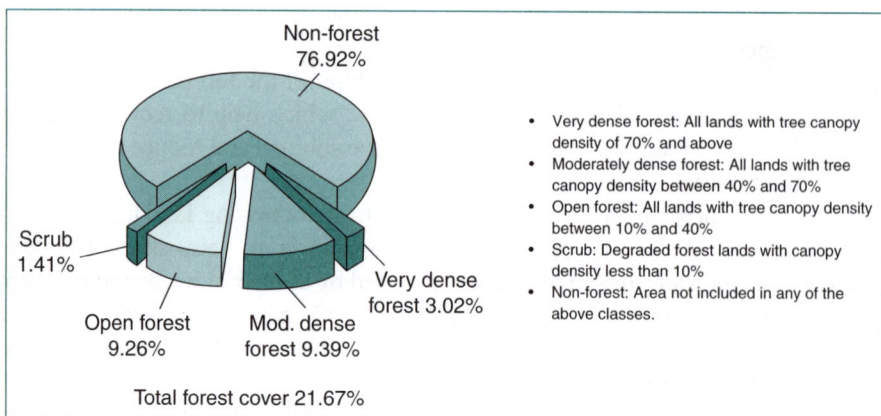

- Very dense forest: All lands with tree canopy density of 70% and above
- Moderately dense forest: All lands with tree canopy density between 40% and 70%
- Open forest: All lands with tree canopy density between 10% and 40%
- Scrub: Degraded forest lands with canopy density less than 10%
- Non-forest: Area not included in any of the above classes.

Fig. 3.2 Forest cover in India

Threats to forest resources: Scientists estimate that India should ideally have 33% of its land under forests. Today, we only have about 12%. Thus, we need to not only protect our existing forests but also to increase our forest cover.

Deforestation and degradation: Deforestation involves a loss in the afforested area. Degradation, on the other hand, refers to the condition of a forest involving a reduction in its quality. One or more of the components such as soil, vegetation and/or fauna of a forest are affected, thereby impacting the overall functioning of a forest ecosystem. Those civilisations that looked after forests by using forest resources cautiously have prospered, whereas those that destroyed forests have gradually become impoverished. Today, expansion of agricultural land, rapid industrialisation, urbanisation, illegal logging and mining are serious causes of loss and degradation of forests in our country and all over the world. Further, dams built for hydroelectric power or irrigation have submerged large tracts and have displaced tribal people whose lives are closely knit to the forest. Realising this, the Ministry of Environment and Forest (MoEF) formulated the National Forest Policy of 1988 to give added importance to Joint Forest Management (JFM), which co-opts local village communities and the Forest Department to work together to sustainably manage our forests. Another resolution in 1990 provided a formal structure for community participation through the formation of Village Forest Committees (VFCs). Based on these experiences, new JFM guidelines were issued in 2000. This stipulates that at least 25% of the income from the area must go to the community. From the initiation of the programme until 2002, there were 63,618 JFM committees managing over 140,953 sq. km of forest in 27 states in India.

Various states have tried a variety of approaches to JFM. The share of profits for the VFCs ranges from 25% in Kerala to 50% in Gujarat, Maharashtra, Orissa and Tripura and 100% in Andhra Pradesh. In many states, 25% of the revenue is used for village development. In many states, non-timber forest products (NTFPs) are available to the people free of cost. Furthermore, some states have stopped grazing while others have rotational grazing schemes that have helped in forest regeneration.

> **Case Study 3.1   Joint Forest Management (JFM)**
>
> The need to include local communities in forest management has become a growing concern. Local people will only support greening of an area if they can see some economic benefit from conservation. An informal arrangement between local communities and the Forest Department began in 1972, in the Midnapore District of West Bengal. JFM has now evolved into a formal agreement which identifies and respects the local community's rights and benefits that they need from forest resources. Under JFM schemes, Forest Protection Committees (FPCs) from local community members are formed. They participate in restoring the green cover and protect the area from being over exploited.
>
> While the causes of deforestation and forest degradation such as timber extraction, mining and dams are invariably some of the needs of a developing country, there needs to be a strategy that looks at long-term ecological gains that cannot be sacrificed for short-term economic gains.

3.3.1   Causes and Impacts of Deforestation

Land use change, agriculture, urbanisation and industrialisation are the primary causes. Fuelwood, timber extraction, grazing and repeated fires also lead to forest degradation.

◆ **Mining:** The land on which mining for coal, iron ore, bauxite and other minerals is carried out is invariably under forest tracts in West Bengal, Bihar, Orissa and the Western Ghats. Mining concessions lead to extensive deforestation. Mining operations are considered one of the

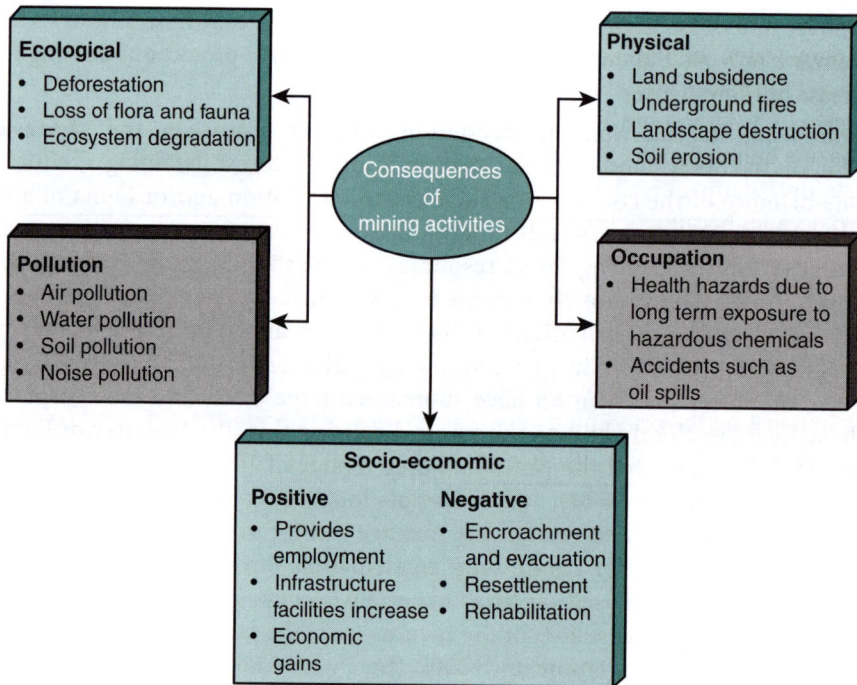

Fig. 3.3   Consequences of mining

main sources of environmental degradation. The extraction of all these products from the lithosphere has a variety of side effects. The depletion of available land due to mining, waste from industries, conversion of land to industry and pollution of land, water and air by industrial waste are the environmental side effects of the use of these non-renewable resources. There is global public awareness of this problem and government actions to stem the damage to the natural environment have led to numerous international agreements and laws directed toward the prevention of activities and events that may adversely affect the environment (**Fig. 3.3**).

❖ **Dam building on environment:** River catchments of all our major rivers and catchment slopes are often covered by thick forests as they have been inaccessible. These forests get submerged when dams are built leading to serious loss of forest cover. Managing a river system is best done by leaving its course undisturbed. Dams and canals lead to major floods during the monsoon.

Problems caused by dams:

❖ The fragmentation and physical transformation of rivers leads to ecological disasters.
❖ Social consequences of large dams due to the displacement of people.
❖ Serious impacts on riverine ecosystems its flora and fauna.
❖ Water-logging and salinisation of the surrounding lands.
❖ Dislodging animal populations, damaging their habitat and cutting off their migratory routes.
❖ Disruption of fishing and water way traffic.
❖ The emission of greenhouse gases from reservoirs due to rotting vegetation and carbon inflows from the catchment is a recently identified impact.

Large dams have had a serious impact on the lives, livelihoods, cultures and spiritual existence of indigenous and tribal people. What they have suffered from the negative impact of dams has often been overshadowed by sharing the benefits from the dam water. In India, of the 16–18 million people displaced by dams, 40%–50% are tribal people, who account for only 8% of our nation's one billion people.

Conflicts over dams have heightened over the last two decades because of their social and environmental impact and failure to achieve targets for adhering to costs as well as achieving promised benefits. Recent examples show how the failure to provide a transparent process that

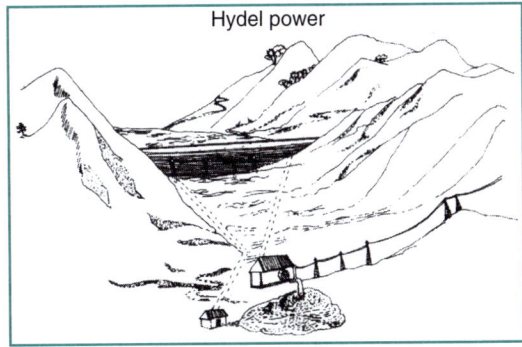
Hydel power

includes the effective participation of local people has prevented these people from playing an active role in debating the pros and cons of the project and its alternatives. The loss of traditional and local controls over equitable distribution remains a major source of conflict.

CASE STUDY 3.2   Sardar Sarovar project

This project is a giant multi-purpose dam project built on the Narmada river valley. It has impacted the lives and households of millions of people living in the river valley. The farmers downstream will get water for their irrigated agriculture at the cost of the tribal folk, tribal farmers along the river banks and fishermen at the estuary, who have lost their homes and their means of livelihood. The question is, why should the local tribals be rendered homeless and displaced, and be relocated to benefit other people down river who are reasonably well-off farmers? Why should the less fortunate be made to bear the costs of development for wealthier farmers? It is a question of social and economic equity as well as the enormous environmental impacts, including loss of the biological diversity of the inundated forests in the Narmada valley.

◈ **Biodiversity:** Development projects not only destroy forest habitats but disrupt the biodiversity of forests and their wildlife. This leads to a fall in the population of several faunal and floral species.
◈ **Tribal populations:** Tribal folk cultures are closely linked to their forest homes as they are foragers who are completely dependent on forest resources. Changes in land use patterns and deforestation disrupt their traditional way of life and their livelihood.

Habitat preservation: what you can do

The rapid destruction of forests and the growth of human habitations and activities have reduced the natural habitats of animals and birds. Loss of habitat is one of the major pressures on several species and has led to the extinction of several rare and endemic species. Many others are seriously threatened. We thus have the responsibility of preserving the remaining habitats and their inhabitants. The following are some 'dos and don'ts' that can help preserve threatened ecosystems.

Do

◈ Visit forests responsibly. Remember to bring out everything you take in and clean up any litter left by others. Stay on marked trails, and respect the fact that wildlife needs peace and quiet. Study the ecosystem; it gives one a greater sense of responsibility to conserve it.
◈ Be kind to animals. Stop friends from disturbing or being cruel to wild creatures such as birds, frogs, snakes, lizards and insects. They all perform vital functions for a healthy ecosystem.
◈ Learn about birds and identify the birds that are common in your area. Understand their food requirements and feeding habits.

◈ Construct and fit artificial nesting boxes for birds. This will encourage birds to stay in your neighbourhood, even if their nesting habitat is scarce.

◈ Birds need water to drink and to keep their feathers clean. You can make a birdbath. Having birds around your home, school or college can even help increase species diversity in the area.

◈ Attract wildlife like small mammals such as squirrels, to your garden by providing running or dripping water. Make a hole in the bottom of a bucket and poke a string through to serve as a wick. Hang a bucket on a tree branch above your birdbath to fill it gradually with water throughout the day.

◈ Protect wildlife, especially birds that are insectivorous and live in your neighbourhood by eliminating the use of chemicals in your garden. Instead, use vermicompost and introduce natural pest predators.

◈ Do your gardening and landscaping using local plants to control the pests in your garden.

Don't

◈ Never bother, tease, hurt or throw stones at animals in a Protected Area and stop others from doing so. Do not feed or befriend the birds and animals. If you see an injured animal contact the forest officials.

◈ Do not kill small animals and insects like dragonflies and spiders, as they act as biological pest-control mechanisms.

◈ Do not use any wildlife products.

◈ Never bring home animals or plants collected in the wild. You could be seriously harming wild populations and the natural ecosystems where they were collected.

◈ Never buy products made from ivory; it is illegal. Elephants are killed for their tusks, which are used to make a variety of ivory products.

◈ Don't use any wild animal or plant products that are collected from the wild which have dubious medicinal properties. You may be endangering a species and even damaging your own health.

Soil conservation: what you can do

Soil degradation affects us all in some way, either directly or indirectly. There are many ways that each of us can help in solving environmental problems due to the loss of soil. Given below are some dos and don'ts for conserving soil.

Do

◈ Cover the soil in your farm or garden with a layer of mulch to prevent soil erosion in the rains and to conserve soil moisture. Mulch can be made from grass-clippings or leaf-litter.

◈ If you plan to plant on a steep slope in your farm or garden, prevent soil erosion by first terracing the area. Terraces help in slowing the rain water running downhill so it can soak into the soil rather than carry the soil away.

◈ Help prevent soil erosion in your community by planting trees and ground-covering plants that help hold the soil in place. You might organise a group of citizens to identify places that need planting, raise funds, work with the local government to plant trees, shrubs and grasses and maintain them over the long term.

◈ Add organic matter to enrich your garden soil; for example, compost from kitchen scraps and manure from poultry and cows are good sources of nutrients. Make sure the manure is not too fresh and that you do not use too much. Healthy soil grows healthy plants, and it reduces the need for chemical fertilisers, insecticides and herbicides.

◈ In your vegetable garden, rotate crops to prevent the depletion of nutrients. Legumes like peas and beans put nitrogen back into the soil.

◆ Set up a compost pit in your college or garden, so that you can enrich your soil with the organic waste from the kitchen and cut down on the amount of waste it sends to a landfill. Set up buckets in your college or lunchroom where fruit and left-over food can be put. Empty the buckets daily into a compost pit and use the rich compost formed in a few weeks to enrich the soil around the college.

◆ Encourage your local zoo, farms and other organisations or people that house a large number of domestic livestock to provide your community with bio-fertiliser made from animal manure. This can be composted to make a rich fertiliser and it forms an additional source of income for the animal owners.

◆ Buy organically-grown produce to help reduce the amount of toxic pesticides used in farms that harm soil organisms. Look for organically-grown produce in your grocery shop or try growing some yourself if you have the space.

Don't

◆ Remove the cut grass. Leave it on the lawn or put the clippings in your potted plants. Cuttings serve as moisture-retention mulch and a natural fertiliser.

◆ Burn leaf litter as it can be carcinogenic. Instead, use leaf litter in composting or as mulch—it is a valuable soil conditioner.

◆ Use toxic pesticides in your garden—they often kill the beneficial organisms your soil needs to stay healthy.

◆ Over-water garden plants, water them only when necessary.

◆ Pollute sources of water or water bodies by throwing waste into them. This is the water you or someone else has to drink!

◆ Throw waste into toilets because finally it goes into water bodies.

3.4 WATER

Water on earth is found in marine, brackish and fresh water ecosystems. Fresh water is found as surface water (rivers, lakes) and as ground water. Wetlands are the intermediate forms between terrestrial and aquatic ecosystems and contain species of plants and animals that are highly moisture dependent. All aquatic ecosystems are used by a large number of people for their daily needs such as drinking water, washing, cooking, watering animals and irrigating fields. However, the world depends on a limited quantity of freshwater. Water covers 70% of the earth's surface, but only 3% of this is fresh water. Of this, 2% is present as polar ice caps and only 1% is usable water in rivers, lakes and subsoil aquifers. Only a fraction of this can be actually used. At a global level, 70% of the water is used for agriculture, about 22% for industry and only 8% for domestic purposes. However, this varies in different countries, and industrialised countries use a greater percentage for industry. India uses 87% of its water for agriculture, 8% for industry and 5% for domestic purposes (National Commission for Water Resources Development Plan, Ministry of Water Resources, 1999).

One of the greatest challenges facing the world this century is the overall management of water resources. They can be overused or wasted to such an extent that they will run dry locally. Water sources can also become so heavily polluted by sewage and toxic substances that it becomes impossible to use the water.

The world population has passed the 6 billion mark. Based on the proportion of young people in developing countries, this will continue to increase significantly during the next few decades. This places enormous demands on the world's limited supply of freshwater. The total annual withdrawals of fresh water today are estimated at an amount twice as much as just 50 years ago (World Commission on Dams, 2000). Studies indicate that a person needs a minimum of 20–40 litres of water per day for drinking and sanitation. However, more than one billion people worldwide have no access to clean water, and to many more, water supplies are unreliable. Local water conflicts are already spreading to states; for example, Karnataka and Tamil Nadu are fighting over the waters of the Cauvery, and Karnataka and Andhra Pradesh over the Krishna.

India is expected to face critical levels of water stress by 2025. At the global level, 31 countries are already short of water and by 2025 there will be 48 countries facing serious water shortages. The UN has estimated that by the year 2050, 4 billion people will be seriously affected by water shortages. This will lead to multiple conflicts between countries over the sharing of water. Around 20 major cities in India face chronic or interrupted water shortages. There are 100 countries that share the waters of 13 large rivers and lakes. The upstream countries could starve the downstream nations, leading to political instability across the world. Examples are Ethiopia, which is upstream on the Nile, and Egypt, which is downstream and highly dependent on the Nile. International accords that will look at a fair distribution of water in such areas will become critical to world peace. India and Bangladesh already have a negotiated agreement on the use of the Ganges.

3.4.1 Use and Over-exploitation of Surface and Ground Water

With the growth in human population, there is an increasing need for larger amounts of water to fulfil everyone's needs. Today in many areas, this requirement cannot be met. The mismanagement of water resources has meant an inequitable distribution of water—some use more water than they need while others do not have access to clean water at all.

Agriculture also pollutes surface water and underground water stores by the excessive use of chemical fertilisers and pesticides. Methods, such as the use of biomass as fertiliser and non-toxic pesticides such as neem products, and using integrated pest management systems, help in reducing the agricultural pollution of surface and ground water. There are many ways in which farmers can use less water without reducing yield, such as, by using the drip irrigation system.

Industries also tend to overlook their environmental impact while maximising their short-term economic gains. Industrial liquid waste is often released into streams, rivers and the sea without treatment. Public awareness may increasingly put pressure on industry to produce products in a manner which minimises the impact on the environment. In the longer term, as people become more conscious of using 'green products', eco-sensitive industries will gain a greater competitive market compared to industries that continue to pollute. As people begin to learn about the serious health hazards caused by pesticides in their food, public awareness can begin putting pressure on farmers to reduce the use of chemicals that are injurious to health.

Global climate change: Changes in climate at a global level, caused by increasing air pollution, have now begun to affect our climate. In some regions, global warming and the El Niño winds have created unprecedented storms. In other areas, they lead to long droughts. Everywhere, the greenhouse effect due to atmospheric pollution is leading to increasingly erratic and unpredictable climatic effects. This has seriously affected regional hydrological conditions.

3.4.2 Floods

Floods have been a serious environmental hazard for centuries. However, the havoc caused by rivers overflowing their banks has become progressively more damaging, as people have deforested catchments and intensified the use of river flood-plains that once acted as safety valves. The wetlands in flood-plains are nature's flood control systems into which swelling rivers could spill over and acted like a temporary sponge holding the water and preventing fast-flowing water from damaging the surrounding land.

Deforestation in the Himalayas has caused floods that kill people, damage crops and destroy homes along the Ganges and Brahmaputra and their tributaries. Rivers change their course during floods and tons of valuable soil is lost to the sea. As the forests are degraded, rain water no longer percolates slowly into the subsoil but runs off down the mountain side, bearing large amounts of topsoil. This soil blocks or silts up the rivers temporarily, but eventually gives way as the pressure mounts, allowing enormous quantities of water to suddenly wash down into the plains below. There, the rivers swell, burst their banks and flood waters spread to engulf peoples' farms and homes.

3.4.3   Droughts

In most arid regions of the world, the rains are very unpredictable. This leads to periods when there is a serious scarcity of water to drink, use in farms or provide for urban and industrial use. Drought-prone areas are thus faced with irregular periods of famine. Agriculturists have no income during such years and as they have no steady income, they have a constant fear of droughts. India has 'Drought-prone Areas Development Programmes', which are used in such areas to buffer the effects of drought. Under these schemes, during years of water scarcity people are given wages to build roads, minor irrigation works and plantation programmes.

Drought has been a major problem in our country, especially in arid and semi-arid regions. It is an unpredictable climatic condition and occurs due to the failure of one or more monsoons. It varies in frequency in different parts of the country. While it is not feasible to prevent the monsoon from failing, good environmental management can reduce the ill effects of water scarcity. The lack of water during drought affects all sectors of society including those dependent on agriculture and industry. It leads to serious food shortage and malnutrition especially among children. Farmers resort to suicide in bad years.

Several measures can be taken to minimise the serious impacts of a drought. This must be done as a preventive measure so that, if the monsoon fails, its impact on local peoples' lives is reduced. In years when the monsoon is adequate, we often use up the supply of water without trying to conserve it or use it judiciously. Thus, in a year when the rains are poor, there is no water even for drinking in the drought-affected areas.

A major factor that aggravates the effect of drought is deforestation. Once the hill-slopes are denuded of forest cover, the rainwater rushes down the rivers and is lost. Forest cover permits water to be held in the area and gradually seep into the ground. This charges the underground stores of water in natural aquifers. This can be used during drought if the aquifers have been filled during a good monsoon. If water from the underground stores is overused, the water table drops and vegetation suffers. Soil and water management, and afforestation are long-term measures that reduce the impact of drought.

Water for agriculture and power generation: India's increasing demand for water, for intensive irrigated agriculture, for generating electricity and for consumption in urban and industrial centres, has been met by creating large dams.

Although dams ensure a year-round supply of water for domestic use and provide water for agriculture, industry and hydropower generation, these development projects are also accompanied by several serious environmental problems. They alter river flows, change nature's flood-control mechanisms such as wetlands and flood-plains, destroy the lives of local people, and the habitats of wild plant and animal species. Intensive irrigation to support water-hungry cash crops like sugarcane produces an unequal distribution of water. Large landholders on the canals get the lion's share of water, while smaller farmers get less and are adversely affected.

'Save water' campaigns are essential to make people everywhere aware of the dangers of water scarcity. A number of measures need to be taken for better management of the world's water resources. These include measures such as:

◆ building several small reservoirs instead of a few megaprojects,
◆ developing small catchment dams and protecting wetlands,
◆ soil management, micro catchment development and afforestation, which enables recharging of underground aquifers, thus reducing the need for large dams,
◆ treating and recycling municipal waste water for agricultural use,
◆ preventing leakages from dams and canals,
◆ preventing loss in municipal pipes,
◆ effective rainwater harvesting and recharging of groundwater in urban environments,
◆ water conservation measures in agriculture, such as using drip irrigation,
◆ pricing water at its real value, which makes people use it more responsibly and efficiently and reduces wastage, and
◆ in deforested areas where land has been degraded, soil management, by making bunds along the hill-slopes and making nalla plugs, can help retain moisture and make it possible to re-vegetate degraded areas.

Sustainable water management is a crucial step towards the world's looming water crisis.

Let's do it!

How much water does one person need a day? Several international agencies and experts have proposed that 50 litres per person per day covers the basic human water requirements for drinking, sanitation, bathing and food preparation. Estimate your average daily consumption by guesswork. Then quantify it across 24 hours by measuring how much you used in a day. Was your estimate done initially correct or incorrect? If so why?

3.4.4 Water Conservation, Rainwater Harvesting, Water Management

Water conservation: Conserving water has become a prime environmental concern. Clean potable water is becoming increasingly scarce. With deforestation, surface run-off increases and the sub-soil water table drops as water has no time to seep slowly into the ground. As many areas depend on wells, it has become necessary to dig deeper and deeper wells. This adds to the cost and further depletes underground stores of water. The water table could take years to recharge even if the present rate of extraction is reduced, which seems hardly possible in most situations.

When we waste water, we do not realise that it affects all of us in so many different ways. Water has to be equitably and fairly distributed so that – household, agriculture and industry – all get an equitable share of the water. The overuse and misuse of water due to various activities and the resulting pollution has led to a serious shortage of potable drinking water. Thus, water conservation is linked closely with overall human wellbeing.

Traditional systems of collecting water and using it optimally have been practiced in India for many generations. These have been forgotten in the recent past. Conserving water in multiple small percolation tanks and *jheels* was an important feature of traditional forms of agriculture. Villages all over the country had one or more common *talabs* or tanks from which people collected or used water carefully.

As carrying water to their homes over long distances was time-consuming and laborious, water was not wasted. Many homes had a kitchen garden that was watered by the waste water. Conservation of water was done in traditional homes through a conscious effort.

During the British period, many dams were built to supply water to growing urban areas. After independence, India's policy on water changed towards building large dams for expanding agriculture to support the green revolution. While this reduced the need to import food material and mitigated food shortages in India, the country began to see the effects of serious water shortages and problems related to its distribution. The newer forms of irrigated crops, such as sugarcane and other water-hungry cash crops, required enormous quantities of water. After a while, such irrigated areas become waterlogged and unproductive. Thus, the ill effects of the poorly-conceived management of water at the national and local level have made it mandatory to consider a new water policy for the country.

Saving water in agriculture: Drip irrigation supplies water to plants near their roots through a system of tubes, thus saving water. Small percolation tanks and rainwater harvesting can provide water for agriculture and domestic use. Rainwater collected from rooftops can be stored or used to effectively recharge sub-soil aquifers.

Saving water in the urban setting: Urban people waste large amounts of water. Leaking taps and pipes are a major source of water loss. Canals and pipes carrying water from dams to the consumer contribute nearly 50% to water loss during transfer. Implementing water distribution infrastructure that reduces such distribution losses as well as reducing the demand for water by saving it, is more appropriate than trying to meet growing demands.

Rainwater harvesting

As we face serious water shortages, every drop of water we can use efficiently is of great value. One method is to manage rainwater in such a way that it is used at source. If as much water as possible is collected and stored, this can be used after the rainy season. In many parts of the world, especially in very dry areas, this has been the traditional practice. However, the stored water has to be kept pollution-free and clean so that it can be used as drinking water. Stored water can grow algae and zooplankton (microscopic animals) which can be pathogenic and cause infections. Thus, keeping the water uncontaminated is important.

Current technologies of rainwater harvesting require that all roof and terrace water passes down into a covered tank (below or above ground) where it can be stored for use after the monsoon (**Fig. 3.4**). This practice is most advantageous in arid areas where clean water is very scarce. However, there are practical difficulties such as the expense of constructing large storage tanks.

Another way of using rooftop rainwater harvesting is to collect the rain water so that it percolates into the ground to recharge wells instead of flowing over the ground into rivers. Thus, by recharging the groundwater, the water table rises and the surrounding wells retain water throughout the year.

Watershed management

Rivers originate in streams that flow down mountains and hill-slopes. A group of small streams flow down hillsides to meet larger streams in the valley, which form the tributaries of major rivers. The management of a single unit of land with its water drainage system is called watershed management. As this makes clean water available throughout the year, the health in the community also improves. Watershed management enhances the growth of agricultural crops and even makes it possible to grow more than one crop in a year in dry areas.

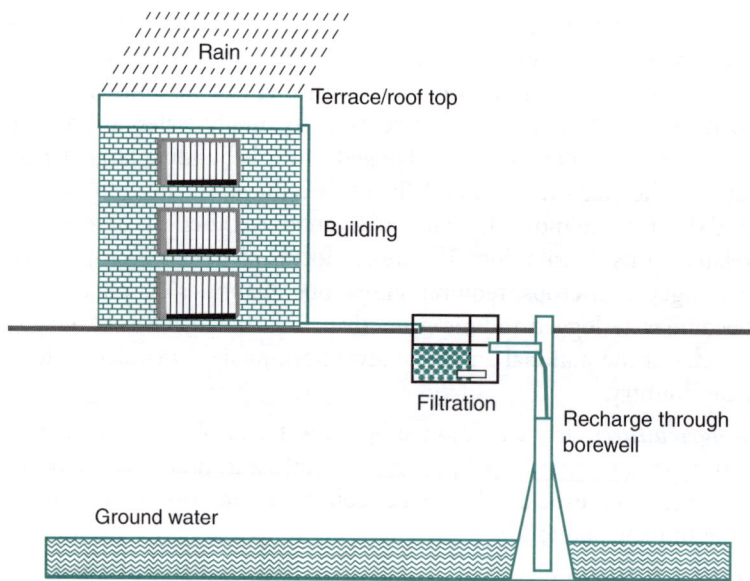

Fig. 3.4   Rainwater harvesting: Recharge of groundwater through borewell

Watershed management begins by taking control of a degraded site through local participation. People must appreciate the need to improve the availability of water both in quantity and quality for their own use. Once the people are adequately sensitised, the communities begin to understand the project and they begin to work together to promote good watershed management.

Technical steps to take appropriate soil conservation measures for a sound watershed management setup are given below.

- Construct a series of long trenches and mounds along the contours of hills to hold the rainwater which allow it to percolate into the ground. This ensures that underground stores of water are fully recharged. This can be enhanced by growing grasses and shrubs, and planting trees (mainly local species) which hold the soil and prevent it from being washed away in the monsoon.

(*Note:* Local grass cover can only increase if free grazing of domestic animals is regulated or replaced by stall feeding. Afforestation should use only local species of trees and shrubs.)

- Make *nalla* plugs in the streams, so that the water is held in the stream and does not rush down the hillside.
- In selected sites, several small check dams should be built, which together can hold back larger amounts of water.

3.4.5 Conflicts Over Water—International and Interstate

International water conflict in the Middle East: Three countries – Ethiopia, Sudan and Egypt – use most of the water that flows in Africa's river Nile, with Egypt being the last in line along the river. To meet the water needs of its rapidly growing population, Ethiopia plans to divert more water from the Nile. Sudan also wants to divert more water. Such upstream diversions would reduce the amount of water available to Egypt. The country cannot exist without irrigation from the Nile. Egypt could go to war with Sudan and Ethiopia if there are more water cuts. Syria plans to build dams and withdraw more water from the Jordan river decreasing the downstream water supply to Jordan and Israel. Israel has given a warning that it may destroy the largest dam that Syria plans to build.

Emerging water shortages in many parts of the world is the most serious environmental problem the world is facing this century. Water is the heart of any country and all economic activities depend on it.

CASE STUDY 3.3   Krishna river water dispute

This dispute is linked with four states—Karnataka, Andhra Pradesh, Telangana and Maharashtra. The main basis of the dispute is associated with the Almati dam constructed on river Krishna in the Bijapur district of Karnataka. It was built in 1968. Karnataka wishes to raise the height of the dam which would adversely affect the other states.

CASE STUDY 3.4   International river water dispute between India and Bangladesh on river Ganga

The Ganga water dispute is connected with the Kolkata port. The flow of water of the Ganga from the steep slopes of Himalayas is very rapid within the boundaries of Uttar Pradesh and Bihar. Other rivers also flow into the Ganga. After its entry into Bangladesh, the flow of water slows and its delta divides into two rivers—the Padma and the Ganga. In 1974, the Farakka barrage was constructed on the river to make water available for the Kolkata port, so that ships can sail up to the dock even during the dry season and the dock remains free of silt.

Bangladesh had objections to this as India diverts most of the water during the dry season towards the Kolkata port which results in insufficient quantity of water in the Meghna and Padma rivers and the gates of Farrakka barrage are opened during the rainy season which results in floods in Bangladesh.

Conserving water: what you can do

Most of India has a good average annual rainfall; however, we still face water shortages nearly everywhere. This is one of the major environmental problems in our country. Conservation of this precious natural resource is very important and is the need of the hour. It should start with every individual. It must start with you! The following are some of the things you can do to conserve this precious natural resource.

Do

◆ Reduce the amount of water used for daily activities; for example, turn off the tap while brushing your teeth to save water.

◆ Use a bucket to wash your clothes instead of using running water; use a mop to clean your bathroom floors instead of pouring buckets of water to do the same.

◆ Reuse the rinsing water for house plants. Collect and reuse the water that vegetables are washed in to water the plants in your garden or your potted plants.

- Always water plants in your garden early in the morning to minimise evaporation.
- Soak the dishes before washing them to reduce water and detergent usage.
- Look for leaks in the toilet and bathroom, to save several litres of water a day.
- While watering plants, water only as rapidly as the soil can absorb the water.
- When you need to drink water, take only as much as you need to avoid wastage.
- Saving precious rainwater is very important. Harvest rainwater from rooftops and use it sustainably to recharge wells to reduce the burden on rivers and lakes.
- Replace chemicals like phenyl, strong detergents, shampoo, chemical pesticides and fertilisers used in your home, with environment-friendly alternatives, such as neem and bio-fertilisers. Groundwater contamination by household chemicals is a growing concern.
- For the *Ganesh festival*, bring home a clay idol instead of a plaster of Paris idol and donate it instead of immersing it in the river to reduce river pollution.

Don't

- Forget to turn off your tap or maintain a slow flow.
- Don't have long showers; minimise them to 4 minutes. A 10-minute shower wastes many litres of water in comparison to using water from a bucket.

3.5 ENERGY RESOURCES

3.5.1 Renewable and Non-renewable Energy Sources

Energy is defined by physicists as the capacity to do work. Energy is found on our planet in a variety of forms, some of which are immediately useful, while others require a process of transformation.

Types of energy: There are three main types of energy: those classified as non-renewable; those that are said to be renewable; and nuclear energy, which uses such small quantities of raw material (uranium) that supplies are, to all effect, limitless. However, this classification is inaccurate because several of the renewable sources, if not used 'sustainably', can be depleted more quickly than they can be renewed.

Non-renewable energy sources

To produce electricity from conventional non-renewable resources, the material must first be ignited. The fuel is placed in a secured area and set on fire. The heat thus generated turns water to steam, which moves through pipes, to turn the blades of a turbine. This converts it into electricity, which we use in various appliances.

Mineral resources: A mineral is a naturally occurring substance of definite chemical composition and identifiable physical properties. An ore is a mineral or combination of minerals from which a useful substance, such as a metal, can be extracted and used to manufacture useful products.

Minerals and their ores need to be extracted from the earth's interior so that they can be used; this process is known as mining. Mining operations generally progress through four stages.

- *Prospecting:* Searching for minerals.
- *Exploration:* Assessing the size, shape, location and economic value of the deposit.
- *Development:* The work of preparing access to the deposit so that the minerals can be extracted from it.
- *Exploitation:* Extracting the minerals from the mines.
- *Restoration:* Returning mined-out lands to as natural a land cover as possible.

Mines are of two types—surface (open castorstrip mines) or deep (or shaft) mines. Coal, metals and non-metal ferrous minerals are all mined differently depending on the above criteria. The method chosen for mining will ultimately depend on how the maximum yield may be obtained under existing conditions at the minimum cost, with the least danger to the mining personnel.

Most minerals need to be processed before they become usable. Thus, 'technology' is dependent on both the presence of resources and the energy necessary to make them 'usable'.

Mine safety: Mining is a hazardous occupation, and the safety of mine workers is an important environmental consideration. Surface mining is less hazardous than underground mining, and metal mining is less hazardous than coal mining. In all underground mines, rock- and roof-falls, flooding and inadequate ventilation are the greatest hazards. Large explosions have occurred in coal mines, killing many miners. More miners have suffered from disasters due to the use of explosives in metal mines.

Environmental problems: Mining operations are considered one of the main sources of environmental degradation. The extraction of all these products from the lithosphere has a variety of side effects. The depletion of available land due to mining, waste from industries, conversion of land to industry and pollution of land, water and air by industrial waste are the environmental side effects of the use of these non-renewable resources. There is global public awareness of this problem and government actions to stem the damage to the natural environment have led to numerous international agreements and laws directed toward the prevention of activities and events that may adversely affect the environment.

> ### Learning through discussion
>
> A coal mining company is to commence operations in a small farming town in Maharashtra. Form two discussion groups in your class: Group A is the mining company that will list out the positive impacts of having this operation in town. Group B represents the concerned public that will list out the negative impacts of having this operation. Discuss and come up with a decision/solution that is sustainable.

Impacts of non-renewable energy sources: These consist of the mineral-based hydrocarbon fuels – coal, oil and natural gas – that were formed from ancient prehistoric forests. These are called *fossil fuels*, because they are formed after plant life is fossilised. At the present rate of extraction, there is enough coal for a long time to come. Oil and gas resources, however, are likely to be used up within the next 50 years. When these fuels are burnt, they produce waste products that are released into the atmosphere as gases such as carbon dioxide, oxides of sulfur, nitrogen and carbon monoxide. All these gases are the causes of air pollution. This leads to respiratory tract problems in an enormous number of people all over the world. They have also affected historic monuments like the Taj Mahal and destroyed forests and lakes due to acid rain. Many of these gases contribute to the greenhouse effect, letting sunlight in and trapping the heat inside the atmosphere. This leads to global warming (a rise in global temperature), an increase in droughts in some areas and floods in other regions. The melting of polar icecaps leads to a rise in sea level, which is slowly submerging coastal belts all over the world. The warming of the seas also leads to the death of sensitive organisms such as coral.

Oil and its impact on the environment: India's oil reserves, which are being used at present, lie off the coast of Mumbai and in Assam. Most of our natural gas is linked to oil and, because there is no distribution system, it is just burnt off. This means that nearly 40% of the available gas is wasted. The processes of oil and natural gas drilling, processing, transport and utilisation have serious environmental consequences, such as leaks in which air and water are polluted and accidental fires that may go on burning for days or weeks before they can be put out. While refining oil, solid waste like salts and grease are produced, which also damage the environment. Oil slicks are caused at sea from offshore oil wells, cleaning of oil tankers and ship wrecks. The most well-known disaster occurred when the huge oil-carrier, the *Exxon Valdez* sank in 1989 and birds, sea otters, seals, fish and other marine life along the coast of Alaska was seriously affected.

Oil-powered vehicles emit carbon dioxide, sulfur dioxide, nitrous oxide, carbon monoxide and particulate matter that are a major cause of air pollution, especially in cities with heavy traffic density. Leaded petrol leads to neurological damage and reduces attention span. Petrol vehicles can be run with unleaded fuel by attaching catalytic converters to all cars, however unleaded fuel contains benzene and butadiene which are known to be carcinogenic compounds.

Delhi, which used to have a serious smog problem due to traffic, has been able to reduce this health hazard by changing a large number of its vehicles to CNG, which contains methane.

This high dependence on dwindling fossil fuel resources, especially oil, results in political tension, instability and a potential cause of future wars between nations. At present, 65% of the world's oil reserves are located in the Middle East.

Coal and its impact on the environment: Coal is the world's largest single contributor of greenhouse gases and is one of the most important causes of global warming. Many coal-based power generation plants are not fitted with devices such as electrostatic precipitators to reduce emissions of suspended particulate matter (SPM), which is a major air polluter. Burning coal also produces oxides of sulfur and nitrogen which, combined with water vapour, lead to *acid rain*. This destroys forest vegetation, damages architectural heritage sites, pollutes water and affects human health. Thermal power stations that use coal produce waste in the form of fly-ash. Large dumps are required to dispose of this waste material; some efforts have been made to use it for making bricks. The transport of large quantities of fly-ash and its eventual dumping are costs that have to be included in calculating the cost benefits of thermal power.

Impacts of renewable energy sources

Renewable energy systems use resources that are constantly replaced and are usually less polluting. Some examples are: hydropower, solar, wind and geothermal (energy from the heat inside the earth). We also get renewable energy from burning trees and even garbage as fuel and processing other plants into biofuels.

One day, in the near future, all our homes may get their energy from the sun or the wind. Your car's fuel tank will probably use biofuel, and your garbage might contribute to your city's energy supply. Renewable energy technologies will improve the efficiency and the cost of energy will reduce. We may reach a point when we may no longer rely mostly on fossil fuel energy.

Hydroelectric power: This uses water flowing down a natural gradient to turn turbines to generate electricity known as *hydroelectric power* by constructing dams across rivers. Between 1950 and 1970, hydropower generation worldwide increased seven times. The long life of hydropower plants, the renewable nature of the energy source, relatively low operating and maintenance costs, and the absence of inflationary pressures, as in fossil fuels, are some of its advantages.

Although hydroelectric power has led to economic progress around the world, it has created serious ecological problems.

◇ To produce hydroelectric power, large areas of forest and agricultural lands are submerged when dams are built in valleys. These lands traditionally provided a livelihood for local tribal people and farmers causing conflicts over land use.

- The silting of the reservoirs (especially as a result of deforestation) reduces the life of the hydroelectric power installations.
- Water is required for many purposes besides power generation. These include domestic, agricultural and industrial requirements. Multiple uses by different users gives rise to conflicts over the equitable allocation of water.
- The use of rivers for navigation and fisheries becomes difficult once the water is dammed to generate electricity.
- The resettlement of displaced people is a problem for which there is no ready solution. The opposition to large hydroelectric schemes is growing, as most dam projects have been unable to resettle or adequately compensate the affected people.
- In certain seismically sensitive regions, large dams can induce increased seismic activity, resulting in earthquakes and the consequent loss of lives and property. There is a great possibility of this occurring around the Tehri Dam in the Himalayan foothills. Shri Sunderlal Bahuguna, the initiator of the Chipko Movement, has fought against the Tehri Dam for several years.

With large dams causing so many social problems, an attempt has been made to develop small hydroelectric generation units. Multiple small dams have less impact on the environment.

3.5.2 Use of Alternate Energy Sources

Solar energy: In one hour, the sun pours as much energy onto the earth as we use in a whole year. If it were possible to harness this colossal quantum of energy, humanity would need no other source. Today, we have developed several methods of collecting this energy for heating water and generating electricity.

Solar heating for homes: Modern houses that use air conditioning and/or heating are extremely energy dependent. A passive solar home or building is designed to collect the sun's heat through large, south-facing glass windows. In solar-heated buildings, *sun spaces* are built on the south side of the structure and act as large heat absorbers. The floors of sun spaces are usually made of tiles or bricks that absorb heat throughout the day and then release heat at night when it is cooler.

In energy-efficient architecture, the sun, water and wind are used to heat a building when the weather is cold and to cool it in summer. This is based on good environment-friendly design and building material. Thick walls of stone or mud were used in traditional architecture as insulators. Small doors and windows kept direct sunlight and heat out. Deeply-set glass windows in colonial homes, on which direct sunlight could not reach, permitted the use of glass without creating a greenhouse effect; a veranda also served a similar purpose. Traditional bungalows also had high roofs and ventilators that permitted the hot air to rise and leave the room. Cross-ventilation where wind can drive the air in and out of a room keeps it cool. Large overhangs or eaves over windows prevent the glass from heating the room inside. Double walls are used to prevent heating and shady trees around the house help reduce the temperature.

Solar water heating: Most solar water-heating systems have two main parts—the solar *collector* and the *storage tank*. The solar energy collector heats the water, which then flows to a well-insulated storage tank. A common type of collector is the *flat-plate collector*, a rectangular box with a transparent cover that faces the sun, usually mounted on a flat roof. Small tubes run through the box, carrying the water or any other fluid such as antifreeze, to be heated. The tubes are mounted on a metal *absorber plate*, which is painted black to absorb the sun's heat. The back and sides of the box are insulated to hold in the heat. Heat builds up in the collector and as the fluid passes through the tubes, it heats up as well.

Solar water-heating systems cannot heat water when the sun is not shining. Thus, homes must also have a conventional backup system. About 80% of homes in Israel have solar water heaters.

Solar cookers: The heat produced by the sun can be harnessed directly for cooking using solar cookers. A solar cooker is a metal box, which is black on the inside to absorb and retain heat. The lid has a reflective surface to reflect the heat from the sun into the box. The box contains black vessels in which the food to be cooked is placed.

India has the world's largest solar cooker programme and an estimated 2 lakh families that use solar cookers. Although solar cookers reduce the need for fuelwood and pollution from smoky wood fires, they have not yet become popular in rural areas as it is felt that they are not suitable for traditional cooking practices. However, they have great potential if marketed well.

Other solar-powered devices: Solar desalination systems (for converting saline or brackish water into pure distilled water) have been developed. In future, they should become important alternatives for our future economic growth in areas where freshwater is not available.

Photovoltaic energy: The alternate energy technology that has the greatest potential for use throughout the world is that of solar photovoltaic cells, which directly produce electricity from sunlight using photovoltaic energy (PV) (also called solar cells or solar panels). PV cells use the sun's light, not its heat, to make electricity. PV cells require little maintenance, have no moving parts and essentially no environmental impact. They work cleanly, safely and silently. They can be installed quickly in small modules, in any place where there is sunlight. Solar cells are made up of two separate layers of silicon,

Rooftop solar panels

each of which contains an electric charge. When light hits the cells, the charges begin to move between the two layers and electricity is produced. PV cells are wired together to form a module. A module of about 40 cells is enough to power a light bulb. For more power, PV modules are wired together into an array. PV arrays can produce enough power to meet the electrical needs of a home. Over the past few years, extensive work has been done in decreasing PV technology costs, increasing efficiency and extending cell lifetimes. Many new materials, such as amorphous silicon, are being tested to reduce costs and automate manufacturing.

PV cells are environmentally benign; that is, they do not release pollutants or toxic material into the air or water, there is no radioactive substance or possible risks of catastrophic accidents. Some PV cells, however, do contain small quantities of toxic substances such as cadmium, and these can be released into the environment in the event of a fire. PV cells are made of silicon (although the most abundant element in the earth's crust), which has to be mined. Mining creates environmental problems.

PV systems only work when the sun is shining and thus need batteries to store the electricity.

Solar thermal electric power (STE): Solar radiation can produce high temperatures, which can generate electricity. Areas with low cloud cover with little scattered radiation, as in the desert, are considered the most suitable sites.

Mirror energy: During the 1980s, a major solar thermal electrical generation unit was built in California, containing 700 parabolic mirrors, each with 24 reflectors, 1.5 m in diameter, which focussed the sun's energy to produce steam to generate electricity.

Biomass energy: When a log of wood is burned, we use biomass energy. Because plants and trees depend on sunlight to grow, biomass energy is a form of stored solar energy. Although wood is the largest source of biomass energy, we also use agricultural waste, sugarcane waste and other farm by products to make energy.

There are three ways to use biomass. It can be burned to produce heat and electricity, changed to a gas-like fuel such as methane or changed to a liquid fuel. Liquid fuels, also called *biofuels,* include two forms of alcohol—*ethanol* and *methanol.* Because biomass can be changed directly into liquid fuel, it could someday supply all our transportation fuel needs for cars, trucks, buses, airplanes and trains by replacing petrol and diesel fuel to *biodiesel* made from vegetable oils. In the US, this fuel is now being produced from soybean oil. Researchers are also developing algae that produce oils, which can be converted to biodiesel. New ways have been found to produce ethanol from grasses, trees, bark, sawdust, paper and farming wastes.

Organic municipal solid waste includes paper, food waste and other organic non-fossil fuel-derived materials such as textiles, natural rubber and leather, that are found in the waste of urban areas. Like any fuel, biomass creates some pollutants, including carbon dioxide, when burned or converted into energy. In terms of air pollutants, biomass generates less pollution in comparison to fossil fuels. Biomass is naturally low in sulfur and therefore, when burned, generates a lower level of sulfur dioxide emissions. However, if burned in the open air, some biomass feedstock would emit relatively high levels of nitrous oxides (given the high nitrogen content of plant material), as well as carbon monoxide, and particulates.

Biogas: Biogas is produced from plant material, animal waste, garbage, waste from households and some types of industrial waste such as fish processing, dairies and sewage treatment plants. It is a mixture of gases which includes methane, carbon dioxide, hydrogen sulfide and water vapour. In this mixture, methane burns easily.

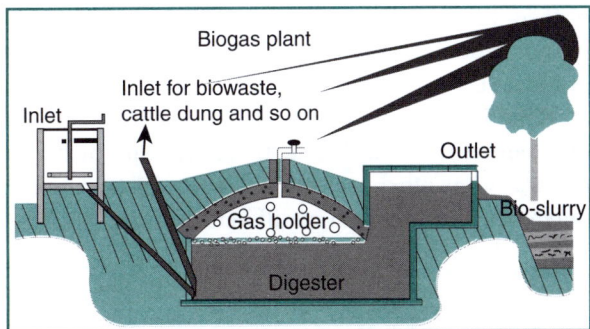

Biogas plants have become increasingly popular in India in the rural sector. The biogas plants use cow dung that is converted into a gas which is used as a fuel. The reduction in kitchen smoke by using biogas has reduced lung problems in thousands of families.

Ethanol produced from sugarcane molasses is a good automobile fuel and is now used in a third of the vehicles in Brazil. The National Project on Biogas Development (NPBD) and Community/Institutional Biogas Plant Programme promote various biogas projects.

Learning by doing—Action learning

Household waste to electrical energy

What you throw out in your garbage today could be used as fuel for someone else. Municipal solid waste has the potential to be a large energy source. Garbage is an inexpensive energy resource. If garbage is collected and delivered to the power plant on a payment basis, it will cover the cost of turning garbage into energy. Garbage is also a unique resource because we all contribute to it.

Keep a record of all the garbage that you and your family produce in a day. What proportion of it is in the form of biomass? Weigh this. How long would it take you to gather enough waste biomass to make a tank-full (0.85 cum) of biogas? (1 ton of biomass produces 85 cu m of biogas.)

Wind power: Wind was the earliest energy source used for transportation by sailing ships. About 2000 years ago, windmills were developed in China, Afghanistan and Persia to draw water for irrigation and to grind grain. Most of the early work on generating electricity from wind was carried out in Denmark at the end of the last century. Today, Denmark and California have large wind turbine cooperatives, which sell electricity to the government grid. In Tamil Nadu, there are large wind farms producing 850 mW of electricity. At present, India is the third largest producer of wind energy in the world.

Wind power

Wind power is a function of the wind speed and therefore, the average wind speed of an area is an important determinant of economically feasible power. Wind speed increases with height. At a given turbine site, the power available 30 m above ground is typically 60% greater than at 10 m.

Environmental impact: Wind power has little environmental impact, as there are virtually no air or water emissions, radiation or solid waste production. The principal problems are bird kills, noise, effect on TV reception and aesthetic objections to the sheer number of wind turbines that are required to meet electricity needs. Although large areas of land are required for setting up wind farms, the amount used by the turbine bases, the foundations and the access roads is less than 1% of the total area covered by the wind farm. The rest of the area can also be used for agricultural purposes or for grazing livestock. Siting windmills offshore reduces their demand for land and produces less visual impact.

Wind is an intermittent source of energy and the changes in wind depend on the geographic distribution of wind. Wind, therefore, cannot be used as the sole resource for electricity and requires some other backup or standby electrical source. In our monsoonal climate there is a great difference in wind speed in different seasons.

Tidal and wave power: The energy of waves in the sea that crash on the land of all the continents is estimated at 2–3million mW of energy. From the 1970s onwards, several countries have been experimenting with technology to harness the kinetic energy of the ocean to generate electricity. Tidal power is tapped by placing a barrage across an estuary and forcing the tidal flow to pass-through turbines. In a one-way system, the incoming tide is allowed to fill the basin through a sluice, and the water so collected is used to produce electricity during low tide. In a two-way system, power is generated from both incoming as well as outgoing tides.

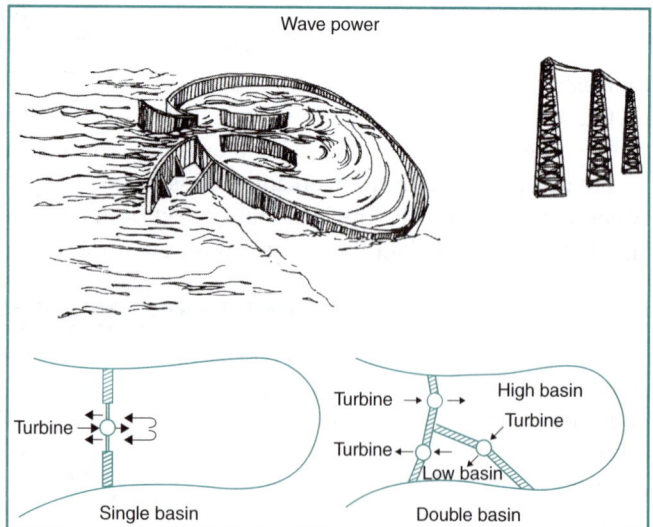

Wave power

Turbine

Single basin

Turbine → → High basin

Turbine

Turbine ← Low basin

Double basin

However, tidal power stations bring about major ecological changes in the sensitive ecosystem of coastal regions and can destroy the habitats and nesting places of water birds and interfere with fisheries.

Wave power converts the motion of waves into electrical or mechanical energy. For this, an energy extraction device is used to drive turbo-generators. Electricity can be generated at sea and transmitted by cable to land. This energy source has yet to be fully explored. The largest concentration of potential wave energy on earth is located between 40° and 60° latitude in both the northern and southern hemispheres, where the winds blow most strongly.

Another developing concept that harnesses energy due to the differences in temperature between the warm upper layers of the ocean and the cold deep-sea water is called Ocean Thermal Energy Conversion (OTEC). This is a high-tech installation, which may prove to be highly valuable in the future. At present, the Department of Ocean Development (DOD) has one plant in Tiruchendur in Tamil Nadu, producing 1 mW a day.

Geothermal energy: This is the energy stored within the earth ('geo' for earth and 'thermal' for heat). Geothermal energy is initiated by hot, molten rock (called *magma*) deep inside the earth, which surfaces at some parts of the earth's crust. The heat rising from the magma warms underground pools of water known as *geothermal reservoirs.* If there is an opening, hot underground water comes to the surface and forms hot springs or it may boil to form geysers. With modern technology, wells are drilled deep below the surface of the earth to tap into geothermal reservoirs. This is called 'direct' use of geothermal energy, and it provides a steady stream of hot water that is pumped to the earth's surface.

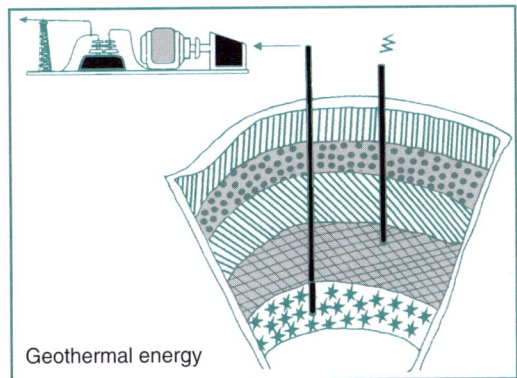
Geothermal energy

Learning by action

Remember that even a single electrical light that is burning unnecessarily is a contributor to environmental degradation. What energy saving actions have you taken in the last 24 hours?

Nuclear power: In 1938, two German scientists, Otto Hahn and Fritz Strassman, demonstrated nuclear fission. They found they could split the nucleus of a uranium atom by bombarding it with neutrons. As the nucleus split, some mass was converted to energy. The nuclear power industry was born in the late 1950s. The first large-scale nuclear power plant in the world became operational in 1957, in Pennsylvania, USA.

As of 2008, India and the US have an agreement facilitating nuclear cooperation in energy and satellite technology between the two countries. Homi Bhabha is considered the father of nuclear power development in India. The Bhabha Atomic Research Centre (BARC) in Mumbai studies and develops modern nuclear technology. India has ten nuclear reactors at five nuclear power stations that produce 2% of India's electricity. These are located in Maharashtra (Tarapur), Rajasthan, Tamil Nadu, Uttar Pradesh and Gujarat. India has uranium from mines in Bihar. There are also thorium deposits in Kerala and Tamil Nadu. This forms our raw material for our nuclear power generation units.

The nuclear reactors use uranium-235 to produce electricity. The energy released from 1 kg of U-235 is equivalent to that produced by burning 3,000 t of coal. U-235 is made into rods that are fitted into a nuclear reactor. The control rods absorb neutrons and thus adjust the fission, which releases energy due to the chain reaction in a reactor unit. The heat energy produced in the reaction is used to heat water and produce steam, which drives turbines that produce electricity.

The drawback is that the rods need to be changed periodically. This has an adverse impact on the environment due to the need for disposal of nuclear waste. The reaction releases very hot waste water that potentially damages aquatic ecosystems, even though it is cooled by a water system before it is released.

The cost of nuclear power generation must include the high cost of disposal of its waste and the decommissioning of old plants. These have high economic as well as ecological costs that are not taken into account when developing new nuclear installations. For environmental reasons, Sweden decided to become a nuclear-free country by 2010, due to the risks of nuclear disaster.

Although the conventional environmental impact from nuclear power is negligible, what overshadows all the other types of energy sources is that an accident can be devastating and the effects last for long periods of time. While it does not pollute the air or water routinely (like oil or biomass), a single accident can kill thousands of people, make many others seriously ill, and destroy an area for decades by its radioactivity which leads to death, cancer and genetic deformities in new born babies. Land, water, vegetation are destroyed for long periods of time. The management, storage and disposal of radioactive waste resulting from nuclear power generation are the largest expenses of the nuclear power industry.

There have been horrifying nuclear accidents at Chernobyl in USSR and at the Three-Mile Island in the USA. The radioactivity unleashed by such accidents can affect humans for generations.

3.5.3   Growing Energy Needs

Energy has always been closely linked to man's economic growth and development. The present strategies for development focussed on rapid economic growth, have used energy utilisation as an index of economic development. This index, however, does not take into account the adverse long-term effects of excessive energy utilisation on society.

Between 1950 and 1990, the world's energy needs have increased four-fold. The world's demand for electricity has doubled over the last 22 years! The world's total primary energy consumption in 2000 was 9096 million tons of oil; a global average per capita that works out to be 1.5 tonnes of oil. Electricity is at present the fastest growing form of end-use energy worldwide. By 2005, the Asia–Pacific region is expected to surpass North America in energy consumption, and by 2020 is expected to consume some 40% more energy than North America!

For almost 200 years, coal was the primary energy source, fuelling the Industrial Revolution in the 19th century. At the close of the 20th century, oil accounted for 39% of the world's commercial energy consumption, followed by coal (24%) and natural gas (24%), while nuclear (7%) and hydro/renewable power sources (6%) accounted for the rest.

Among the commercial energy sources used in India, coal is a predominant source, accounting for 55% of energy consumption estimated in 2001, followed by oil (31%), natural gas (8%), hydro (5%) and nuclear power (1%). In India, biomass (mainly wood and dung) accounts for almost 40% of the primary energy supply. While coal continues to remain the dominant fuel for electricity generation, nuclear power has been used increasingly since the 1970s and 1980s, and the use of natural gas increased rapidly in the '80s and '90s.

Urban problems related to energy: In the past, urban housing in India required relatively smaller amounts of energy than we use at present. Traditional housing in India required very little temperature adjustments as the materials used, such as wood and bricks, handled temperature changes better than the current concrete, glass and steel of ultra-modern buildings.

> ### Embodied energy
> Materials like iron, glass, aluminium, steel, cement, marble and burnt bricks, which are used in urban housing, are very energy intensive. The process of extraction, refinement, fabrication and delivery are all energy consuming and add to the pollution of the earth, air and water. The energy consumed in the process is called 'embodied energy'.

Until the 1950s many urban kitchens were based on fuelwood or charcoal. This was possible and practical when homes had chimneys and kitchens were isolated from the rest of the house. Smoke became a problem once this changed to apartment blocks. Kerosene thus became a popular urban fuel. This changed to electrical energy and increasingly to natural gas by the 1970s, in most parts of urban India.

Urban centres in hot climates need energy for cooling. The early systems of fans changed into air-conditioning, which consumes enormous quantities of energy, thus increasing natural resource use. New buildings in our country have taken to using large areas covered by glass. While in cold climates this uses the greenhouse effect to trap the warmth of the sun inside, in our hot climate this adds several degrees to the temperature inside. Thus, it requires even more energy to run large central air-conditioning units. High-rise buildings in urban centres also depend on energy to operate lifts and require an enormous number of lights.

Urban transport depends on energy—mainly from fossil fuels. Most urban people use their own individual transport rather than public transport systems for a variety of reasons. Urban transport in different cities and even different parts of a city are either inefficient or overcrowded, so even middle-income groups tend to use their own private vehicles. This means more and more vehicles on the road, which leads to traffic congestion, waste of time for all the commuters, and a great load of particulate matter and carbon monoxide from the exhaust of vehicles. This, in turn, causes a rise in the number of people having serious respiratory diseases. Thus, there is a need to develop a more efficient public transport system and discourage the use of individual vehicles in all our urban areas.

Each of us as environmentally conscious individuals must reduce our use of energy. An unnecessary light left on carelessly adds to a large amount of wasted energy. If we learn to save electricity, we would begin to live a more sustainable lifestyle.

Energy conservation

Conventional energy sources affect nature and human society in different ways. India needs to rapidly move towards a policy to reduce energy needs and use cleaner energy production technologies. Approaches towards energy conservation in India's include:

- A shift to alternative energy use and renewable energy sources that are used judiciously and equitably would bring about environmentally friendly and sustainable lifestyles. This would also reduce India's dependency on imported oil.
- Electricity losses in India during transmission and distribution are extremely high—approximately 30%–45%. This occurs even before we turn on our electrical appliances at home! Minimising these losses is critical to effective energy conservation in India.

◇ Small hydrogeneration units are environment-friendly. They do not displace people, destroy forests or wildlife habitats or kill aquatic and terrestrial biodiversity. They can be placed on several hill-streams, canals or rivers. The generation depends on flowing water due to gravity. However, this fails if the flow is seasonal. An estimated potential of about 15,000 MW of small hydrogeneration projects exist in India. Andhra Pradesh, Arunachal Pradesh and Assam are examples of some states where small hydro-projects have been implemented.

◇ Enhancing fuelwood plantations and managing them through JFM.

◇ Using energy-efficient cooking stoves or chulhas which help the movement of air through them, makes the wood burn more efficiently. Additionally, these stoves have a chimney to minimise indoor air pollution and thus reduce respiratory problems.

◇ Biomass generated from firewood, cattle dung, crop residue such as rice husk, coconut shells or straw can be converted into biogas or liquid fuels such as ethanol and methanol. This can provide a low-cost fuel option for heating purposes such as cooking. Biogas can be compressed and used to power motor vehicles.

◇ Unplanned and inefficient public transport systems, especially in cities, waste a large amount of energy. Providing for bicycle paths during the town planning stage would encourage more people to conserve energy by using bicycles.

◇ In the agricultural sector, irrigation pumps to lift water are energy intensive; these are either electrical or run on fossil fuels. Alternative energy sources such as solar-powered irrigation pumps can be used instead.

◇ Large-energy consumers include chemical industries, especially petrochemical units, iron and steel, textiles and paper. Efforts must be made by these industries to be more energy efficient.

Conserving energy: what you can do

Many of the mineral resources such as coal, petroleum and oil are non-renewable sources of energy. At the current rate of fossil fuel consumption, the present oil reserves on the earth will last only for the next 30–50 years. Crores of rupees are being spent to extract, process and distribute coal, petroleum and electricity. Experiments are being carried out to generate energy from wind and photovoltaic cells. At an individual level, every one of us should try to conserve energy. Given below are some of the things you can do to conserve energy.

Do

◇ Turn off the lights fans and air-conditioning when not necessary.
◇ Use low voltage lights.
◇ Use tube lights and energy-saving bulbs (LEDs and/or CFLs) as they consume less electricity.
◇ Use alternative sources of energy like solar power for heating water and for cooking food.
◇ Cut down on the use of electrical appliances.
◇ In summer, shut the windows, curtains and doors early in the morning to keep the house cool.
◇ Use a pressure cooker as much as possible to save energy.
◇ Turn off the stove immediately after use.
◇ Plan and keep things ready before you start cooking.
◇ Keep the vessels closed while cooking and always use small, narrow-mouthed vessels to conserve energy.
◇ When the food is almost cooked, switch off the gas stove and keep the vessel closed. It will get completely cooked with the steam already present inside.
◇ Soak rice and pulses before cooking to reduce cooking time and save fuel.
◇ Get your family to eat together, it will save re-heating fuel.

◆ Select a light shade of paint for walls and ceilings as it will reflect more light and reduce electrical consumption.
◆ Position your reading tables near the window and cut down on your electricity bill by reading in natural light.
◆ Use a bicycle—it occupies less space, releases no pollutant and provides healthy exercise.
◆ Try using public transport systems like trains and buses as far as possible.
◆ Walk rather than drive wherever possible.
◆ Get your vehicles serviced regularly to reduce fuel consumption and reduce pollution levels.

Don't

◆ Don't use unnecessary outdoor decorative lights.
◆ Do not use a geyser during summer; instead, heat water naturally with the help of sunlight.
◆ Don't use halogen lamps as they consume a lot of electricity.
◆ Do not put food in the refrigerator when it is still hot.

Learning by doing

It is easy to waste energy but cheaper to save it than generate it. We can conserve energy by preventing or reducing energy waste and by using resources more efficiently. People often waste energy because the government subsidises it. If the real cost was levied, people would not be able to afford to waste it carelessly.
Do an energy audit of your home/institution.

CASE STUDY 3.5 Oil-related disasters

The Deepwater Horizon oil spill, also known as the Gulf of Mexico oil spill or the BP oil spill, is considered amongst the largest oil spills in history. On 10 April 2010, an oil well blowout occurred about 5,000 feet below the ocean surface causing a catastrophic explosion on the Deepwater Horizon offshore oil drilling platform. Hundreds of millions of litres of oil caused severe pollution of the sea. Work to drill relief wells to permanently close the oil well continues. The best estimate of the spill so far is approximately 12,000 to 19,000 barrels of oil per day. Petroleum toxicity is expected to affect the habitats of thousands of marine and bird species. This ongoing environmental disaster has already impacted the fishing and tourism industries in the Gulf of Mexico.

CASE STUDY 3.6 Solar power

◆ In 1981, a plane called 'The Solar Challenger' flew from Paris to England in 5 hours, 20 minutes. It had 16,000 solar cells glued to its wings and tail, and they produced enough power to drive a small electric motor and propeller. Since 1987, every three years there is a World Solar Challenge for solar-operated vehicles in Australia, where the vehicles cover 3000 km!
◆ Space technology requires solar energy and the space race spurred the development of solar cells. Only sunlight can provide power for long periods of time for a space station or a long-distance spaceship.
◆ Japanese farmers are substituting PV-operated insect killers for toxic pesticides.
◆ In recent years, the popularity of Building Integrated Photovoltaics (BIPVs) has grown considerably. In this application, PV devices are designed as part of the building materials (that is, roofs and siding) both to produce electricity and reduce costs by replacing the costs of normal construction materials.

CASE STUDY 3.7 Hydel power in the Western Ghats

In 1882, the first hydroelectric power dam was built in Appleton, Wisconsin. In India, the first hydroelectric power dams were built in the late 1800s and early 1900s by the Tatas in the Western Ghats of Maharashtra. Jamshedji Tata, a great visionary who developed industry in India in the 1800s, wished to have a clean source of energy to run cotton and textile mills in Bombay as he found people were getting respiratory infections due to coal-driven mills. He therefore asked the British Government to permit him to develop dams in the Western Ghats to generate electricity. The four dams are the Andhra, Shirowata, Valvan and Mulshi hydel dams. An important feature of the Tata power projects is that they use the high rainfall areas in the hills as storage areas. While the rivers flowing eastwards from the Western Ghats are dammed in the foothills near the Deccan Plateau, the water is tunnelled through the crest of the Ghats to drop several hundred metres to the coastal belt. Large turbines in the power plants generate electricity for Mumbai and its giant industrial belt.

3.6   ROLE OF AN INDIVIDUAL IN THE CONSERVATION OF NATURAL RESOURCES

The critical point

Until fairly recently, humans acted as if we could endlessly exploit the earth's ecosystems and natural resources such as soil, water, forests and grasslands and continue to permanently extract minerals and fossil fuels. However, in the last few decades, it has become increasingly evident that the global ecosystem has the capacity to sustain only a limited level of utilisation. Biological systems cannot go on replenishing resources if they are overused or misused at the current rate of utilisation. At a critical point, increasing pressure destabilises their natural balance. Even biological resources traditionally classified as 'renewable' – such as those from our oceans, forests, grasslands and wetlands – are being degraded by overuse and may be permanently destroyed. And no natural resource is limitless. Non-renewable resources will be rapidly exhausted if we continue to use them as intensively as at present.

The two most damaging factors leading to the current rapid depletion of all forms of natural resources are increasing consumerism on the part of the affluent sections of society and rapid population growth linked to poverty. Both factors are the result of choices *we make as individuals*. As individuals we need to decide,

- What will we leave to our children? (Are we thinking of short-term or long-term gain?)
- Is my material gain someone else's loss?

In general, acquiring goods has become a way of life for the majority of people in the developed world. Population growth and the resulting shortage of resources most severely affect people in the developing countries. In nations such as ours, which are both developing rapidly and suffering from a population explosion, both factors are responsible for environmental degradation. We must ask ourselves if we have perhaps reached a critical flashpoint, at which economic 'development' affects the lives of people more adversely than the benefits it provides.

SUMMARY

- All resources from nature constitute our *natural capital*, which is not factored into routine economic thinking in business and industry. Natural resources are classified as, renewable resources, non-renewable resources and perpetual resources. Perpetual resources are those resources that are everlasting. They are inexhaustible and do not require any form of renewal. Examples of these are solar, wind, tidal wave power, geothermal energy and energy from flowing water.

- A renewable resource, if harvested sustainably, will regenerate after its use. Nature can replenish these resources if given enough time.

- A non-renewable resource can take millions of years to regenerate and is therefore irreplaceable after consumption.

- Pressures of an increasing population, changes in land use, over-consumption and pollution are some of the major threats to our fast depleting natural resources.

- Conservation and natural resource use efficiency along with an equitable distribution of natural resources is the key to a sustainable future.

QUESTIONS

1. Name any two natural renewable resources and two advantages and two disadvantages of each.
2. What is a non-renewable resource? Give two examples of non-renewable resources.
3. In terms of resource use, what are some of the steps we can take to lead a more sustainable lifestyle?
4. What is the role of an individual in the conservation of mineral resources?
5. Describe different ways of conservation of water.

Biodiversity and Conservation

In pushing other species to extinction, humanity is busy sawing off the limb on which it perches.

—Paul Enrich

Biodiversity conservation is a lifeline to a better world future.
Break the web of life and we are doomed forever.
Preserve it so we can have a brighter future.
It is our generation's most critical choice.
Only we can make it happen.

—Erach Bharucha

I will argue that every scrap of biological diversity is priceless, to be learned and cherished, and never to be surrendered without a struggle.

—E O Wilson

Learning Objectives

In this chapter you will learn,

- What the levels of biodiversity are
- What the biogeographical zones of India are
- Why India is a mega-biodiversity nation
- What the threats to biodiversity are
- What ecosystem and biodiversity services are

Purpose

Biological diversity in all forms of life is the basis of all living things and we depend on the processes that occur in nature. Modern civilisation has a great negative impact on many forms of life. If these impacts go on increasing, these varied species of plants and animals will be driven to extinction.

Our Role

The first step is to appreciate the enormous diversity, complexity and importance of biodiversity in our lives. Care for nature must lead us into positive actions in our own surroundings and by demonstrating this to others. Joining a conservation action group and organisations that lobby with the government to support and preserve biodiversity is a primary duty of each one of us.

4.1 LEVELS OF BIOLOGICAL DIVERSITY

Biological diversity is nature's variety in the biosphere—the living (biotic) part of our world. This variety can be observed at three levels—the genetic variability within a species, the variety of species within a community and the organisation of species of distinctive plants and animals in various ecosystems.

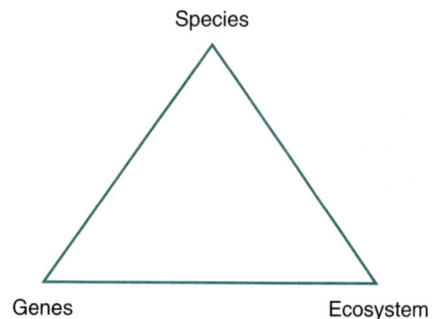

- ◆ Genetic diversity: Biological diversity, also referred to as biodiversity, is that part of nature which includes the differences in genes among the individuals of a species. (*How does each of us differ?*)
- ◆ Species diversity: The variety of all the plant and animal species in an area locally, in a region, in the country, or the world. (You can see so many different species of plants and animals around you).
- ◆ Ecosystem diversity: The types of ecosystems, both terrestrial and aquatic, within a defined area. (Look around and observe the differences in ecological systems—the forests, grasslands, wetlands, hills, rivers, the sea, the agricultural fields and villages and towns. Wherever you live, there are natural or human-modified ecosystems such as wilderness areas, agricultural areas or urban areas.)
- ◆ Landscapes: All that you see around you is organised into these ecosystems of interconnected species of plants and animals; they form different landscapes.

You are only one individual, of one species, among 1.9 million species living on the earth.

If one appreciates the richness of species in the world, one begins to realise that a large number are disappearing due to human activities. We break down nature's food chains and the interlinked chains that form complex food webs in the thousands of ecosystems in the world. We use dangerous chemicals indiscriminately which is not sustainable. We even alter genes which carries several environmental risks.

4.1.1 Genetic Diversity

Each individual of any animal or plant species differs widely from other individuals in its genetic makeup. Due to the large variety of combinations possible in its genes, every individual has its own unique specific set of characteristics. Genes are made of several complex combinations of chemicals.

For example, each human being is different from another, each tiger has unique stripes, every dog looks different. This genetic variability is essential for a healthy breeding population of a species. If the number of breeding individuals is reduced, the dissimilarity of genetic makeup is lost and in-breeding occurs. This leads to genetic anomalies and eventually to the extinction of the species. For example, individuals of the African cheetah have little genetic variability, thus they do not breed successfully and may become extinct in the near future. The diversity in wild species forms a *gene pool* for successful multiplication.

This genetic variability has been used for cultivating crops and breeding domestic animals that have been developed over thousands of years. Today, the variety of nature's bounty in the wild relatives of crops is being used to create new varieties of productive and disease-resistant crops. It is also used to breed superior domestic animals. Modern biotechnology manipulates genes to develop better medicines, new agricultural, horticultural and industrial products. The variability in genes is the foundation on which biological diversity is based. Humankind's future depends on the variability and diversity in the genes of each species and even each individual. If we disrupt this genetic diversity in living organisms by endangering wildlife or inappropriate genetic manipulation through biotechnology, the damage will be irreversible and lead to extinction of species. Even with today's high-end technology, our scientists are unable to successfully reverse extinction.

Introspection

Can you imagine a world without the thousands of species that live around us in our environment? It would be bleak and lifeless even if we lose little bits of it. Animals and plants are wondrous creations of the evolutionary processes that have occurred over millions of years. Species once lost from an area can destabilise the whole ecosystem as they are all linked together in food chains, food webs and food pyramids. Humans are one little strand in that web—a single cog in the wheel of nature.

4.1.2 Species Diversity

The number of species of plants and animals present in a region constitutes its species diversity. This diversity is seen both in natural ecosystems and in agricultural ecosystems. Some areas are richer in species than others. For example, natural undisturbed tropical forests have a much greater richness of species than monoculture plantations developed by the Forest Department for timber production.

Due to their high diversity, natural forest ecosystems provide a large number of non-timber forest products (NTFPs) that local people depend on. This includes fruit, fuelwood, fodder, fibre, gum, resin and medicines. Timber plantations do not provide the large variety of goods essential for local consumption. In the long term, the economic sustainable returns from NTFPs is said to be greater than the economic returns from felling a forest for its timber. Thus, felling a natural forest is a short-term benefit but a long-term loss. The commercial value of a natural forest, with all its species richness and its important ecological services is much greater than a plantation. The natural forest with its rich biodiversity provides a much larger service component than a plantation. Clearing up air and holding water and soil is much more efficiently done by natural forests than plantations. Pollinators and seed dispersers are great service providers in nature.

Modern intensive agricultural ecosystems have a relatively lower diversity of crop varieties than traditional agro-pastoral farming systems, where multiple crops were planted as a mixed crop. Each farmer, by exchanging seeds with others, maintained a high crop-diversity. India once had 10–30 thousand varieties of rice. Now we have only a few left.

Aquatic ecosystems are rich in species—the most species-rich are shallow seas in small patches of coral reefs. Fisher-folk are dependent for their livelihood on both marine and fresh water fish. Fish are an important part of their diet.

At present, conservation scientists have been able to identify and categorise about 1.9 million species on earth. However, this is only a fraction of what really exists. Many species are still being identified by research scientists, especially among flowering plants and insects.

Critical thinking

Areas that are rich in species diversity are called *hotspots* of diversity. India is among the world's 15 nations that are exceptionally rich in species diversity. Our country has very important globally recognised hotspots. This includes the Himalayas, the North-eastern states and the Western Ghats. Isn't this incredible!! It gives us a great sense of pride, but it is also a great responsibility towards our future citizens, who will be dependent on the wealth of diversity that we have. Think about it ... what can you do as a good citizen?

4.1.3 Ecosystem Diversity

A large variety of ecosystems exist on earth, each with their own complement of distinctive interlinked species based on the differences in their habitats. This is based on thousands of years of evolutionary processes. It is related to interactions between species and their habitat, as well as interactions between different species. Ecosystems have thus evolved and continue to change over long periods of time as species adapt to their changing habitat. Ecosystem diversity can be described for a specific geographical region, or a political entity such as a country, a state, a taluka, a village or a city. Distinctive natural ecosystems or biomes include forests, grasslands, deserts and mountains as well as aquatic ecosystems such as rivers, lakes, ponds, wetlands and the sea. Each eco-region has areas modified by human society such as farmland, or grazing pastures for their livestock, which are referred to as cultural landscapes. This forms a mosaic of elements that constitute different landscape elements. The landscapes are thus formed of natural wilderness ecosystems and cultural (human-made) ecosystems.

An ecosystem is referred to as 'natural' when it is relatively undisturbed by human activities, or 'modified' when it is converted to other types of uses, such as farmland, tree plantations or urban and industrial areas. Ecosystems are most natural in the wilderness where human impacts are low. If natural ecosystems are overused or misused, their productivity eventually decreases and they are then said to be degraded ecosystems.

Some ecosystems are very rich in species (hotspots of biological diversity) while others are important as they form a unique habitat for endemic species which are not found anywhere else in the world. Some ecosystems are fragile and are known as eco-sensitive areas. Others are more resilient and can revert to their original state if human impacts are minimised. In today's world, it is the modified irrigated mechanised farmland that is spreading as our human population requires enormous quantities of food. The highly human dominated ecosystems – cities and industrial areas – are spreading very rapidly across the world as well as in India. This change in the landuse patterns occurs at the cost of natural ecosystems, and our earth's incredible biological diversity.

> **Reflective thinking**
>
> It is ecosystems, species and genetic diversity that together form our living world. Can human beings survive without biodiversity?

4.2 BIOGEOGRAPHIC ZONES OF INDIA

India is divided into ten major regions based on the geography, climate, the pattern of vegetation and the communities of mammals, birds, reptiles, amphibians, insects, other invertebrates and microorganisms, that live in an area (**Colour Plates 4.1, 4.2, Table 4.1**). Each of these large eco-regions contains a variety of ecosystems that have specific communities of plant and animal species. All plants and animals interact with their habitat, as well as with each other to form food chains, food webs and food pyramids, which are all a part of various biogeochemical cycles based on energy from the sun, the water cycle, climate, geography and soil characteristics.

The important subgroups include,

- the *terai*—the lowland where the Himalayan rivers flow into the plains with tall grasslands, wetlands and sal forests,
- Kutch—the Great and Little Rann which change seasonally from desert to wetland in the monsoon, and
- the Central Highlands of the Vindhya and Sathpura ranges—belt of hills where teak and sal forests meet, which is rich in biodiversity.

> **Critical thinking—India's species richness**
>
> India has 47,513 plant and 91,000 animal species. India has 410 mammal species, 526 reptile species, 1266 bird species and 405 amphibian species. This makes our country a globally recognised mega diversity nation. Isn't this remarkable!!

4.2.1 Biodiversity Patterns and Global Biodiversity Hotspots

Biomes are large regions across the earth that are similar in structure, composition and functioning ecosystems. This is characterised by the types of naturally occurring vegetation in the terrestrial

Table 4.1 Biogeographic zones, specific impacts, distribution across Indian states

Zone	Ecosystems and biomes	Impacts and threats	Distribution across states
1. Trans Himalaya	High altitude cold desert plateau, wetlands, glaciers, alpine scrub vegetation	Climate change	Jammu, Kashmir, Ladakh
2. Himalaya	Hotspots, perennial peaks, snow covered steep slopes, river valleys conifers and broad-leaved forests at lower elevation	Climate change, snow melt, receding glaciers, large pilgrimage tourism, deforestation, rivers impacted by dams	Jammu, Kashmir, Ladakh, Himachal Pradesh, Uttarakhand, Assam, North-eastern states
3. Desert	Sand dunes, scrubland, low rainfall, sporadic downpours	Intensive irrigated agriculture impacts desert flora and fauna	Rajasthan
4. Semi-arid	Scrubland, dry grassland, seasonal wetlands (Rann)	Conservation to intensive agriculture, loss of dry land species and crop varieties	Punjab, Rajasthan, North Gujarat
5. Western ghats	Hotspot, ancient hill range, basaltic and lateritic plateaus, high rainfall. high biodiversity, evergreen/semi-evergreen forests, ground endemic plants on plateau tops	Fragile eco-sensitive area which has very high biodiversity value, threatened with landuse change, urbanisation, tourism	Maharashtra, Goa, Karnataka, Kerala
6. Deccan plateau, semi-arid zone	Ancient plateau of Gondwana land, semi-arid grassland and scrubland, low rainfall, central highland has unique fauna and flora, deciduous forest	Urbanisation, industry, irrigated farmland, change from grassland pastures to other uses due to industrialisation	Madhya Pradesh, Maharashtra, South Gujarat, West Bengal, Jharkhand, Odisha, Chhattisgarh, Andhra, Telangana, Karnataka, Tamil Nadu
7. Gangetic plain	Gangetic valley flood plains of Ganga and tributaries, ancient food bowl of India's early civilisation	Agriculture and industrialisation, river pollution, urbanisation	Punjab, Haryana, Uttarakhand, Uttaranchal, Bihar, West Bengal
8. Coast	Coastal plains, river deltas, mangrove forests, coastal lagoons, sandy shores, rocky shores, mudflats (salt pans), coral reef hotspots	Very fragile region impacted with ports, industries, thermal power, plants, urbanisation and industry	Western and eastern coastal belts of the Arabian Sea and Bay of Bengal, off-shore coral reefs in Gujarat, Maharashtra, Tamil Nadu, mangroves of West Bengal
9. North-east	Biodiversity rich forests of endemic flora and fauna, carnivores and broad-leaved forests, riverine vegetation, rich local cultures based on hunter–gatherer communities	Shifting agriculture, hunting pressure, dams on rivers	Meghalaya, Mizoram, Manipur, Nagaland, Assam, Arunachal Pradesh, Tripura, Sikkim
10. Islands	Super bio-rich for endemism, contains both South Asian and South East Asian flora and fauna, orchid ferns, subspecies of avifauna, marine ecosystem with extensive coral reefs with high biodiversity of marine vertebrate and invertebrate fauna, cultural diversity of tribal folk	Logging in the past, excessive tourism with rapidly growing impacts	Andaman and Nicobar, Lakshadweep

part of the world. The terrestrial biomes are created by a combination of climate—the range of precipitation, its seasonality and temperature variations. This influences vegetation at an ecosystem level. Each biome has its own component of plants, animals, fungi and microorganisms.

Aquatic biomes are influenced by the physical environment. They include marine and freshwater systems. The large oceans and their temperature variations affect global climate and wind. Marine algae and bacteria produce oxygen and take up CO_2. Water depth influences sunlight penetration. It creates superficial zones where algae grow, and a deep zone or benthic zone at the bottom of the ocean.

Biodiversity is linked to abiotic factors such as climate, altitude, rainfall, soil, water and seasonality (regimes which determine its vegetation characteristics and its dependent fauna). This is an outcome of millions of years of evolutionary processes. These evolutionary changes have led to differences in biodiversity adapted to varied types of living spaces as communities of plants and animals. Thus, at the global level, there are biogeographic realms which have an evolutionary basis.

Major terrestrial biomes include:
- Tropical forests
- Savanna grasslands
- Deserts
- Temperate grasslands
- Temperate broadleaf forests
- Northern coniferous forests
- Tundra
- High mountains
- Polar ice

Major aquatic biomes (zone) include:
- Marine oceans and benthic zone
- Intertidal zone
- Estuaries (deltas)
- Coral reefs
- Fresh water
- Rivers and streams
- Lakes
- Wetlands

The ancient continental drift: As the world's tectonic plate movements separated the ancient continent called Pangea into smaller land masses with large oceans between them, species of plants and animals were sequestrated within each of these major terrestrial realms. Several times through the long history of evolution, great upheavals in climate gradually altered these species and led to extinctions on the one hand and speciation (formation of new species) on the other. Currently we have more species than ever before, but we are also losing species faster than ever before due to human activities. This is why our era is known as the '*anthropocene*'. This extinction is a great loss to our future generations.

India is a hotspot of biological diversity, rich in endemic plants and animals. India has a special status as we are an emerging economy. We have very rich local knowledge of biodiversity in our traditional communities. India has not only a high level of diversity but a great diversity of traditional crop varieties and breeds of livestock. We also have high biotechnological expertise. Thus, India can use its own biological diversity for economic development by creating new medicines and industrial products. However, biodiversity has its own intrinsic values related to ethical, spiritual and aesthetic aspects of our lives.

Global hotspots of biodiversity

The earth's biodiversity is distributed in specific ecological regions. There are thousands of eco-regions in the world. Of these, 200 are said to be the richest, rarest and most distinctive natural areas. These areas are referred to as the 'Global 200'.

A majority of species are yet to be scientifically named and studied. Most of the world's bio-rich areas are in the developing nations. The bio-rich regions include South America, South Asia and South East Asia. In contrast, the majority of the countries capable of exploiting biodiversity through biotechnology are the industrially advanced nations, in the economically developed industrialised world. These rich nations, however, have lower levels of floral and faunal biodiversity.

It has been estimated that 50,000 endemic plants, which comprise 20% of global plant life, probably occur in only 18 hotspots in the world. The three globally recognised regions which are rich in biological diversity are South America, South East Asia and South Asia (in which India is a major area for its biological assets). This is distributed in both natural and human-modified ecosystems. Countries which have a relatively large proportion of these biodiversity hotspots such as India, are referred to as mega-diversity nations. India is thus a major mega-diversity hotspot. Within India, the N-E states, Western Ghats and Himalayas are considered to be very rich in species.

The rate at which extinction of species is taking place throughout the world and in our country remains obscure. It is likely to be extremely high, as our wilderness areas are shrinking rapidly.

India's globally recognised biodiversity hotspots

The globally accepted national hotspots which are included in the world's most bio-rich areas are in the forests of the Eastern Himalayas, the North-eastern states and the Western Ghats. The Andaman and Nicobar Islands are also extremely rich in species and many subspecies of different animals and birds have evolved on these small islands separated by the sea from our mainland. The islands are exceptionally rich in orchids and ferns.

Mangrove river deltas are highly productive ecosystems and rich breeding grounds of marine and brackish water species. This makes their conservation of great value especially as they prevent erosion and reduce the destructive force of storms, cyclones and tsunamis in coastal areas.

Among the endemic species, that is, those found only in India, a large proportion are concentrated in these bio-rich hotspots. The Andaman and Nicobar Islands alone have as many as 2200 species of flowering plants and 120 species of ferns. Out of 135 genera of land mammals in India, 85 (63%) are found in the North-eastern states of India. The North-eastern states have 1,500 endemic plant species. A major proportion of amphibian and reptile species, especially snakes, are concentrated in the Western Ghats. There are 1,500 endemic plant species in the Western Ghats which makes this range of hills a globally important area for conserving plant life.

Coral reefs in Indian waters surround the Andaman and Nicobar Islands, the Lakshadweep Islands, the gulf areas of Gujarat, shallow seas off the coast of Tamil Nadu and the Angria bank, a submerged coral reef which lies along the coast of Maharashtra. These sites are super rich in species, and are now considered as important and species rich as tropical evergreen forests! At several of these hotspots, species are threatened due to human activities. The disastrous long term effects of climate change are irreversible and will lead to extinction of species.

In the 1980s, early conservation biologists rated tropical forests of South America, Africa, South Asia and South East Asia as having an exceptionally high diversity of flora and fauna. By 2000, however, conservation biologists began to rate coral reefs as being extremely species rich in various taxa as they provided specialised breeding habitats for fish, crustaceans, molluscs, starfish, sea horses and many marine mammals. Hotspots of biodiversity have been included in Protected Area Systems – National Parks and Wildlife sanctuaries – to safeguard their flora and fauna. Bio-rich areas such as sacred groves which are ancient mature forests are of great conservation significance. These small cultural hot specks of diversity are protected by local tribal folk across most forest tracts of India. They are now referred to as 'other effective' (area based) conservation measures.

4.3 INDIA AS A MEGA-DIVERSITY NATION

India has an ancient geological history when the earth was covered by a single large continent called Pangea. After Pangea split 70 million years ago, India was linked to the southern part called Gondwanaland. The southern landmass, together with Africa, South America, Australia and the Antarctic split away from the northern part which included Eurasia and North America. Ancient flora and fauna thus have some common elements in India. Later, as tectonic movements shifted India northward across the equator to join the Northern Eurasian continent, the intervening shallow Tethys sea became the meeting place of the two land masses. This led to the gradual lifting up of the Himalayas as the plates collided against each other. Thus, plants and animals that had evolved both in Europe and in the Far East migrated into the Indian subcontinent even before the Himalayas formed a barrier to their dispersal. A final influx came from Africa with Ethiopian species, which were adapted to the savannas and semi-arid regions. This last shift of flora and fauna could in all probability have occurred during the last ice age when land bridges were reformed between India and Africa across the Arabian Sea due to a fall in sea level caused by expansion of polar ice during the last ice age several thousand years ago. Thus, India's special geographical position between three distinctive centres of biological evolution and radiation of species is responsible for the origin of our rich and varied biodiversity today.

Common plant species of india

India is rich in floral species in all its different biogeographic regions. This rich floral diversity gives our country its mega-diversity status (**Colour Plate 4.3**).

Teak: This tree is from the southwest of peninsular India. It is a common tree in deciduous forests, and yields the much-sought-after timber used for making furniture. During the early British period, it was cut down from many forest tracts to build ships. As stocks were diminishing, the British selected areas which they called Reserved Forests where teak was planted for the Government's use. Teak is grown extensively by the Forest Department and is a highly priced wood. The teak tree is identified by its large leaves, which grow to more than 40–50 cm long and 20 cm wide; it has tiny flowers and fruit. In the winter, the trees shed all their leaves. Some teak forests that have exceptional populations of wildlife have been declared as PAs and included in our national parks and wildlife sanctuaries.

Sal: This is a common species found in several types of forests of the North-eastern region of India and central India. It has bright green foliage and its canopy remains green almost throughout the year. Sal wood is hard and durable. Sal trees yield a large number of seeds used in making cosmetics. Sal forests are rich in wild mammals, birds, reptiles and insect life. Several of these areas are included in our network of national parks and sanctuaries.

Mango: This has become one of our most popular horticultural species, with different varieties being grown all over the country. The mango tree is an evergreen species and has small flowers pollinated by insects. In the forest, fruit-dependent animals such as monkeys, squirrels, fruit bats and birds relish its ripe fruit.

Ficus: The *peepal*, banyan and many other ficus species comprise this group of trees. They are all ecologically of great importance as many different species of insects, birds and mammals feed on ficus berries. The flowers are inside the berries. They are pollinated by a specific wasp, which lays its eggs inside the berries on which the larvae feed and grow. The ficus trees bear berries throughout the year, thus supplying nutritious food to several animal species when the other tree species have no fruit. Ficus species are thus known as 'keystone' species and support a major part of the food web in several ecosystems. The *peepal* and banyan are considered sacred and are protected in India.

Neem: Its scientific name is *Azadirachta indica*. It has been traditionally used in indigenous medicine. It has small yellow fruit; the leaves and fruit are bitter in taste. Its leaves are used extensively as an environmentally friendly insecticide. It grows extremely well in semi-arid regions and is planted in afforestation programmes where the soil is poor and rainfall is low.

Tamarind: The tamarind grows to a large size and is known to live for over 200 years. Its fruit is a curved pod with sour pulp, containing a number of squarish seeds. It is commonly grown as a shade tree and for its edible sour fruit, which contains high concentrations of vitamin C. It is used as a preservative and an additive in food to give a tangy flavour. It is valued for its timber as well as for fuelwood.

Babul: This is a thorny species that is characteristic of the semi-arid areas of Western India and the Deccan plateau. The babul grows sparsely in tracts of grassland and around farms, and is used for fodder and fuelwood. It remains green throughout the year even under the driest conditions and is browsed on by wild animals and cattle. It has small leaves and bright yellow flowers and small pods with multiple seeds. Its main characteristic is its long, sharp, straight thorns, which prevent excessive browsing of its older branches.

Ber (Zizyphus): These are the small trees or shrubs typically found in the arid and semi-arid areas of India; it is a favourite of frugivorous birds. The popular fruit is commonly collected and sold in local markets.

Jamun: This tree is an evergreen species, which has a tasty purple fruit. It is a favourite with people and with many wild birds and mammals. It grows in many parts of India and has several varieties with fruits of different sizes.

Tendu: This is a mid-sized, deciduous tree, commonly found in dry deciduous forests throughout the Indian subcontinent. There are around 50 Indian species. The leaves are elliptical and leathery and its young leaves are extensively used for making beedis. The fruit is brownish-yellow and astringent. Tendu-leaf collection necessitates burning the undergrowth and slashing the branches of the trees to get at the leaves. The resulting disturbance to wildlife is a serious issue in PAs.

Flame of the forest (Butea monosperma): This tree grows in many parts of India. It has bright orange flowers when it is leafless, and is thus called the 'flame of the forest'. The flowers are full of nectar, which attract monkeys and many nectar-dependent birds.

Coral tree (Erythrina): A common deciduous tree that is leafless in February, when it gets bright scarlet flowers that attract many birds to their nectar.

Amla (gooseberry): This deciduous medium-sized tree is known for its sour, greenish-yellow fruit, which are rich in vitamin C. They are used as medicine, in pickles, and for dyeing and tanning.

Dipterocarps: This group of trees grows in evergreen forests of the southern part of the Western Ghats and in the North-east parts of India, in high rainfall areas. It grows to an enormous height and has a wide girth. The seed has a pair of wing-like structures which aid in wind dispersal.

Oak (Quercus): It is a large tree and is economically important. Oaks provide the finest hardwoods of great strength and durability used for high-quality furniture. Some of its species are also excellent fodder trees.

Pine: There are five species of true pines found in India in the Himalayan region. The timber of these trees is used in construction, carpentry and the paper industry. Pine resin is used to make turpentine, rosin, tar and pitch. Pine leaves are thin and needle-like. The male and female spores are produced in woody cones.

Cycads: Cycads, along with conifers, make up the gymnosperms. They are among the most primitive seed plants, and have remained virtually unchanged through the past 200 million years (since Jurassic times). There are five species found in India, mostly in high rainfall areas.

Coconut: It produces the coconut, filled with liquid and a soft white edible, jelly-like material that hardens when the fruit ripens. It is a common ingredient of food in India, especially in the southern states. It is extensively cultivated along the coastal regions and islands of India. Most parts of the tree yield useful products, such as broomsticks from its leaves and fibre from the husk of dried coconuts.

Orchids: This is the largest group of flowering plants in the world, with over 18,000 known species. Of these, 1500 species are found in India of which 700 species are found in the North-eastern states. These plants are terrestrial or epiphytic herbs. The flowers show a range of bright colours and great variation in structure. A large number of orchid species are found in the Western Ghats, the North-east and the Andaman and Nicobar Islands.

Drosera: This is a small insectivorous plant, usually 5–6 cm in height, which has tiny hairs that secrete a sticky droplet of fluid on which insects get stuck. The leaf winds around the struggling insect, which is then slowly digested. The plant has pretty flowers. It grows in shallow, poor-quality soil. It is a rare plant and is found in small patches.

Lotus: The lotus is the national flower of India. This is an aquatic floating plant. Its leaves are circular, flat and covered with a waxy coating. The flower grows on an erect stalk with several petals ranging from pink-violet to white. It is widely distributed in wetland habitats and shallow parts of lakes and marshy areas. The rhizome, stalks of the leaves and seeds are considered delicacies. The flower has been a traditional motif in Indian art.

Grasses: Grasses form the second-largest group of flowering plants in the world. They are a very important group of fodder species and are used for various other purposes, such as making fibre, paper, thatching material for roofs, oil, gum and medicines. The economically important grasses include sugarcane, bamboo and cereals like rice, wheat, millets and maize.

Bamboo: This is a group of large grass-like species that grow in clumps to great heights in many forests of India. They are used for constructing huts and making several useful household articles in rural areas, such as baskets, farm implements, fences and mats. The young shoots are used as food. They are extensively used in the pulp and paper industry as raw material. The bamboo plants flower after more than two decades; the plant then dies. The flowering produces thousands of seeds, which results in the slow regrowth of the bamboo. Bamboo is a favourite food of elephants and other large herbivores of the forest, such as gaur and deer.

Wild relatives of crop plants: All our present-day cultivated varieties of rice, which are grown for food, come from wild varieties of rice, many of which have originated in India, China and Indonesia. Rice is one of the staple foods of the world. Although the wild varieties are not used as food crops, they are important as they contain genes that can be used to develop disease or pest resistant crop varieties. Many local varieties of rice have already been lost, as most farmers now grow only high-yielding varieties. India once had thousands of rice varieties grown by traditional farmers. Most of these are not grown now as high yielding varieties are preferred. This leads to a loss of genetic diversity.

Common animal species of india

(a) Mammals [Colour Plate 4.4: (a)]

The common deer species found in India include the *sambar, chital, barasingha* and barking deer. *Sambar* live in small family parties especially in hilly forested areas and feed mainly on shrubs and leaves of low branches. They are dark brown in colour and have large thick antlers, each having three branches. *Chital* or spotted deer live in large herds in forest clearings where they graze on the grass. They have a rust-brown body with white spots, which camouflages them in the forest. Each antler has three branches.

The rare *hangul* deer is found only in Kashmir. It has a magnificent spread of antlers with six branches on each antler. The *barasingha* or swamp deer has wide hoofs that enable this beautiful animal to live in boggy areas of the *terai*. Each antler has six or more branches. The tiny barking deer lives in forests all over India. It has two ridges on its face and short antlers with only two branches. Its call sounds like the bark of a dog.

The blackbuck is the only true antelope found in India. It lives in large herds. The males are black on top and cream below and have beautiful spiral horns that form a 'V' shape. The *chinkara*, also known as the Indian gazelle, is a smaller animal, pale brown in colour, with beautiful curved horns. The rare *chausingha* (four-horned antelope) is the only animal in the world that has four horns. The *nilgai* is the largest of the dryland herbivores. The males are blue-grey in colour. The *nilgai* have white markings on the legs and head, and short strong spike-like horns.

A very special rare species is the Indian wild ass, endemic to the Little Rann of Kutch. The Himalayan pastures support several species of wild goat and sheep, many of them restricted to the region, like the goral and the Himalayan *tahr* or mountain goat. The Nilgiri *tahr* is found in the Nilgiri and Anamalai hills in South India.

The one-horned rhinocerous is now restricted to Assam, but was once found throughout the Gangetic plains. The *gaur* is now also restricted to the *terai*. The Indian elephant is distributed in the North-eastern and Southern states. It is threatened by habitat loss and poaching for ivory. The gaur is found in patches in several well-wooded parts of India. The best known predator of our forests is the tiger. Its gold and black stripes hide it perfectly in the forest undergrowth. It preys on herbivores, such as the *sambar* or *chital,* and on domestic animals. Its numbers had declined due to poaching for its skin, and for the supposed magical value of its teeth, claws and whiskers. It has been extensively killed for the supposed medicinal properties of its bones that are used in Chinese medicine. In the recent past, its population has grown.

The Asiatic lion is now found only in the Gir forests of Gujarat. The leopard is more adaptable than the tiger and lives both in thick forests and degraded forest areas. The small jungle cat and the leopard cat are found in forested areas. Their populations are small. The most typical predator of the Himalayas is the snow leopard, which is very rare and is poached for its beautiful skin, which is pale grey with darker grey ring-like markings.

The wolf, jackal, fox and the wild dog or *dhole* form a group called canids. The wolves are now highly threatened as they have become increasingly dependent on domestic sheep. Thus, the shepherds constantly devise ways to kill the wolves.

The *Indian pangolin* lives mainly in burrows. It has a small triangular shaped head and large, overlapping scales to protect its body. Indian pangolins are insectivores that feed on ants and termites. It is extensively poached for its scales.

One of the common monkey species of the forest is the bonnet macaque, which has a red face, a very long tail and a whorl of hair on the scalp which looks like a cap. The rhesus macaque, which is smaller has a shorter tail than the bonnet. A rare macaque is the lion-tailed macaque found only in a few forests of the southern Western Ghats and Anamalai ranges. It is black in colour, has long hair, a grey mane and a tassel at the end of its tail that looks like a lion's tail. The common *langur* has a black face and is also known as the Hanuman langur. The rare golden langur is golden-yellow in colour and lives along the banks of the river Manas in Assam. The capped langur is an uncommon species of North-east India. The rare black Nilgiri langur lives in the southern Western Ghats, the Nilgiris and Anamalais. The hoolock is a rare primate of North-eastern India.

(b) Birds [Colour Plate 4.4: (b)]

There are over 1200 bird species found in India in different habitats. Most of our forest birds are specially adapted to life in certain forest types. There are several species of hornbills that live on fruit. They have heavy curved beaks.

Frugivores such as parakeets, barbets and bulbuls live on fruit. Insectivorous birds of many species live on forest insects. They include various species of flycatchers, bee-eaters, babblers and thrushes. The male Asian paradise flycatcher is a small beautiful white bird with a black head and two long white trailing tail feathers. The female is brown and does not have the long tail feathers. There are several birds of prey such as eagles, falcons and kites, many of which are now endangered.

The grasslands support many species of birds. The most threatened species is the Great Indian Bustard, a large brown stately bird with long legs, which struts about through grasslands looking for locusts and grasshoppers. Another rare group of threatened birds of grasslands are the floricans. There are many species of quails, partridges, larks, munias and other grain-eating birds that are adapted to the grasslands.

Several species of aquatic birds such as waders, gulls and terns, live along the seashore and go out fishing many kilometres to the sea. Many of these birds have lost their coastal habitats due to pollution. The freshwater aquatic birds have long legs and are known as waders—such as stilts, egrets and sandpipers. Several species of ducks and geese are found in wetlands across India. Different species of storks, cranes, spoonbills, herons, flamingoes and pelicans are linked to aquatic habitats. Many aquatic species are migrants. They breed in Northern Europe or Siberia and come to India in the thousands during the winter.

(c) Reptiles [Colour Plate 4.4: (c)]

India has a wide variety of lizards, snakes and turtles and tortoises, with a high level of endemism. The lizards include the common garden lizard, the fan-throated lizard, chameleon, skink, common monitor and water monitor. Most of these are threatened due to trade in reptile skins. The common Indian snakes include the rock python, the grass snake and the vine snake. We rarely appreciate the fact that only a few species of snakes such as the king cobra, cobra, krait and Russell's viper are venomous while most other snakes are harmless. The star tortoise is found in arid lands and is illegally exported for pet trade.

The olive ridley is a marine turtle and comes to nest in large numbers on the coast of Odisha; this unique phenomenon is called the *arribada*. The freshwater turtles include the flap-shell turtle. Many turtles are becoming increasingly rare due to the poaching of adults and eggs. The crocodile is our largest reptile, which is poached for its prized skin. The *gharial* is a species of fish-eating crocodile, endemic to India and is highly threatened.

(d) Amphibians [Colour Plate 4.4: (d)]

Most of the amphibians found in India are frogs, toads and skinks. These include several species such as the Indian bullfrog and tree frog. These amphibians are mostly found in the hotspots in the North-east and the Western Ghats. It is now believed that global warming and increasing levels of UV radiation may be seriously affecting amphibian populations in some areas.

(e) Invertebrates [Colour Plate 4.4: (e)]

Invertebrates include a variety of taxa that inhabit both terrestrial and aquatic ecosystems. Microscopic animals such as protozoa and zooplankton form the basis of the food chain in aquatic habitats. Coral is formed by colonies of polyp-like animals. Worms, mollusks (snails), spiders, crabs, jellyfish and octopus are a few of the better-known invertebrates found in India.

There are more than a million known insect species on earth. They include grasshoppers, bugs, beetles, ants, bees, butterflies and moths. India is rich in its butterfly and moth species.

(f) Marine life

Marine ecosystems are the habitat for a large variety of fish and crustaceans, such as crabs and shrimp, which are used as food. The other endangered species in our waters include marine turtles and whales (killed for their fat). There are a large number of species of freshwater fish found in Indian rivers and lakes that are now threatened by the introduction of fish from abroad as well as due to being introduced from one river into another. Fish are also now seriously affected by pollution. Marine fisheries are being over-harvested in our coastal waters and the fish catch has decreased seriously over the last few years. Mechanised boats with giant, small-meshed nets are a major cause for the depletion. There are many endangered fish such as the *mahseer*, which once grew to over a metre in length. Many species of marine animals, such as whales, sharks and dolphins that live in the Indian Ocean, are now threatened by extinction due to fishing in the deep sea.

For further details see:

◆ *The Wonders of the Indian Wilderness*, The biodiversity of India, Mapin Publications, Ahmedabad.
◆ S H Prater, *The Book of Indian Animals*, BNHS.
◆ S Ali, *The Book of Indian Birds*, BNHS.
◆ Vivek Menon, *Indian Mammals*, a field guide.
◆ Johnsingh A J T and Nima Manjrekar, *Mammals of South Asia*, Volumes 1 (2012) & 2 (2015), Universities Press.

4.3.1 Endangered and Endemic Species of India

Human actions have, in the last few generations, led many floral and faunal species down a path that ended in extinction. Species such as the cheetah, the Siberian crane and the pink-headed duck, that were a part of India's fascinating glamour species, will never be seen in our country again. We do not know how many other floral and faunal species are threatened and may soon go down the same path unless they are carefully protected. Many undiscovered species may have already gone down the irreversible pathway of extinction. Insects, molluscs, coral and other less studied organisms, and many wild plants may have become extinct even before being studied by zoologists and botanists.

While species are being protected in our national parks and sanctuaries, the extensive changes in landuse and a variety of human activities is a major factor in the loss of abundance of many species. Data on abundance of species is available for only a few glamour species, the current status of less known or rare plants and animals are merely conjectural estimations. Thus, many of the less glamorous little known species could well be on the brink of extinction without us knowing about their threatened or even critically endangered status. We may be losing species that are still not even named or studied. The IUCN classifies endangered species into 9 categories as follows:

1. Extinct (EX)
2. Extinct in the wild (EW)
3. Critically endangered (CR)
4. Endangered (EN)
5. Vulnerable (VU)
6. Near threatened (NT)
7. Least concerned (LC)
8. Data deficient (DD)
9. Not evaluated (NE)

India's biodiversity
Faunal species 91,000
Floral species 47,513

Among the bio-rich nations, India is in the top 10 or 15 countries for its great variety of plants and animals, many of which are not found elsewhere. India has 91,000 species of animals—410 different mammals (rated the eighth highest in the world), 1,250 species of birds (eighth in the world), 450 species of reptiles (fifth in the world).

There are 69,903 known species of insects, including 13,000 butterflies and moths in India. India has 47,513 plant species with 17,527 flowering plants, most of which are angiosperms (fifteenth highest in the world). These include especially high species diversity of ferns (1022 species) and orchids (1082 species). It is estimated that the number of unknown species could be several times higher.

Reflective learning

We still need taxonomists in the fields of botany and zoology, biologists to study flora and fauna, to know the number and status of our plant and animal kingdoms. We need specialists in IT and GIS to deal with the wealth of species and ecosystem data. We need conservation scientists to assist foresters to manage Protected Areas better. We need lawyers to punish offenders of wildlife crime. We need the judiciary to interpret environmental laws. We require trained police, forest guards and coast guards to catch offenders who are part of a growing illegal trade in flora and fauna. Landuse planners and architects must be sensitive to the needs of biodiversity conservation. We need to be aware of social issues and the equitable solutions for marketing the goods and services present in biodiversity. Thus, every field in natural or social science can contribute to conserving biodiversity. This is India's natural capital. Please reflect on what you can do in your future career to preserve our biological assets.

Around 18% of Indian plants are endemic with the flowering plants (angiosperms) showing a much higher degree of endemism. A third of these flowering plants are not found elsewhere in the world. Among the amphibians that are found in India, 62% are unique to this country. Of the 153 species of lizards recorded, 50% are endemic. High endemism has also been recorded

Table 4.2 India's biodiversity status

Species	India's world ranking	Number of species in India
Mammals	8	410
Birds	8	1266
Reptiles	5	450
Amphibians	15	405
Angiosperms	15–20	17527

for various groups of insects, marine worms, centipedes, mayflies and freshwater sponges.

Apart from the high biodiversity of wild plants and animals in India (**Table 4.2**), there is also a great diversity in cultivated crops and breeds of domestic livestock. This is a result of several thousand years during which India's farmers and traditional folk have reared their livestock breeds and have grown and treasured thousands of indigenous varieties of crops.

Conservation can only happen if people everywhere value biological diversity and ecosystem services. It is therefore important to disseminate knowledge on issues related to sustainable economic, societal and ecological development through conservation of our bio-resources.

It is now evident that for long term sustainable development, the world's people must alter the consumerist pattern of life through a balance between economic, societal and environmental concerns. Since these three pillars of sustainable development have different sets of goals, a complex balance has to be achieved so that conservation and use of natural resources are achieved at the same time. This means, reducing our footprint on earth and supporting and preserving our resources for the future is our duty, as enshrined in our country's constitution.

The science of *conservation biology* includes strategies and actions that can prevent extinction, or further loss of the abundance of several species. It helps in preventing degradation of natural capital and preventing the genetic erosion of threatened species. However, this growing scientific expertise must lead to positive actions to save biodiversity, especially as climate change looms over the earth as a result of human ignorance and carelessness.

Conservation biology is based on the taxonomic documentation of floral and faunal species. While it has grown out of classic taxonomy it even uses traditional knowledge of plants and animals. This is known as trans-disciplinary studies and includes the knowledge of traditional communities such as tribal folk, agriculturists and fisher-folk. It looks at the inter-relationships of species as communities of living things, thus defining the role of different ecosystems and landscape typologies in our country.

India's natural landscapes have been modified by human activities over several thousands of years. This has created cultural landscapes at the cost of the wilderness ecosystems. This has disrupted the integrity of the habitats of wild species. Human interference has changed the behavioural patterns of wildlife in relation to environmental alterations of their habitats made by people in the recent past. Conservation biology is the basis of mitigating and reversing the effects of climate change on species. Finding appropriate eco-restoration strategies for providing wildlife with their specific habitat requirements is now the greatest challenge.

Mammal species of conservation significance:	Species present in only one natural population which needs translocation to another PA:
• Indian wild ass	• Indian wild ass
• Phyre's leaf monkey	• Markhor
• Malabar civet	• Hangul
• Malay sun bear	• Sangai
• Markhor	• Dryland swamp deer
• Argali/nayan	• Asiatic lion
• Pallas cat	• Urial
• Rusty spotted cat	
• Tibetan antelope	
• Tibetan gazelle	
• Pig-tailed macaque	
• Stump-tailed macaque	

Case Study 4.1   The Great Indian Bustard

The Great Indian Bustard (*Ardeotis nigriceps*) is listed as a critically endangered species with around 150 individuals left exclusively in India. It lies within Schedule I of the Wildlife Protection Act 1972, which gave this species the highest protection status. GIBs are facing an imminent extinction risk as their population has steadily declined by 75% in the last 30 years. The GIB was distributed throughout the western half of India, but currently they are present in only a few grasslands in Rajasthan. Currently the largest population of GIB is present in the Thar landscape of Rajasthan.

Direct and indirect human exploitation is one of the reasons for the decline of the species. However, the loss of habitat (dry grasslands) is a major factor. Changed cropping patterns have led to food scarcity due to loss of insects killed by pesticides. Being low and heavy flyers, they face a high risk of fatal collisions with power lines. Also, an increased numbers of feral dogs as well as wild predators such as foxes, mongooses and cats have increased predation pressure on nests and chicks. Effective conservation of the bustard aims at integrating all these components into a holistic conservation plan for priority bustard landscapes. Bustards are now to be bred in captivity through a scientifically managed ex situ conservation programme in Rajasthan, so that the birds can be released into their natural habitat in future.

Conservation biology provides strategies for protecting various species from extinction and preserving our ecosystems and the genes of various species. This not only needs good science, but also requires knowledge of human behaviour, the sustainable use of nature's resources and ecosystem services. Advocacy towards a more conservation-based governance is part of our duty towards our future citizens.

Colour Plate 4.1: Biogeographic zones of India

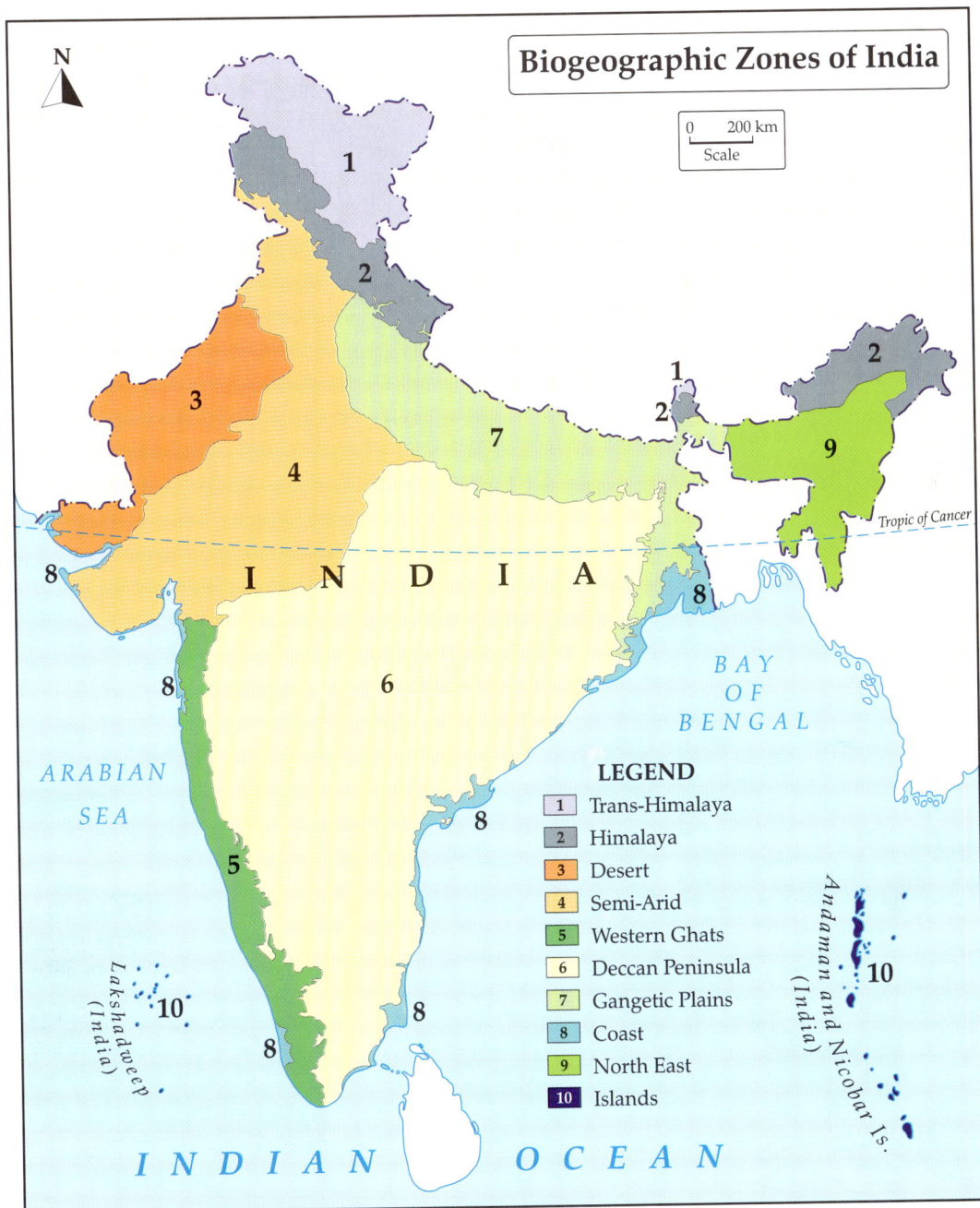

Biogeographic Zones of India

N

0   200 km
Scale

1

2

3

4

5

6

7

8

9

10

2

1
2

9

Tropic of Cancer

BAY
OF
BENGAL

ARABIAN
SEA

Lakshadweep
(India)

Andaman and Nicobar Is.
(India)

LEGEND

1	Trans-Himalaya
2	Himalaya
3	Desert
4	Semi-Arid
5	Western Ghats
6	Deccan Peninsula
7	Gangetic Plains
8	Coast
9	North East
10	Islands

I N D I A

I N D I A N   O C E A N

1. ©Government of India, Copyright 2020.
2. The responsibility for the correctness of internal details rests with the publisher.
3. The territorial waters of India extend into the sea to a distance of twelve nautical miles measured from the appropriate base line.
4. The external boundaries and coastlines of India agree with the Record/Master Copy certified by Survey of India.
5. The spellings of names in this map, have been taken from various sources.

Colour Plate 4.2: Biogeographic zones of India

Trans Himalaya

Himalaya

Desert

Semi-arid

Western Ghats

Deccan plateau

Gangetic plain

Coast

Mangroves (Coastal)

North-east region

North-East

Islands

Colour Plate 4.3: Common plant species of India (flora)

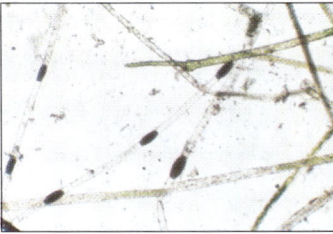
Algae

Fern

Bamboo

Drosera

Pine

Coniferous trees

Ficus

Teak

Indian coral tree

Strobilanthus shrub

Orchid

Lotus

Colour Plate 4.4: Common animal species of India

Colour Plate 4.4: (a) Mammals

Leopard

Tiger

Lion

Wolf

Gaur

Nilgai

Elephant

Rhinocerous

Chital

Sambar

Blackbuck

Porcupine

Hare

Macaque

Hanuman langur

Colour Plate 4.4: (b) Birds

Peafowl

Pheasant

Jungle fowl

Greater flamingoes

Vultures

Barn owl

Colour Plate 4.4: (c) Reptiles

Garden lizard

Crocodile

Monitor lizard

Vine snake

Turtle

Star tortoise

Colour Plate 4.4: (d) Amphibians and fish

Toad

Tree frog

Fish

Colour Plate 4.4: (e) Invertebrates

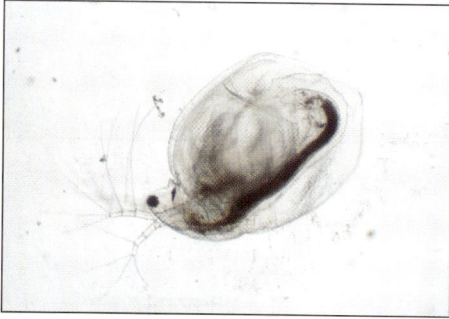

Zoo plankton

Coral

Beetle

Blue mormon butterfly

Dragonfly

Crab

Snail

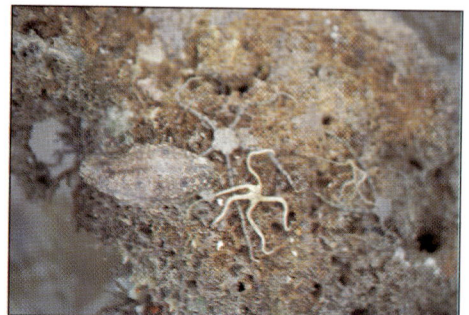

Brittle star

Colour Plate 4.5: Ecosystems

Coniferous forest

Deciduous sal forest

Deciduous teak forest

Thorn Forest

Grassland

Wetland

River

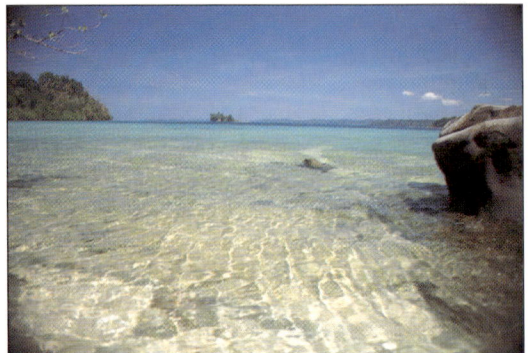
Marine

To appreciate the endangered, endemic and rare species of India, it is important to have an overview of the wide variety of plant and animal species found in the natural ecosystems left in the country. Of the well-known species that are seen across India, in the 10 biogeographic zones, many species have been restricted to small ranges and some are found only in a single or a few Protected Areas. For example, the Asiatic lion is found only in the Gir National Park, the hangul (Kashmir stag) is now seen only in Dachigam in Kashmir. The hard land barasingha is restricted to Kanha National Park. The golden langur is found mainly in Manas and a few in other areas. The rhino is seen in Kaziranga and a few have been translocated to Manas in Assam. Several species of avifauna such as the Great Indian Bustard and florican, that were once widely distributed in our semi-arid grasslands, now occur in isolated small pockets of Rajasthan and thus cannot breed successfully due to the loss of its grassland habitat which has been changed to irrigated farmland or plantation of fast growing trees. Hundreds of coral species in the shallow seas of Gujarat, Tamil Nadu, Andaman and Lakshadweep Islands are being severely bleached and are dying as a result of climate change. Species of fish such as the mahseer are now rarely seen in our rivers. There are several floral species in small areas of the Himalayas and Western Ghats, many of which are not yet researched for their medicinal properties are becoming rare due to over harvesting.

Case Study 4.2   Siberian crane

The critically endangered Siberian crane (*Leucogeranus leucogeranus*), also known as the Siberian white crane or the snow crane, is the third rarest and the most threatened species of crane in the world. This species breeds in Russia and winters in China, and formerly was seen in India in the Keoladeo, Ghana (Bharatpur) bird sanctuary. It used to migrate over 5,000 km through seven other countries. It was found in wetland habitats throughout the year along its migratory route and at wintering and breeding sites. It is highly dependent on wetlands, where it was killed along its migratory route.

Historic records from India suggest a wider winter distribution in the past including records from Gujarat, near New Delhi, and even as far east as Bihar. In 1974, as many as 75 birds wintered in Bharatpur and this declined to a single pair in 1992 and the last bird was seen in India in 2002. India has unfortunately lost this bird forever.

Case Study 4.3   Gujarat's gentle giant: The whale shark

The whale shark (*Rhincodon typus*) is known to be the largest fish on the earth. This plankton-feeding, ovoviviparous and highly migratory shark species is found along the Indian coastline. Prior to 2001, catch statistics and anecdotal reports across the shores of Gujarat, revealed that this unique species was declining severely due to the unsustainable and unregulated capture to meet international trade demands for shark fins, liver oil, skin and meat. Additional threats included accidental entanglement in trawl nets, collision with boats and extensive coastal pollution.

In 2001, the Wildlife Trust of India (WTI) urged the Ministry of Environment and Forests (MoEF) for legal protection of this species by placing it in Schedule I of the Wildlife (Protection) Act, 1972. In 2002, the fish was included under Appendix II of the Convention on International Trade in Endangered Species (CITES). WTI conducted a survey along the coastal town of Veraval, the hub of whale shark slaughter on the coast of Gujarat, which showed that there were low awareness levels on poaching and protecting the species among local people. A *whale shark awareness campaign* was launched by WTI in 2004 to change the perception of local people and local protection of the species. In 2008, WTI with support from TATA Chemicals, launched the Whale Shark Conservation Project. A religious leader, Morari Baba, also supported the conservation of the whale shark. The objective of the project was to rescue and release the animal if caught in a net. Photo identification and geotagging of the whale sharks was done. In 2013, 372 whale sharks had been rescued and voluntarily released by the fishermen. WTI has extended the project beyond the Gujarat coast to other western coastal states of India and collaborates in the global efforts in whale shark conservation, research and management.

CASE STUDY 4.4   Olive ridley turtles

The olive ridley turtle (*Lepidochelys olivacea*) is recognised by its olive coloured carapace, which is heart-shaped and rounded. These turtles are known to spend their entire lives in the ocean and migrate thousands of kilometres for feeding and mating in certain sites. The female turtle returns to the same beach where she was born to lay her eggs. To do that, she sometimes swims for thousands of kilometres. She lays her eggs in a pit on the beach, covers them with sand, and returns to the sea. The eggs hatch after 50–60 days. Their number is decreasing and they are classified as 'Endangered' under the Wildlife (Protection) Act, 1972.

The olive ridleys face serious threats across their migratory route, habitat and nesting beaches, due to human activities, development and exploitation of nesting beaches for ports, and tourist centres, extensive poaching for their meat and eggs, shell and leather. Accidental killing of adult turtles through entanglement in trawl nets of fisher-folk and gill nets due to uncontrolled fishing during their mating season especially around nesting beaches, kills thousands of turtles each year.

To avoid such threats WWF-India, along with the fishermen community, has been involved in protecting the olive ridley rookery at the mass nesting site at Rushikulya, in Odisha, by fencing off the nesting area and patrolling it till hatching and ensuring a safe passage for the hatchlings to the sea. The Odisha government has made it mandatory for trawls to use Turtle Excluder Devices (TEDs), a net specially designed with an exit cover which allows the turtles to escape while retaining the fish catch. The villagers of Anjarle and Velas in the Maharashtra coast (which is a famous nesting site of the olive ridley turtle) have taken it upon themselves to protect the turtles. The local people spot the nests and measure the depth and move the eggs to a barricaded nesting site on the same beach, where they are transferred to similar sized nests. The nests are marked with the date on which they were found, and the estimated date of hatching. A person is assigned to check on the eggs every morning and evening to help the young turtles to make their way to the ocean.

The Indian cheetah and the pink headed duck are globally extinct. The Siberian crane that used to be seen only in the Bharatpur wetland in India, has not been seen there for about twenty years and now flies from its breeding grounds in Russia only to China. On the brink of early extinction are our six species of vultures that are dying due to the use of diclofenac in veterinary medicine. When vultures feed on carcasses of livestock that have been treated with diclofenac, they die due to kidney failure. India has lost over 90% of the population of vultures which are an important species of several ecosystems as they clean up dead material.

Elephants need corridors to move between feeding areas in different seasons. As the corridors are increasingly used for farmland and urbanisation, there is a serious conflict between elephants and local people. However, the population of species such as tigers and elephants which were falling 50 years ago are slowly recovering in well managed Protected Areas.

Several species are endangered by human activity. The endangered species are categorised as *vulnerable, rare, indeterminate* or *threatened*. Some species are endemic and are at a high risk of extinction if they are wiped out in India. Illegal poaching of animals for medicines and wild medicinal plants take them to the brink of extinction.

Learning by critical thinking

◆ Understanding the diversity of flora is thus crucial to human wellbeing.
◆ What would you feel if the tiger became extinct, or if elephants are killed by poachers, or are affected by loss of habitat or corridors they need to go from one feeding area to another?
◆ How would you feel if any other species you know and love were to disappear forever? Think of a few examples.
◆ So, why don't you feel the same sadness and loss for the other lesser-known species that are likely to become extinct in the near future? They are out of sight and out of mind.
◆ Although they are all a part of the same great diversity of plants and animals and are an equally important part of nature, we don't give them their due importance.
◆ Think about it, we need to be equally sensitive to all living creatures.

India's common species of flowering plants and conifers are distributed across our 10 biogeographic zones. The broad leaved plants are found in the whole of peninsular India, while coniferous trees are found mainly in the Himalayas. Some are seen naturally in a single zone, while others are found across many different zones. There are species such as pine and oak which are found only in the Himalayas and North-eastern states. The thorny species are found in the Thar desert, semi-arid zone and the dryer parts of the Deccan plateau (neem, babul, ber). The Western Ghats, North-eastern states, and the Andaman and Nicobar Islands, are rich in evergreen trees and endemic species, especially the rare orchids and ferns. There are species that grow only along our coastal belt. Mangrove species favour brackish areas and deltas of rivers—the Sunderbans being the largest mangroves.

4.3.2 Keystone Species and Ecosystems

Each species requires several other species that are linked to it in the food chains and web of life of its ecosystem. Thus, a community of species using the same habitat with inter-linking food chains forms a feeding 'guild' in which the species have similar food habits and share the same niche or microhabitat. Among these many inter-linkages, a few species are used by a relatively large number of species as food. Species that anchor many species with itself in the food chain and food pyramid are called *keystone* species.

Banyan trees are a keystone species as they provide food for many different species of birds and mammals that feed on its figs. These also fruit in succession over several months, thus providing food to frugivorous species when other species of trees do not have any fruits. Loss of a keystone species produces a serious food shortage for many species linked to it.

Some predators can also be considered as keystone species as they have a major controlling influence on herbivore populations. Large herbivores such as elephants and gaur that require large quantities of food also play a major role in modifying their own habitat by selectively feeding on a favoured tree species.

If the abundance of a keystone species is decreased in an ecosystem, or substituted by an alternate species due to human activities, or by the accidental invasion of an imported exotic species, a cascade effect may result in the abundance of different species in the ecosystem. The decrease in the population of several species which are dependent on the keystone species may eventually lead to extinctions.

Another group of keystone species are those that tend to modify their environment, called *ecosystem engineers*. A classic example is the termite colonies in a forest that modify soil characteristics by breaking down leaf and fallen wood into smaller fragments. Insects increase the surface area of dead material on which fungi and bacteria act, help reform and refresh the soil with nutrients for plants to grow on. Soil fauna such as earthworms, insects and other soil invertebrates play this important role of controlling engineering functions in the ecosystem.

Snags and fallen dead wood

Old trees frequently develop hollow interiors and holes where branches have fallen off. These hollows are an important part of the habitat of many faunal species. Hornbills, woodpeckers, barbets, squirrels and civets use hollows for resting and nesting year after year. Bats use caves for their colonies. Fish depend on perennial pools of water.

A fallen dead tree gradually disintegrates by the effects of the physical environment, weathering due to wind and rain or chemical degradation. The dead tree is the food and home for insect life, worms and many invertebrates that are dependent on old standing dead wood, fallen dead logs and uprooted trees. Thus, removing these dead trees from a habitat has several serious consequences for the integrity of a natural ecosystem.

Some species have highly specialised feeding habits. If the abundance of such feeding trees is reduced, it affects the population of the specialised species that use it for food. If pollinators such as insects are destroyed by pesticides, the dependent trees do not get fruit. There are various interlinked mechanisms in nature which must remain intact for retaining these resources. This is a serious problem for agriculture and horticulture as these insects pollinate our crops and orchards.

4.3.3   Keystone Resources and Ecosystem Linkages

Just as some species act as keystone species in a community, a highly specialised *hot speck* or habitat of ecosystem diversity of a relatively small size, can harbour a large population of faunal or floral species. This includes species that are rare, endangered or highly endemic to the particular habitat. These habitats are considered as *key resource areas*. Many species can be linked to these highly specialised areas which are often small. A good example is the sacred groves of tribal communities which are hundreds of years old and are patches of old growth forest. Other good examples in the Indian context are the lateritic plateaus of the Western Ghats that harbour a large number of endemic monsoonal ground flora. Waterfalls, hot springs, caves and hollows in trees that act as nesting sites for birds and dens for mammals are specialised key resource areas.

Certain species form important link species in a food chain. These are highly specialised pollinators or seed dispersal agents. Many berries which are eaten by birds must pass through an animal's gut to be able to germinate. Fireflies have a complex life cycle. The females require soft bodied worms and molluscs (snails) to feed on, which are only found near pools of water. The male mates and dies a week later as it does not have an intestinal tract. Many species of spiders have large females and very small males. Once mating occurs, the female eats the male.

In a dry summer there may only be a few waterholes or streams for wildlife to drink from. These may be surrounded by the only green grass left in a large tract of forest. Herbivores inevitably move into such key areas in summer. Giant squirrels are essentially arboreal and rarely descend to the ground. Thus, they require a continuous canopy for their specific habitat needs.

India must also protect and grow its specialised crop varieties. The traditionally grown Indian cultivars included 30,000 to 50,000 varieties of rice and a number of other cereals such as millets and wheat as well as vegetables and fruits. The highest diversity of cultivars (cultivated food crops) is concentrated in the high-rainfall areas of the Western Ghats, Eastern Ghats, lower Himalayas and the North-eastern hills. Many of these areas are the farmlands of tribals who have continued to use these rare and genetically useful traditional crop varieties.

Indian National Gene Bank Collection (NBPGR) has collected a large number of species and varieties of indigenous crops. Modern gene banks have collected over 67,000 cereals and 26,542 pulses and 16585 millets grown in India. The people who have been traditionally keeping livestock have bred and nurtured 27 indigenous breeds of cattle, 40 breeds of sheep, 22 breeds of goats and 8 breeds of buffaloes. Many indigenous breeds are increasingly rare today, or are on the verge of extinction due to our misguided use of exotic foreign or mixed breeds. Jerseys and Holsteins have largely replaced Indian cattle like the Vaichur, Geer (Brahma bull) and Khillar cattle. Cross breeding with rare local breeds is leading to genetic erosion.

High-yielding cultivars and genetic manipulations have eaten away into centuries-old species of crops. Cash crops have replaced many specialised traditional ancient varieties of food crops. In the forestry sector, eucalyptus and wattle plantations have replaced the mixed *shola* forests. Species such as subabul, acacia, auriculoformis, glyricidia planted to green degraded forest areas has a negative impact on the biodiversity of wild flora and fauna which is indigenous to our dry areas. The Indian landscape is slowly beginning to lose its individuality.

4.4   THREATS TO BIODIVERSITY

Biodiversity is threatened across the world. After centuries of over exploitation of the diversity of plant and animal life, the nations of the world have come to realise that biodiversity must be conserved for the wellbeing of humans.

Some of the countries with high levels of biodiversity are located in South America, for example, Brazil and Costa Rica, Southeast Asian countries, for example Malaysia and Indonesia, and South Asian countries which include India, Pakistan, Bangladesh and Srilanka. The species diversity found in the bio-rich countries of South Asia is, however, different from those of South America and South East Asia. This makes it imperative to preserve our own biodiversity as a major economic resource for the future. While a few of the other mega-diversity nations have developed the technology to use their wild species (which have specific genetic properties) through biotechnology and genetic engineering, India is capable of applying its inherent strength in biotechnology to use the bioresources of wild and domestic species. This is used for manufacturing new drugs and a large number of industrial products.

Throughout the world, the value of biologically-rich natural areas is now being increasingly appreciated as being of unimaginable economic value. International agreements, like those made at the World Heritage Convention, attempt to protect and support such wilderness areas. India is a signatory to the World Heritage Convention and has included several Protected Areas as Natural World Heritage Sites. These include Manas on the border between Bhutan and India, Kaziranga in Assam, Bharatpur in UP, Nandadevi in the Himalayas, the Sunderbans in the Ganges delta in West Bengal and a chain of Protected Areas in the Western Ghats.

India is a signatory to the Convention on Biological Diversity that ensures that our bioresources are considered as our own sovereign national resources. No other foreign nation has access to our resources without the permission of India's National Biodiversity Authority.

India has also signed the Convention on International Trade on Endangered Species (CITES), which is intended to reduce the utilisation of endangered plants and animals by controlling trade in their products and in the pet trade.

The Union Ministry of Environment Forests and Climate Change (MoEF&CC) is the nodal agency for implementing provisions of the Convention of Biological diversity (CBD) in India. India has a National Biodiversity Authority in Chennai and has developed a National Biodiversity Action Plan under the Biodiversity Act of 2002.

Many plant species are increasingly threatened due to changes in their habitat induced by human activity. Apart from the major trees, shrubs and climbers that are extremely habitat specific and thus at greater risk, thousands of small herbs are also severely threatened by habitat loss. Orchids are a group under threat due to the illegal export market. Many plants are threatened due to over-harvesting for ingredients used in ayurvedic medicines in far eastern countries and even modern allopathic medicinal products. Cosmetics also use a large number of raw materials from plants.

The trade in Non-Timber Forest Products (NTFP) is overexploiting several plants. The Biodiversity Act of 2002 is a legal tool to prevent over harvesting by strengthening local village Biodiversity Management Committees to preserve their own village biodiversity.

To protect endangered species, India has created the Wildlife Protection Act. This includes a list of plants and animals categorised according to the level of threat to their survival. It ensures that Protected Areas are carefully managed and poachers are caught. In 2002, India passed the Biological Diversity Act in response to signing the Convention on Biological Diversity. It prevents illegal export of plants or animals and their products for commercial and other use abroad. It empowers local Biodiversity Management Committees at the village Panchayat level to manage their own wild flora and fauna. The Act also recognises that local people have preserved the biological diversity that is present in traditional cultivars (crops) and domestic livestock breeds, for many centuries. This is their intellectual property and they should derive economic benefits as they have consciously preserved genetic and species diversity of crop varieties and livestock breeds over many generations. Currently foreign companies are trying to use this valuable resource by exploiting the knowledge of the traditional folk cultures.

Experiential learning

How can we appreciate endangered or endemic species that we may never even see? The root to this understanding can only come about by observing and learning about some of our more common species of plants, animals, birds and aquatic species in nature that are present around us.

Threats

The threats to biodiversity include habitat loss, poaching, human–wildlife conflicts and biological invasions. Economically unsustainable development frequently has multiple ill effects. Some threats are permanent and cannot be reversed even after decades. Others are reversible through good pro-active conservation management.

4.4.1 Habitat Loss

Humans are overusing/misusing most of the natural ecosystems. Due to this unsustainable use of resources, once-productive forests and grasslands have been turned into deserts and wastelands all over the world. Mangroves have been cleared for fuel wood and prawn farming. This has led to a decrease in the habitat essential for the breeding of marine fish. Wetlands have been drained to increase agricultural land. These changes have grave economic implications in the longer term.

Worldwide, the current destruction of the remaining large areas of wilderness habitats, especially in the diverse tropical forests and coral reefs, is the most palpable threat to biodiversity. Scientists have estimated that human activities are likely to eliminate approximately 10 million species by the year 2050.

There are about 1.9 million species of plants and animals in the world that are known to science at present. A number of species are not studied or identified. The number is likely to be at least ten times greater. Plants including flowering plants and insects, as well as other forms of life not known till date, are continually being identified in the world's hotspots of diversity. Unfortunately, at the present rate of extinction, about 25% of the worlds' species will undergo extinction fairly rapidly. This may occur at the rate of 10,000 to 20,000 species per year. This is 1000 to10,000 times faster than the expected natural rate! Much of this mega-extinction is related to human population growth, industrialisation and changes in land-use patterns. A major part of these extinctions will occur in bio-rich areas such as tropical forests, wetlands and coral reefs. The loss of wild habitats, due to rapid human population growth and short-term gains from economic development, is a major contributor to the rapid global destruction of biodiversity.

4.4.2 Poaching of Wildlife

Specific threats to certain animals are related to large illegal economic benefits. The skin and bones of tigers, ivory of elephants, horns of rhinos and perfume of the musk deer are extensively used abroad. Bears are killed for their gall bladders which are used in medicine in the Far Eastern countries. Coral and shells are collected for export and are illegally sold on the beaches of Chennai, Kanyakumari, and the Andaman and Nicobar Islands. Tortoises, birds and other small animals are packed into tiny containers and smuggled abroad for pet trade.

A variety of wild plants with real or sometimes dubious medicinal value are being over-harvested. The commonly collected plants include rauwolfia, *Nux vomica*, datura and several other species from the Himalayas and Western Ghats. The garden plants collected for illegal trade include orchids, ferns and mosses. The Convention on International Trade in Endangered Species has been signed by many countries to prevent illegal trade of endangered species. India is a signatory of the convention and in spite of attempts to reduce poaching, it is still rampant and carried out by a highly destructive poachers network. However, the major criminals are the traders and exporters, while the exploited poacher who is apprehended is a poverty stricken local individual living in a bio-rich area.

Introspection: Action learning

- ❖ Who is the largest criminal—the major exporter and illegal trader, the local functionary who helps poachers, the transporter or the poor individual who lays traps for the wildlife?
- ❖ Have you ever used a plant or animal product that is illegally sold?
- ❖ What should we do if we see the sale of an endangered species?
- ❖ Will you inform the police and/or Forest Department?

4.4.3   Human–Wildlife Conflicts

There are three major concerns when such conflicts occur—crop damage by elephants, deer; livestock lifting by tigers and leopards; rare occurrence of predators killing humans. The most important causes are (i) lack of wilderness on the periphery of PAs (buffer zones); (ii) land use changes into intensified canal-fed farms prompting predators and herbivores to move into human landscapes for food and cover, thus coming into direct contact with people; (iii) conversion of elephant migration corridors into tea and coffee plantations; (iv) damage of crops by Nilgai and wild boar causing serious economic loss. All these lead to local resentment to the presence of PAs.

Conflict resolution is highly complex. (i) Each PA should maintain ecologically sensitive buffer zones between farmlands, settlements or industrial areas that are at the fringes of the National Parks and sanctuaries. (ii) Resettlement and rehabilitation of people who live inside core areas of Tiger Reserves and National Parks must be adequately done. (iii) Crop damage, livestock predation and loss of human life due to wildlife must be prevented and compensated by the country when it occurs, just as it takes responsibility when disasters such as floods, droughts and earthquakes occur. Local people cannot be expected to pay the price for conserving wildlife. (iv) Eliciting local participation for management of the PA by enhancing alternative income by capacity building, education and skill development programmes, marketing of NTFP at a sustainable level, ecotourism through homestay facilities and eco-restoration of degraded land outside the PA. This will ensure that locals do not resent the presence of the PA.

More people die from road accidents, unhygienic air and water, and vector borne preventable diseases than from wild animals. Tigers and leopards are killed by poachers for trade of skin and bones, elephants for ivory and smaller wild animals for pet trade; these should be a part of conflict management. While we have policies, laws and compensation schemes for human–wildlife conflict, our efforts at implementing conflict management at ground level fall way short of what is required.

Learning through critical thinking

Loss of wild species occurs due to the destruction of natural ecosystems, either for conversion to agriculture or industry. Over-extraction of resources leads to severe changes in ecosystems. Pollution of air, water and soil degrades the environment. Poaching and illegal collection of medicinal plants is an ever present negative impact. What do you think should be done to change to a more nature oriented society?

CASE STUDY 4.5 Project Tiger

Project Tiger was launched by the Government of India with the support of WWF-International in 1973. It was aimed at protecting this key species and all its habitats. Project Tiger was initiated in nine tiger reserves in different ecosystems of the country. In 2020, the number of tiger reserves has increased to 50. Project Tiger recognised the fact that tigers cannot be protected in isolation and that to protect the tiger, its habitat needs to be preserved and carefully managed. The sudden disappearance of the tigers from the Sariska exposed a network of illegal poachers who were exporting the skin and bones of tigers for Chinese markets. Tiger estimations are done using camera traps placed in the tiger reserve and a complete programme can identify each individual based on the pattern of its stripes.

The National Tiger Conservation Authority (NTCA) has been set up to further the aims and objectives of tiger reserves. The 2018 tiger estimation report stated that there are 2967 tigers.

CASE STUDY 4.6 Project Elephant

Project Elephant was launched in 1992 by the MoEF to provide financial and technical support to wildlife management efforts by the states for their free-ranging populations of wild Asian elephants. The project aimed to achieve long term survival of the populations of elephants in their natural habitats by protecting the elephants, their habitats and migration corridors. The project also supported research of the ecology and management of elephants, conducted public education and awareness programmes for conservation among local people and provided improved veterinary care for captive elephants. Through the project, measures for mitigating human–elephant conflicts and opposing the undue pressures of human interference and domestic stock and the impact on crucial habitats has been addressed.

The 2018 elephant estimation report stated that there are 29964 elephants across the three major habitats in India. The WTI has identified 100 corridors through which elephants migrate seasonally and it is imperative that these areas be provided with a 'right of passage'. However, conflicts with local people have to be addressed in a sensitive manner.

CASE STUDY 4.7 Silent Valley National Park

The Silent Valley National Park is located in the North-east corner of the Palakkad district, Kerala. Silent Valley was named a National Park only in 1985. Before that there was a plan for a hydroelectric project that threatened the park's rich wildlife. This initiated a strong environmental-social movement in the 1970s, known as the 'Save Silent Valley' movement, which resulted in cancellation of the project and creation of the park in 1985. Dr Salim Ali's intervention with the then Prime Minister Indira Gandhi, led to the creation of the protected area and cancellation of the dam in Kerala. Initially there was only 89.52 sq. km area under the division which forms the core zone of the National Park. In 2007, an area of 148 sq. km was added to this division as buffer zone. The National Park is home to 41 mammals, 211 birds, 49 reptiles, 47 amphibians, 12 fish, 164 butterflies and 400 species of moths. Apart from this, the park harbours a viable population of the lion tailed macaque which also is the flagship species of the Park.

The tribal communities living around the National Park include the Kurumba, Muduga, Irula and Kattu Naiken tribes. These people are mainly agriculturists who usually follow shifting cultivation (Panchakkad). Six Eco Development Committees (EDCs) around the park facilitate different outreach programs and welfare activities by the park and help in management. Due to this, Silent Valley has evolved into an ecological haven and the local community now feel that their environmental needs are taken care of by the presence of the valley forests and environmental threats will be mitigated by its presence.

4.4.4 Biological Invasion

Invasive species: An important factor that disrupts forest biodiversity is the introduction of exotic weeds (invasive species), which are not a part of the natural vegetation. Some common examples in India are lantana (*Lantana camara*), eupatorium (*Eupatorium perfoliatum*), water hyacynth and parthenium (*Parthenium hysterophorus*) or congress grass. These have been imported into the country from abroad and have invaded large tracts of our natural forests. These invasive shrubs and weeds spread at the expense of the diverse range of natural indigenous forest undergrowth species. This has a serious negative impact on the diversity of insects, birds and other wildlife species, which depend on a large diversity of local plant species. This is a serious and increasing impact on our biodiversity.

Island flora and fauna, which have high endemism in small isolated areas surrounded by the sea on all sides, have been most seriously affected by human activity. This has already led to the extinction of many island plants and animals (the dodo in Madagascar is a well-known example). Bringing species into our country from other regions rapidly depletes biodiversity, especially of our islands.

4.4.5   Conservation of Biodiversity (In situ and Ex situ)

National initiatives in implementing conservation areas: Conservation areas are not new to India. Several ancient indigenous cultures dedicated an area to a favoured or often feared animistic deity. This led to conservation of the habitat and its species in sacred groves or other hallowed sites. In the more recent historical context, conservation areas were implemented for a wide variety of reasons. The earliest conservation efforts in India that are well-documented and are an evidence of the protection of sites and specific species, were implemented by the great ruler Ashoka. He not only promulgated the Buddhist philosophy of *ahimsa*, but gave up eating wild animals. Ashokan edicts in the 3rd century BC specified the species that had to be protected in a well-documented list. Today we refer to our modern lists as scientifically selected 'scheduled' species. In ancient India, elephant reserves were forests that were left undisturbed for breeding of wild elephants. This was one of ancient India's earliest forest management strategies. The rulers required elephants for their war machinery and had perfected traditional methods to catch and domesticate them centuries ago.

Most monarchs who were powerful in the ancient Indian states created repositories of wildlife and refuges for animals for their own royal shikhars. This prevented their subjects from foraging and/or hunting for food, or killing wild animals from the forests that were considered the rulers' personal royal reserves.

The Mughal rulers, who were especially fond of large mass killing of animals for sport (often by enclosing the unfortunate creatures in corrals), killed thousands of animals. They however created large reserves to protect animals and birds from being killed by local people. The British in India followed the same practice as far back as the early period of the East India Company. They even had well managed 'shooting blocks' where some form of sustainable hunting was allowed with permits for shikhar and gun licences. Hunting was however not closed in the breeding season in earlier times. Later there were closed seasons in the breeding period of fauna. Prior to independence, the Maharajas of Indian states continued to have private hunting reserves and had made fairly strict rules to punish poachers. Many of their well looked after hunting areas became National Parks and Wildlife Sanctuaries in independent India. In the recent past, Conservation Reserves and Community Reserves have been added through an Amendment of the Wildlife (Protection) Act in 2003. India currently, in 2019 has 869 Protected Areas.

The Biological Diversity Act of 2002 has brought in the concept of conservation of biodiversity along with sustainable use of resources. This is midway between a strict conservationist approach and a new strategy which takes into account the rights of local people for the use of their livelihood resources.

In situ conservation

In situ biodiversity conservation is done by protecting natural habitats and all the species that live in it. This is the major objective of our National Parks and Wildlife Sanctuaries. Biodiversity at all levels – genetic, species and intact ecosystems – is preserved in situ by setting aside an adequate representation of wilderness as protected areas in each of our 10 biogeographic zones. These consist of a network of Protected Areas under the State forest departments.

The Protected Areas – national parks and sanctuaries – are notified to preserve major wildlife species such as tigers, lions, elephants, gaur and deer, as well as birds, reptiles, fish, amphibians and invertebrates. They also preserve all species of trees, shrubs, climbers and ground flora. These relatively intact protected areas preserve all the microscopic unicellular plants and animals. Thus, each ecosystem is preserved in its natural state. This includes all the terrestrial and aquatic freshwater ecosystems, marine protected areas and all their associated species.

Observational learning

A species cannot be protected individually as all species are interdependent. Hence the whole ecosystem must be protected. The biologist's view point deals with areas that are relatively species-rich, or those where rare, threatened or endangered species are found. Those areas that have endemic species require special protection. As rare or endemic species are usually found only in a small area, these species can easily become extinct due to human activities. Such areas must be given an added importance as their biodiversity is a special feature of the region.

Think how important these Protected Areas are for the future generations who have a right to enjoy seeing and learning about the 1.9 million species on earth. Can we neglect them?

Reflection: The need for Protected Area corridors

Animals such as elephants require different types of habitats to feed in during different seasons. They utilise open grasslands after the rains when the young grass shoots are highly nutritious. As the grasses dry, the elephants move into the forest to feed on foliage from the trees and bamboos. Thus corridors between protected areas have to be kept intact for elephants to migrate seasonally.

India's Protected Areas

In India we have 769 Protected Areas (2018). There are 103 national parks, 544 wildlife sanctuaries, 76 conservation reserves and 46 community reserves. India has 8% of the world's species in only 2% of the world's terrestrial land. This gives India a mega diversity status at the global level. Thus the 5% of India's land in which the government has notified Protected Areas is of great importance to our nation as well as for the world.

Protected Areas which help preserve our natural resources provide vital services for people who live outside their boundaries. Water from streams in the Protected Area flows out of forests, and insects pollinate croplands. The natural habitat provides clean air and provides water for drinking and irrigation. Ecosystem services include preserving good soil with its nutrients.

Biodiversity provides opportunities for tourism, peace and quiet and a feeling of appreciation of the many beautiful things present in our world. Thus, while Protected Areas restrict environmental goods from being taken out for unsustainable use (both for consumptive purposes by local people and productive goods for sale and income generation) the economic value of services from the Protected Areas provides innumerable and unaccountable services for surrounding people which far outweighs the potential value of restricting the use of its saleable goods.

In today's world, natural areas are under threat due to changes in landuse. This decreases the natural habitats which affects the diversity of flora and fauna.

We cannot as citizens protect species on our own. But we can do many things to support our governments to set up and fund the management of Protected Areas. Peoples' action groups and NGOs help in creating a lobby force to protect our biological resources both within and outside the network of PAs.

While 50 years ago this was an ad hoc selection of areas that had major mammal or bird populations, it was evident that this could not be expected to effectively protect ecosystems, and all our flora and fauna with an adequate representation of the great variations in genetic differences in individuals.

In 1988, Rodgers and Panwar of the Wildlife Institute of India designed a scientific base for selecting and notifying an effective Protected Area System in India. This was based on geography, climate (rainfall and temperature), soil, vegetation patterns and fauna distribution in the ten different biogeographic zones. The Protected Areas are now selected according to the needs of the individual biogeographic zones.

Wildlife sanctuaries and National Parks of India

Among the 763 Protected Areas in India, some have been created in order to protect highly endangered or endemic species of wild plants or animals found nowhere else in the world (**Table 4.3**). There are 50 Tiger Reserves which are the most critically important Protected Areas. Tiger reserves include a variety of ecosystems and habitats.

The Protected Area in Ladakh protects the high altitude plateau for the snow leopard and several species of wild sheep and goats. The rare black necked crane nests in its wetlands.

The Great Himalayan National Park is the largest sanctuary in this fragile Himalayan ecosystem and is one of the last homes of the beautiful snow leopard. The Dachigam sanctuary is the only place where the rare *hangul* (Kashmir stag) is found. There are several sanctuaries in the low lying *terai* region south of the Himalayas and in the Indo–Gangetic plain. Of these, the Kaziranga National Park is the most famous as it is the home of the one-horned rhinocerous. In the last few decades, some of them have been translocated to Manas and other PAs. The terai Protected Areas have elephant, gaur, wild boar, swamp deer and hog deer, and are known for tigers, leopards and rare small wildcats. The bird life is extremely rich and includes ducks, geese, pelicans and storks in the wetlands. The Manas sanctuary, in addition to the common *terai* species, also includes the rare golden *langur* and the very rare pygmy hog, which is the smallest wild boar in the world. The florican is found only in a few undisturbed grasslands in the *terai* sanctuaries.

The sal forest PAs of Madhya Pradesh includes Kanha which offers a wonderful opportunity to observe wildlife and is the habitat of the rare barasingha. It is the only protected area in which a sub-species of the barasingha is found.

Keoldev Ghana national park—Bharatpur, is one of the most famous water-bird sanctuaries in the world. Thousands of ducks, geese, herons and other wading birds can be seen here. In the past, this was the only home of the very rare Siberian crane. It used to migrate to India every winter. Siberian cranes have not been seen in India after 2002.

In the Thar desert, wildlife is protected in the Desert National Park. Here, large numbers of blackbuck, nilgai and chinkara can be seen. The last few Great Indian Bustards still live in these arid lands. Ranthambore is the most well-known Protected Area for observing tigers in thorn forests and scrubland in the semi-arid zone. The Great Rann which is the only breeding area of the flamingo in India, and the Little Rann of Kutch wild ass sanctuary are unique ecosystems. The star tortoise, the desert fox, the caracal and several other arid area birds and reptiles are found in this unique ecosystem.

In Gujarat, the Gir sanctuary protects the last population of the majestic Asiatic lion. This thorn forest with patches of deciduous forest is also the home of large herds of chital, sambar, nilgai and wild boar. Forest birds, reptiles and insect life are abundant in this sanctuary. The sanctuaries of the Western Ghats and associated hill ranges in the Nilgiris and Agasthyamalai hills protect some of the most diverse evergreen and shola forests. The Western Ghats are considered a global biodiversity hotspot and is recognised as a UNESCO world heritage site. The sanctuaries of the Western Ghats are the habitat of highly threatened species including the Malabar giant squirrel, the flying squirrel and a variety of hill birds. Several species of amphibians, reptiles and insects are found in the high rainfall evergreen tracts of the ghats, which are also rich in endemic plant life. A unique aspect of the ghats is its tree-less lateritic

Table 4.3 Important National Parks in India

No	Name	State	Biogeographic zones
1.	Hemis NP	Jammu and Kashmir	TH
2.	Valley of Flowers NP	Uttarakhand	H
3.	Dachigam NP	Jammu and Kashmir	H
4.	Namdapha NP	Arunachal Pradesh	H
5.	Desert NP	Rajasthan	D
6.	Sariska NP	Rajasthan	SA
7.	Gir NP	Gujarat	SA
8.	Ranthambore NP	Rajasthan	SA
9.	Mudumalai NP	Tamil Nadu	WG
10.	Eravikulam NP	Kerala	WG
11.	Periyar NP	Kerala	WG
12.	Silent Valley NP	Kerala	WG
13.	Bandipur NP	Karnataka	WG
14.	Nagarhole NP (Rajiv Gandhi NP)	Karnataka	WG
15.	Bandhavgarh NP	Madhya Pradesh	DP
16.	Kanha NP	Madhya Pradesh	DP
17.	Pench NP	Maharashtra	DP
18.	Tadoba NP	Maharashtra	DP
19.	Dudhwa NP	Uttar Pradesh	GP
20.	Corbett NP	Uttaranchal	GP
21.	Sundarbans NP	West Bengal	M/C
22.	Gulf of Kachchh Marine NP	Gujarat	M
23.	Borivili NP (Sanjay Gandhi NP)	Maharashtra	M/C
24.	Kaziranga NP	Assam	NE
25.	Manas NP	Assam	NE
26.	Mount Harriett NP	Andaman and Nicobar Islands	I

plateaus with very little soil cover, on which a large number of rare and endangered monsoon plants grow. They form carpets of multicoloured flowers many of which are very rare. An example is the Kass plateau near Satara.

Protected Areas such as Bhimashankar, Koyna, Chandoli and Radhanagari preserve the rich flora in Maharashtra; Bandipur, Bhadra, Dandeli and Nagarhole which are important Protected Areas in Karnataka; Eravikulam and Periyar Protected Areas, and the Silent Valley in Kerala are important habitats of a rich compliment of endemic flora and fauna. In the Nilgiri hills, the rich forest sanctuaries protect some of the important habitats of the Indian elephant in South India. Examples include Bandipur, Mudumalai, Wayanad and Bhadra. During the last several years, a large number of the tuskers have been ruthlessly killed for their ivory. The Eravikulam sanctuary protects the last pocket of the Nilgiri tahr, the only wild goat species south of the Himalayas.

Two important estuarine bird sanctuaries meant for preservation of coastal ecosystems are the Chilika lake in Odisha and Point Calimere in Tamil Nadu. The Sunderbans protect the largest mangrove delta in India. The Marine national park in Gujarat protects the shallow areas of the sea. Its islands, coral reefs and extensive mudflats are home to a wide range of marine life which include mammals such as whales and dolphins, fish such as the whale shark, marine invertebrates, starfish, mussels, molluscs and coral.

Over a hundred Protected Areas have been created in the Andaman and Nicobar islands to preserve their very special island ecosystems in different forest types. The marine ecosystem with its wealth of coral reefs is of high biological importance.

The need for an integrated Protected Area system

Protected Areas, to be effective, must be established in every biogeographic region. It is expected that at least 5% of every biogeographic zone should be protected for its flora and fauna. A relatively larger representation must be included in highly fragile ecosystems, or areas of high species diversity, or high endemism. Protected Areas must also be integrated with each other, by establishing corridors between adjacent areas wherever possible, so that wildlife can move between them. This is important to maintain a healthy breeding population so that inbreeding does not occur and the genetic diversity is maintained.

In our country, which has a rapidly growing human population, it is not easily feasible to set aside more and more land to create new Protected Areas. The need to provide a greater amount of land for agricultural, industrial, mining and urban needs has become an increasing cause of concern in a land whose natural resources are being used at unsustainable levels. This forms a major impediment for creating new Protected Areas as it creates conflicts with local people who require the use of ecosystem goods and services. Having said this, there is an urgent need to add to our Protected Areas to preserve our very rich biological heritage. Much of the natural wilderness has already undergone extensive changes. The residual areas that have high levels of species richness, endemism or habitats of endangered plants and animals must be notified as National Parks and Wildlife Sanctuaries. Other smaller areas can be made into Conservation Reserves or Community Reserves (Community Conserved Areas) which are managed by local people. It is now also essential to identify 'other effective (area based) conservation measures'. Wherever necessary, local people must be compensated adequately if the Protected Area poses a threat to their livelihood.

The International Union for Conservation of Nature and Natural Resources (IUCN) states that it is essential to include at least 10% of all ecosystems as Protected Areas for the long term conservation of biodiversity. Other scientists suggest a minimum of 5% of each biogeographic region. India currently has around 5% of land in the Protected Areas. However, this includes old plantations of sal or teak, which were developed for timber and thus are relatively poor in biological diversity with a low level of 'naturalness'.

There are only a few good grasslands left in our country which have been notified as Protected Areas. Some are over-grazed wastelands in areas that were once flourishing grasslands. Most of these areas have a low biological value and need careful management to allow them to be restored to a more natural state, with their full complement of plants and animals.

Only a few wetlands have been made into Protected Areas. These require better management as they occur in land under revenue, irrigation, forest and other departments. We have very few marine sanctuaries. The Gulf of Kutch and the sea around the Andaman Islands are very important coral reefs which protect marine invertebrates, fish and marine mammals such as dolphins and whales.

Isolated PAs cannot protect all the genetic diversity of species. Thus, there is a need to develop corridors so that wild species can transit from one PA to another safely. Such corridors are vital for animals such as elephants.

A major strategy to reduce impacts on the biodiversity of the Protected Areas is to provide a sustainable source of income for local people living around them on their fringes and buffer areas. Protected Areas curtail traditional grazing practices and access to fuelwood sources. These resources must be provided by developing alternate sources of income in the buffer areas. Plantations of fuelwood and good grassland management in areas outside Protected Areas can help reduce the pressure on the habitat of wildlife within the protected area. Effective management practices must ensure that local people around Protected Areas derive a direct economic benefit from it. Involving local people in Protected Area management and developing tourist facilities that support income generation for the local people helps in enlisting their support for the protected area.

A carefully designed management plan which incorporates an 'eco-development' component aimed at providing a source of fuelwood, fodder and alternate income generation for local people is an important aspect of Protected Area management. Alternative income for local people is encouraged through ecotourism and homestay facilities for tourists. However this should not exceed the carrying capacity of tourists to the Protected Area.

A major drive for conservation of biological diversity can only be implemented through mass environmental education programmes on the value of protecting our dwindling biological resources and our wild plants and animals.

Ex situ conservation

There are situations in which an endangered species is so close to extinction that in situ measures have failed to protect it. The species may be rapidly driven to extinction unless ex situ measures are initiated outside Protected Areas. Conserving the species outside its natural habitat in a carefully controlled situation such as a botanical garden for plants, or a zoological park for animals, where expertise is created to multiply the species under artificially managed conditions, is known as ex situ conservation. A successful ex situ breeding programme must ensure that the species, after its recovery to a safe level, is reintroduced into its original wild habitat. This requires rehabilitation of the degraded habitat and removal of causes such as changes in landuse, poaching, or other disturbances that are a threat to the ecosystem. Eliminating the primary cause of reducing the population of the species must be adequately addressed before its reintroduction into its wild habitat. These breeding programmes for rare plants and animals are however more costly than managing a protected area through in situ conservation strategies. Thus, they act as a last refuge for threatened wild species of plants and animals.

A few of the successful breeding programmes include the centres for breeding the endangered vultures by Bombay Natural History Society at Pinjore, Haryana, pygmy hog of Assam, and pheasants at Chail in Uttarakhand. The rhino has been translocated and released in Manas in Assam and a few other PAs from their original last home in Kaziranga. In situations such as floods or accidents where animals are at risk, there must be more lifetime care facilities to prevent their untimely death. This has been successful in Kaziranga through Wildlife Trust of India (WTI) as the Brahmaputra floods and kills elephants and rhinos.

The Botanical Survey of India (BSI) and Botanical gardens preserve collections of plant life that is endangered. Many of these are used in traditional and Ayurvedic medicine. These plants may soon become extinct due to overexploitation. Another form of preserving an endangered plant or animal species is by preserving its genes as germ plasm in a gene bank. This preserves its genetic properties which can be used to develop better productivity, or disease resistance, when required in the future. However, this is expensive and needs major laboratory facilities for different eco-regions.

When an animal is on the brink of extinction, it must be carefully bred so that in-breeding does not lead to a loss of its genetic diversity. Breeding from the same stock can lead to poorly adapted progeny, or even an inability to get enough offspring. Modern breeding programmes are undertaken in specialised zoos that provide for all the animal's needs, including enclosures that stimulate their wild habitats. There may also be a need to assist artificial breeding. Thus, while most zoos are meant to provide visitors with a visual experience of seeing a wild animal at close quarters and provide them with information about the species, a modern zoo has to go beyond these functions to include the breeding of endangered species as a conservation measure in off-visitor areas of the zoo.

In India, successful ex situ conservation programmes have been carried out for several species. The Darjeeling zoo has successfully bred several endangered Himalayan mammals. Three species of crocodiles have been reared highly successfully in ex situ conditions in The Madras Crocodile Bank Trust and Centre for Herpetology. Another success has been the breeding and rehabilitation of the very rare pygmy hog by the Guwahati zoo. The Delhi zoo has successfully bred the rare Manipur brow-antlered deer. The BNHS has a centre for breeding endangered vultures, which will be releasing birds back into the wild in the next few years.

CASE STUDY 4.8   Vulture breeding programme

Vultures were one of the most common bird species in India. In the last two decades, there has been a drastic decline in the number of vultures across India. Three of the nine species in India are now critically endangered according to the IUCN. Diclofenac is given to sick cattle. Once the animal dies, the carcasses are left out in the open where vultures feed on them. These birds are nature's most efficient scavengers and play a major role in keeping the environment clean. Vultures feeding on these animals get poisoned by diclofenac. This causes renal failure, visceral gout and eventually death. Meloxicam has now been recommended as an alternative to diclofenac. Though the Government of India has banned diclofenac, vetenary practioners still use diclofenac sold for human use for sick cattle. Thus, the ban is still relatively ineffectual.

In 2001, a Vulture Conservation Breeding Centre was established in Pinjore in Haryana by the Bombay Natural History Society (BNHS) and the Haryana Forest Department. The centre houses white-backed vultures, long-billed vultures, slender-billed vultures and the Himalayan griffon. The birds have been bred successfully at Pinjore and will be released gradually over the next few years. The government has now set up several breeding centres for vultures. The success of this programme implemented through the Bombay Natural History Society has made it possible to release ex situ bred vultures back into the wild.

CASE STUDY 4.9   Red panda

The red panda (*Ailurus fulgens*) is a small arboreal mammal found in the forests of India. It is found in Sikkim, western Arunachal Pradesh, Darjeeling district of West Bengal and parts of Meghalaya. It is the state animal of Sikkim. It is listed as 'Endangered' in the IUCN red list of Threatened Species and under Schedule I of the Indian

Wildlife (Protection) Act, 1972. The red panda has the highest legal protection at par with other threatened species. WWF-India has been working since 2005 in the eastern Himalayan region for the conservation of this species.

In Sikkim, an area of 1341 sq. km was identified as potential habitat, and it was found that more than 60% of it fell outside the Protected Area network, making the red panda population vulnerable to several threats. WWF worked with the local communities for conserving this species. In Sikkim, more than 200 individuals were trained in manufacturing bio-briquettes in the areas around Barsey Rhododendron Sanctuary and Khangchendzonga Biosphere Reserve. These briquettes, made from coal produced from agricultural waste and mud, were used for cooking and heating. This helped reduce fuelwood consumption from forest resources. The conservation committee of Community Conserved Areas (CCAs) were trained to look after all red panda conservation initiatives in the area with assistance from WWF-India. They were equipped with camera traps, GPS and field gear to independently conduct surveys and patrol the region to monitor red panda movements and control illegal activities. WWF-India had also collaborated with the Sikkim Anti-Rabies and Animal Health (SARAH).

CASE STUDY 4.10   Sangai

Sangai is a medium-sized deer with uniquely distinctive antlers. It has an extremely long brow tine, which gives it a beautiful shape. The forward protruding beam appears to come out from the eyebrow. This signifies its name, brow-antlered deer. The Sangai is an endemic and endangered deer found only in Manipur in India. It is also the state animal of Manipur. Its original natural habitat is the floating marshy grasslands known as *phumdi* of the Keibul Lamjao National Park, located in the southern parts of the Loktak Lake. This is the largest freshwater lake in eastern India.

Sangai was believed to be almost extinct by 1950, but six individuals were spotted in 1953 and the State of Manipur has protected the species to increase the population to 204. Sangai faces a threat from the steadily degenerating habitat of *phumdi* as a result of continuous inundation and flooding caused due to an artificial reservoir in parts of the Loktak Lake.

Two pairs of brow-antlered deer were brought to the Delhi zoo from Manipur in 1962. Since then their number has been increasing. This is an example of ex situ conservation action. However the animals have not been rehabilitated in the wetlands of Manipur.

CASE STUDY 4.11   Pygmy hog (The first successful ex situ conservation breeding programme in India)

Pygmy hog (*Orcula salvania*) is one of the smallest, rarest and specialised members of the pig family. It was known to occur across a narrow strip of early successional tall grassland plains along the southern Himalayan foothills, in NW Assam, where the species was 'rediscovered' in 1971 after it was long suspected to have become extinct. It is listed as a critically endangered species by the International Union for Conservation of Nature (IUCN); its population size is estimated to be fewer than 250 mature individuals. Today, the Pygmy hog is found in just three places in Assam—Manas, Sonai Rupai and Orang. The recovery programme was launched in 1995, during which the species was reduced to a single declining wild population of a few hundred hogs in the Manas National Park, with no individual in captivity in the world.

With the support of SOS (Save Our Species), Ecosystems–India, in collaboration with the Assam Forest Department, Durrell Wildlife Conservation Trust and the SSC Wild Pig Specialist Group started the ex situ conservation programme for the first time in India. The project maintained about 60 captive hogs at two places in Assam.

CASE STUDY 4.12   Great Indian Bustard

In 2019, MOEF and CC initiated a programme for breeding the endangered GIB in a special centre at Rajasthan. The captive bustards have been successfully bred and several chicks are now growing satisfactorily. However, it will be several years before the captive bred bustards can be released into the wild.

4.5   ECOSYSTEM AND BIODIVERSITY SERVICES

Biodiversity provides a variety of environmental services for human communities from its diverse species ecosystems and genes (**Colour Plate 4.5**). Preserving these various dimensions of biological diversity is essential at the global, regional and local levels, for supporting human life and wellbeing. The production of oxygen, reduction of carbon dioxide, maintenance of the water cycle and protection of the soil are some important ecosystem services. Many ecological processes such as pollination, seed dispersal, germination of seeds and commensalism among species are valuable services of nature on which human communities depend.

4.5.1   Ecological Services

Biological diversity is essential for preserving ecological processes such as fixing and recycling of nutrients, soil formation, circulation and cleansing of air and water, and by providing global life support as plants absorb carbon dioxide and give out oxygen. Biological diversity plays a key function in maintaining the water balance within ecosystems, watershed protection, maintaining stream and river flows throughout the year, erosion control and local flood reduction. For natural forests, grasslands, wetlands, rivers and marine ecology, these functional aspects are all related to biodiversity. These are examples of the hidden services for humanity which are provided by nature.

Our food, clothing, housing, energy needs and medicines are all resources that are directly or indirectly linked to the biological variety present in the biosphere. This is most obvious in the case of tribal communities who directly gather resources from the forest, or fisher-folk who catch fish in marine or freshwater ecosystems. For others, such as agricultural communities, biodiversity is used to select and grow the best suited crops that can grow in different environmental conditions. Urban communities generally use the greatest amount of environmental goods and services, which are all indirectly drawn from the biodiversity inherent in natural ecosystems.

It has become obvious that the preservation of biological resources is essential for the wellbeing and the long-term survival of human society. This diversity of living organisms, which is present in the wilderness as well as in crops and livestock, plays a major role in human sustainable development. The preservation of biodiversity is therefore integral to any strategy that aims at improving the quality of human life.

4.5.2   Economic Value

Consumptive use value: A straight forward example of consumptive uses is the direct utilisation of timber, food, fuelwood and fodder, collected and used from the surrounding ecosystem by local communities. Biodiversity contained in the ecosystem provides forest dwellers with their daily needs of food, building material, fodder, medicines and a variety of other products. They know the qualities and different uses of wood from different species of trees. They collect a large number of local fruits, roots and plant material that they use as food or medicinal products. These are known as Non-Timber Forest Products (NTFP). Rural people use construction material and make artefacts for their households from forests, grasslands, wetlands and marine systems. Fisher-folk are completely dependent on fish, crustaceans and other invertebrates and know where and how to catch fish and other edible aquatic animals and plants.

Thus, the consumptive use of biodiversity is greater than the economic value of forests, grasslands, wetlands or marine resources that are sold for trade.

Man and the web of Life

The biodiversity of a region influences every aspect of the lives of the people who inhabit it. Their living space and livelihoods depend on the type of ecosystem. Even people living in urban areas are dependent on the ecological services provided by the surrounding agricultural landscapes and distant PAs indirectly. Protected Areas are key providers of ecosystem services that are difficult to quantify. We frequently do not see this in everyday life, as it is not necessarily obvious. However, our lives are linked with every service that nature provides. The quality of water we drink and use, the air we breathe, the soil on which our food grows, are all influenced by a wide variety of living organisms, both plants and animals, and the complex ecosystem with which each component species is linked with in the web of life.

While it is well-known that plant life removes carbon dioxide and releases oxygen that we require to breathe, it is less obvious that fungi, small soil invertebrates and even microbes are essential for plants to grow. The functions in natural forests maintain water in the rivers that flow out of the forests in the post-monsoon period. This is rarely appreciated but is a vital component that supports our lives. Ants play a crucial role in transferring nutrients to the soil; we must recognise this in order to understand how we are completely dependent on each of the living things in the 'web of life' on earth.

Reflective thinking

The wilderness is an outcome of a long evolutionary process that has created an unimaginably great diversity of living species and the various ecosystems on earth in which all creatures including human beings live. Think about this and we cannot but want to protect our earth's unique biodiversity!

Productive use value: This category of resources comprises all the marketable goods that are collected and processed from ecosystems. These are goods that are collected from nature and sold as marketable produce. All our food comes from agricultural ecosystems that are indirectly or directly linked to nature's biosphere. Animal husbandry is dependent on pasture land or forests. The clothes we use come from cotton fields. Human civilisations have created millions of items that are sold in every market place which have originated from some natural resource. For the agricultural scientist, the biodiversity found in the wild relatives of crop plants or traditional crop varieties is the basis for developing more productive crops. The biotechnologist uses bio-rich areas to prospect and search for potential genetic properties in plants or animals that can be used to develop better varieties of crops. This is used in modern farming to develop more productive varieties of cultivars, or to develop better livestock.

To the pharmacist, biological diversity is the raw material from which new drugs can be made from wild plant or animal sources. Ayurvedic medicine is totally dependent on plant species. Local forest dwellers collect medicinal plants and sell them in village markets. To industrialists, biodiversity is a rich storehouse from which new industrial marketable products can be developed.

Genetic diversity enables scientists and farmers to selectively develop better crops and domestic animals through careful breeding programmes or through genetic engineering. Earlier, this was done by selecting the best varieties or pollinating crops to obtain a more productive or disease-resistant strain. Today, it is increasingly being done by genetic engineering—selecting genes from one plant and introducing them into another. New crop varieties (cultivars) are being developed using the genetic material found in wild relatives of crop plants through biotechnology.

Even today, species of plants and animals are being constantly discovered in the wild. These wild species are the building blocks for the betterment of human life. Their loss would result in great economic losses in the future. Among the known wild species, only a tiny fraction has been investigated in terms of their food, medicinal or industrial use. New ways of utilising bio-resources are constantly being studied by scientists.

A variety of industries (pharmaceuticals, cosmetics) are highly dependent on identifying compounds of great economic value. A wide variety of undiscovered wild species of plants are still being discovered in undisturbed natural forests. This is called *biological prospecting*.

4.5.3 Social Values of Biological Diversity

Biodiversity has to a great extent, been preserved by traditional societies that valued it as a resource (Table 4.4) and appreciated that its depletion would be a great loss to their society. Thus, apart from the local use or sale of products manufactured from biodiversity, the social aspect must also be considered. For example, many plants and animals are considered sacred in India and are worshipped. For example, tulsi, peepal, tiger, snake and many other wild species are venerated and therefore protected. Traditionally, the cow and various other animals have been venerated in our culture and religious beliefs. The Ramayana and Mahabharata mention many wild animals.

Table 4.4   Commonly used modern drugs derived from plant sources

Drug	Plant source	Use
Atropine	Belladonna	Anticholinergic; reduces intestinal pain in diarrhoea
Bromelain	Pineapple	Controls tissue inflammation due to infection
Caffeine	Tea, coffee	Stimulant of the central nervous system
Camphor	Camphor tree	Rubefacient; increases local blood supply
Cocaine	Cocoa	Analgesic and local anesthetic; reduces and prevents pain during surgery
Codeine	Opium poppy	Analgesic; reduces pain
Morphine	Opium poppy	Analgesic; controls pain
Colchicine	Autumn crocus	Anti-cancer agent
Digitoxin	Common foxglove	Cardiac stimulant used in heart diseases
Diosgenin	Wild yams	Source of female contraceptive; prevents pregnancy
L-DoPa	Velvet bean	Controls Parkinson's disease, which leads to jerky movements of the hands
Ergotamine	Smut-of-rye or ergot	Control of hemorrhage and migraine headaches
Glaziovine	*Ocotea glaziovii*	Antidepressant
Gossypol	Cotton	Male contraceptive
Indicine N-oxide	*Heliotropium indicum*	Anti-cancer agent
Menthol	Mint	Rubefacient; increases local blood supply and reduces pain on local application
Monocrotaline	*Cotolaria sessiliflora*	Anti-cancer agent
Papain	Papaya	Dissolves excess protein and mucus, during digestion
Penicillin	*Penicillium* fungi	General antibiotic, kills bacteria and controls infection by various microorganisms
Quinine	Yellow cinchona	Antimalarial
Reserpine	Indian snakeroot	Reduces high blood pressure
Scopolamine	Thorn apple	Sedative
Taxol	Pacific yew	Anti-cancer (ovarian)
Vinblastine	Rosy periwinkle	Anti-cancer agent; controls cancer in children
Vincristine	*Vinca rosea (Sadaphali)*	Anti-cancer agent; controls cancer in children

Source: E O Wilson, *The Diversity of Life,* Norton Protected Area perback, in association with Harvard University Press, 1993

Valmiki begins his epic with a description of a hunter who kills a crane. Valmiki admonishes him as the crane's partner will pine for its dead mate.

Traditional Knowledge Systems (TKS) on the uses of plants and animals, are best known to tribal and other indigenous groups of folk committees across India. This information is being rapidly lost. Preserving this knowledge on biodiversity amongst tribal people, fisher-folk, farmers, livestock breeders and people who know medicinal plants, is of great importance as it has enormous future economic potential. For example, to respond and adapt to climate change we will require the use of this TKS of biological diversity of our ancient culture. The ethical spiritual and existence values of flora and fauna are important reasons to preserve biodiversity.

CASE STUDY 4.13   Kokkare Bellur, Karnataka (Coexistence of humans and wildlife)

Kokkare Bellur is a village in Karnataka. Every year, hundreds of spot-billed pelicans, painted storks, ibis and other birds establish breeding colonies on the trees in the centre of the village. The local people have protected the birds, believing that they bring luck, good rains and better crops. The villagers collect the droppings of the birds and use it as natural fertiliser. The droppings of these fish-eating birds are rich in nitrates.

The owners of the trees inhabited by the birds dig pits under the trees, in which the guano is collected. Silt from nearby lakes and ponds are mixed with the guano, which is used in their fields and sold as organic fertiliser. The village people have planted trees around their homes to encourage nesting. Young birds that fall out of their nests are rescued and fed till they are ready to fly away.

CASE STUDY 4.14   Living bridges of the Khasi tribes of Meghalaya

The Khasi tribals who live deep inside the valleys of Meghalaya, North-East India, make bridges using live tree roots by joining them across streams. This region receives heavy rainfall, which causes streams to swell thus leaving some of the most remote villages cut off from the rest of the world. To solve this problem, the Khasis of the older generation who were experts in root crafts devised indigenous solutions of constructing these bridges. The technique that is unique to this tribal community is highly sustainable and long lasting.

Currently the tribal villagers grow ficus trees on either side of rivers and direct the roots to join midway to form a living crossing. Some living root bridges are created entirely by using the roots of the trees. The strongest bridges can carry more than 50 people at once and can live to more than 180 years. The living bridges take 15 years to get enough strength to be used on a daily basis. The bridges are now a large tourist attraction and eco-development and homestay facilities have been developed in the villages near the incredible living bridges.

4.5.4   Ethical and Moral Values of Biodiversity

Ethical values related to biodiversity conservation are based on the importance of protecting all forms of life. In many religions all living creatures have the right to live on earth. The Indian Constitution also includes the rights of all living beings.

Reflective learning

Humans are only a small part of the earth's great diversity of species. We do not know if life as we know it exists anywhere else in the universe. Do we have the right to destroy life forms, or do we have a duty to protect all forms of life?

Apart from the economic importance of conserving biodiversity, there are several cultural, moral and ethical values associated with the sanctity of all forms of life. Indian civilisation has, over several generations, preserved nature through local traditions. This has been an important part of the ancient philosophy of many of our cultures. We have in our country a large number of sacred groves or *devrais* preserved by tribal people in several states. These groves of ancient trees are preserved around many sacred shrines or temples. These old forests act as gene banks for wild plants that must be preserved. Thus these spiritual values can be linked to modern science today.

4.5.5 Aesthetic Value

The appreciation of the presence of biodiversity for its inherent value and beauty is enough reason for preserving species. The contribution of biodiversity towards enhancing our knowledge, our aesthetics, imagination and creativity is an important reason to preserve biological diversity. Wildlife is an important tourist attraction that has economic considerations and work options.

Observational learning

Biodiversity is a beautiful and wonderful aspect of nature. Sit in a forest and listen to the birds. Watch a spider weave its complex web. Observe a fish feeding on algae. It is all fascinating in its complexity! This is referred to as *biophilia*—the love for living things.

In India, history and culture is replete with plant and animal imagery. Symbols from wild species such as the lion of Hinduism, the elephant of Buddhism and deities such as Lord Ganesha and the vehicles of several deities that are linked to different animals, have been venerated for thousands of years. The holy basil or tulsi plant has been grown in the courtyards of households for centuries.

4.5.6 Informational Value, Communication Education and Public Awareness (CEPA)

Providing information about biodiversity has a strong educational value. This is done through a variety of strategies for communication, education and public awareness.

4.5.7 Option Value

Keeping possibilities open for the use of bio-resources in the future is called *option value* for preserving biodiversity. It is impossible to predict which of our species, or traditional varieties of crops and domestic animals, will be of use in the future. To continue to improve cultivars and domestic livestock, we need to return to the wild relatives of crop plants and animals. The preservation of biodiversity must also include traditionally used genetic strains already in existence in our indigenous crop varieties and domestic breeds of livestock that have specific properties. This is of great value for using this genetic estate to counter the effects of climate change on our food security.

We need to leave these options open for the betterment of future human societies. If we do not, they will consider our generation as destroyers of nature's bounty. Our generation holds the key to a better future for our children, and their children. We cannot permit extinction through the impacts we make on nature today without thinking about the consequences for the future.

SUMMARY

- Biological diversity refers to the variation of life forms at three levels—genetic variability within a species, the variety of species within a community and the organisation of species into ecosystems.

- Biological diversity provides a variety of environmental goods and services. These can be categorised as having consumptive or productive use values, socio-cultural values, aesthetic values and option values.

- India is a bio-rich nation, it ranks among the top 15 counties that are exceptionally rich in species diversity, many of which are endemic to the country and found nowhere else in the world.

- Areas that are rich in species diversity are called *hotspots*. India's hotspots are concentrated in three areas—the forests of the Himalayas and North-eastern states, the forests of the Western Ghats and the Andaman and Nicobar Islands.

- India's aquatic and marine ecosystems, coral reefs, coastal mangroves and islands are of exceptional value.

- Threats to biodiversity include habitat loss due to rapid urbanisation and industrialisation, poaching of wildlife for short-term economic gain, and finally conflicts arising from the need for human–wildlife coexistence in certain areas and impacts of exotic (foreign) invasive (rapidly spreading) species.

- In situ conservation refers to protecting species in their natural habitat by setting aside an adequate representation of wilderness as Protected Areas, consisting of a network of national parks and wildlife sanctuaries, conservation reserves and community reserves.

- These Protected Areas must be integrated with each other, by establishing corridors between adjacent areas wherever possible so that wildlife can move between them.

- Ex situ conservation refers to protecting species outside their natural habitat, in conditions that can be closely controlled and monitored. The genetic material of endangered species is also preserved in gene banks for future use. The ultimate purpose is to restore ecosystems in the wild and rehabilitate wild species.

QUESTIONS

1. Define biological diversity.
2. What are some of the major causes of species loss in the 21st century? State at least three causes.
3. Discuss any two common plant species, in terms of their description and use.
4. What is in situ and ex situ conservation? Give examples of each.
5. What is a keystone species?
6. Give a few examples of endangered species and endemic species in India. Which are the hotspots of biodiversity in India?
7. Why is India called a megadiversity country?
8. What are the different values of biological diversity?

Environmental Pollution

We spray our elms, and the following spring, the trees are silent of robin song, not because we sprayed the robins directly but because the poison travelled step by step through the now familiar elm–earthworm–robin cycle.   —*Rachael Carson*

This quotation appeared in Rachael Carson's book entitled *Silent Spring* published in 1962. This book inspired a controversy and initiated a major change in the thinking about the safety of using pesticides and other toxic chemicals.

Learning Objectives

In this chapter you will learn,

◆ What environmental pollution is
◆ What air, water, soil, chemical and noise pollution are; their types, causes, effects and control measures
◆ About nuclear hazards
◆ About solid waste management and waste to energy methodologies

Purpose

Unless we all realise that pollution is caused by human activities, the world will soon become unliveable. Our activities affect the air, water and soil in so many diverse ways. We pollute the air in our environment with noxious gases from fuel-based vehicles, industries and the noise we generate. Many of these pollutants impact both the non-living and living parts of the earth. We can reduce this by prudent behaviour.

Our Role

We can only reduce pollution if we have an understanding of what pollutes air, water, soil, and destroys biological systems and (living) creatures. With this deeper understanding we should learn to care for the earth and her resources, so that we use them more and more carefully.

5.1   ENVIRONMENTAL POLLUTION

Pollution refers to the effect of undesirable changes in our surroundings that have harmful effects on plants, animals and human beings. This occurs when development furthers only short-term economic gains at the cost of long-term ecological benefits for humanity. The biggest ecological changes are those created by humans. In the last few decades, we have contaminated the air, water and land (on which life itself depends) with a variety of waste products. Pollution is brought about by the misuse of the earth's resources in unsustainable ways.

Pollutants include solid, liquid and gaseous substances present in greater than natural abundance, produced by human activity that have a detrimental effect on the environment and human health. The nature and concentration of a pollutant determines the severity of its detrimental effects on human health. An average human requires about 12 kg of air each day, which is nearly 12–15 times greater than the amount of food we eat. So, even a small concentration of pollutants in the air becomes more significant in comparison to similar levels present in food. Pollutants that enter the water have the ability to spread to distant places, especially in the marine ecosystem.

Human actions are all too frequently imprudent. This behaviour results in environmental economic and societal losses. Prudent behaviour towards the environment prevents environmental degradation. If the environment is not given due consideration, the cost of restoring the balance in nature is much more than in preventing the damage. Landuse is damaged by disregarding good land and water resource management.

5.1.1 Types of Pollution

From an ecological perspective, pollutants can be classified as follows:

Degradable or non-persistent pollutants: These can be rapidly broken down by natural processes; for example, domestic sewage and discarded vegetables.

Slowly degradable or persistent pollutants: These are pollutants that remain in the environment for many years in an unchanged condition and take decades or longer to degrade; for example, DDT (and other pesticides) and most plastics.

Non-degradable pollutants: These cannot be degraded by natural processes. Once they are released into the environment, they are difficult to eradicate and continue to accumulate; for example, toxic elements such as lead or mercury and nuclear waste.

Observational learning

Observe and reflect on the causes, effects and control measures with respect to different kinds of pollution in your own immediate environment. Do you contribute to it? How can you minimise the damage you are causing?

5.1.2 Causes of Pollution

Pollutants enter our air, water and soil from different sources. However they are frequently interlinked and if they occur simultaneously may exacerbate their impacts on the environment. Primary pollutants affect the environment directly. Secondary pollutants react with primary pollutants and these affect the environment indirectly. Different pollutants degrade our air, water and soil. Thus, most pollutants may affect one or more of our life support systems and affect us in different ways. For example, particulate matter in the air affects the respiratory system, chemicals and sewage affect drinking water leading to gastrointestinal diseases, pesticides sprayed on crops can cause cancer and deplete biodiversity, nuclear waste and disasters cause cancer and congenital defects in future generations.

Natural sources

Air is polluted by strong winds which create large amounts of particulate matter. Water is polluted by natural processes in certain situations where eutrophication occurs due to excess of vegetation or algal growth.

Anthropogenic sources

Air is polluted (atmosphere) frequently by human activities. The exhaust from vehicles and industries which contains gas or smoke as a byproduct, is released into the atmosphere. Burning of agricultural waste also leads to pollution of air. Water is polluted by sewage from households and industrial effluents. They pollute both fresh water and marine ecosystems. Soil is most frequently polluted by the overuse of chemical fertilisers and pesticides. In the long term, this affects agriculture.

Primary air pollutants

Gaseous pollutants include SO_2, Cl_2 and NO_2. However, there are also sulphur, nitrogen, carbon and inorganic compounds of various types that human (anthropogenic) activities release into our atmosphere.

Secondary air pollutants

When primary pollutants are combined with each other they form several secondary pollutants. Secondary air pollutants include those that are formed by reactions between volatile organic compounds (VOCs), oxides and nitrogen (NO_x).

5.1.3 Effects of Pollution

Pollution can have short term, medium term or long term effects on our environment. They affect human health and well being and damage our surroundings. In most situations their effects can be reversed by good environmental management. However, some types of pollution may take many years to reverse, for example, toxic chemicals and radioactive material.

> **Learning by reflection**
>
> It is better and cheaper to prevent pollution than to correct it after it has led to environmental ill health or a disaster.

5.1.4 Control of Pollution

Controlling pollution cannot occur merely by formulating laws and adopting best practices on paper. It requires the will of the people. This positive thinking comes from public awareness and is a democratic process. Laws and the judiciary can punish offenders. However, the fact that laws are broken is a failure of good environmental protection. Preventing pollution is better than trying to fix a damaged environment.

Every 'polluter must pay' if their activities have led to environmental degradation. Water, soil as well as noise polluters affect the lives, livelihoods and habitats of thousands of people. The perpetrators must compensate for these losses. Deforestation must provide tenures under schemes such as Compensatory Afforestation Fund Management and Planning Authority (CAMPA) for reforestation and eco-restoration.

> **Observational learning**
>
> Reflect on the types, causes, effects and controls of air, water, soil, chemical and noise pollution that you experience in your daily life. Each type and cause of pollution must be dealt with through different management systems. However, at a holistic level, each has to be seen as a comprehensive environmental problem.

5.2 AIR POLLUTION

History of air pollution: The origin of air pollution on earth can be traced back to the time when humans started using firewood for cooking and heating. Hippocrates mentioned air pollution in 400 BC. With the discovery and increased use of coal, air pollution became more pronounced, especially in urban areas. It was recognised as a problem 700 years ago in London in the form of smoke pollution, which prompted King Edward I to make the first anti-pollution law to restrict people from using coal for domestic heating in the year 1273. In the year 1300, another Act banning the use of coal was passed in England. Defying the law led to imposition of capital punishment. In spite of this, air pollution became a serious problem in London during the Industrial Revolution due to the widespread use of coal in industries. The earliest recorded major disaster was the 'London smog' that occurred in 1952, which resulted in more than 4000 deaths due to the accumulation of air pollutants over the city for five days.

In Europe, around the middle of the 19th century, a black form of the peppered moth began appearing in industrial areas. In its regular form, the peppered moth is well camouflaged when it settles on a clean, lichen-covered tree. However, this peppered pattern was easily spotted and picked up by birds from the smoke-blackened bark of trees in industrial areas. So, while the normal pepper-patterned moths were successful surviving in clean non-industrial areas, only black moths were successfully camouflaged in industrial areas. Thus, with the spread of industrialisation, an increased incidence of the black form of moths was observed. This genetic adaptation affected not only the peppered moth but also occurred in many other moths. This is a classic case of pollution leading to adaptation in a species over time.

Air pollution began to increase in the beginning of the 20th century with the development of transportation systems and the large scale use of petrol and diesel. Pollution due to auto exhaust remains a serious environmental issue in many developed and developing countries, including India.

Case Study 5.1   Bhopal gas tragedy

The greatest industrial disaster leading to serious air pollution took place in Bhopal, where the extremely poisonous methyl isocyanate gas accidentally leaked from Union Carbide's pesticide manufacturing plant, on the night of 2 December 1984. The ill effects on health have affected hundreds of thousands of people. The ensuing soil and ground water contamination is felt even today. After 35 years, the verdict of this tragedy still continues to raise controversies and exposes the gap in the 'green laws' in India, when they relate to industrial disasters. The accused from the Bhopal disaster were sentenced to two years of imprisonment, the maximum charge for causing death by negligence, along with a fine. An appeal to review this punishment and modify our laws is now under way.

5.2.1   Types of Air Pollution

Undesirable solid or gaseous particles in the air bring about air pollution which is harmful to human health and the environment. The air may become polluted by natural causes such as

| Learning through critical thinking |

What in your opinion would be a more appropriate verdict to this tragedy? What can we do to prevent the recurrence of such an enormous tragedy?

volcanoes which release ash, dust, sulphur and other gases, or by forest fires. Occasionally forest fires are caused by lightning. Most forest fires in India occur either due to human carelessness or are created purposely to elicit a fresh flush of grass growth for grazing livestock. However, unlike pollutants created by human activity, naturally occurring pollutants tend to remain in the atmosphere for a short time and do not lead to permanent atmospheric changes.

Outdoor pollution: Pollutants that are emitted directly from identifiable sources are produced both by natural events (dust storms and volcanic eruptions) and by human activities (emissions from vehicles, thermal power plants that use coal and many industries). These are called *primary pollutants*. There are five primary pollutants that together contribute to about 90% of the global air pollution. These are carbon oxides (CO and CO_2), nitrogen oxides, sulphur oxides, volatile organic compounds (mostly hydrocarbons) and suspended particulate matter.

Carbon monoxide is a colourless, odourless and toxic gas produced when organic materials such as natural gas, coal or wood are incompletely burnt. Vehicular exhaust is the largest single source of carbon monoxide. However, carbon monoxide is not a persistent pollutant. Natural processes can convert carbon monoxide to other compounds that are not harmful. Therefore, the air can be cleared of carbon monoxide if no new carbon monoxide is introduced into the atmosphere.

Sulphur oxides are produced when sulphur containing fossil fuels are burnt. *Nitrogen oxides* are found in vehicular exhaust. These are significant, as they are involved in the production of secondary air pollutants such as ozone, which is harmful at lower levels of the atmosphere but not in the stratosphere.

Hydrocarbons are a group of compounds consisting of carbon and hydrogen atoms. They either evaporate from fuel supplies or are remnants of fuel that has not burnt completely. Hydrocarbons are washed from the air when it rains and get into surface water. They form an oily film on the surface and do not cause problems until they react to form secondary pollutants. Using higher oxygen concentrations in the fuel–air mixture, using valves to prevent the escape of gases and fitting catalytic converters in automobiles, are some of the modifications that can reduce the release of hydrocarbons into the atmosphere.

Particulates are small pieces of solid material (for example, smoke particles from fires, bits of asbestos, dust particles and ash from industries) which disperse into the atmosphere (Table 5.1). The effects of particulates range from soot to the carcinogenic (cancer causing) effects of asbestos and also dust particles and ash from industrial plants that are dispersed into the atmosphere.

Table 5.1 Types of particulates

Term	Meaning	Examples
Aerosol	General term for particles suspended in air	Sprays from pressurised cans
Mist	Aerosol consisting of liquid droplets	Sulphuric acid mist
Dust	Aerosol consisting of solid particles that are blown into the air or are produced from larger particles by grinding them down	Dust storm
Smoke	Aerosol consisting of solid particles or a mixture of solid and liquid particles produced by chemical reactions such as fires	Cigarette smoke, smoke from burning garbage
Fume	Generally is the same as smoke, but often applies specifically to aerosols produced by condensation of hot vapours of metals	Zinc/lead fumes
Fog	Aerosol consisting of water droplets	
Smog	Term used to describe a mixture of smoke and fog	
Plume	Geometrical shape or form of the smoke coming out of a chimney	

Repeated exposure to particulates causes them to accumulate in the lungs and interfere with the ability to exchange gases during respiration.

Lead is a major air pollutant that remains largely unmonitored and is emitted by vehicles. High lead levels have been reported in the ambient air in metropolitan cities. Leaded petrol is the primary source of air borne lead emissions in Indian cities. Unleaded petrol is now used for reducing this pollutant.

Indoor pollution: Pollutants are also found indoors from the infiltration of polluted air from outside and from various chemicals used or produced inside buildings. Both indoor and outdoor air pollution are equally harmful. In India, many homes still use wood or coal for cooking and heating which causes severe respiratory conditions especially in women and children who spend a lot of time indoors.

What happens to pollutants in the atmosphere: Once pollutants enter the troposphere they are transported downwind, diluted by the large volume of air, and transformed through physical or chemical changes. Pollutants are also removed from the atmosphere by rain during which they become attached to water vapour. This subsequently forms rain or snow, which drenches and covers the earth's surface. The atmosphere disperses some pollutants by mixing them in the very large volume of air that covers the earth. This dilutes the pollutants to acceptable levels. The rate of dispersion, however, varies in relation to several aspects such as the topography and meteorological parameters.

5.2.2   Causes of Air Pollution

Air pollution is caused by several human activities. This includes outdoor and indoor pollution. Traffic produces exhaust emissions. All coal- and oil-fired units lead to air pollution. Industry is a major polluter. Thermal power stations based on coal cause serious levels of air pollution.

5.2.3   Effect of Topography on Air Pollution

No breeze

Normally, as the earth's surface becomes warmed by the sun, the layer of air in contact with the ground also gets heated by convection. This warmer air is less dense than the cold air above it, so it rises. Thus, the pollutants produced in the lower layer of air are effectively dispersed.

However, on a still evening, this process is reversed. On a sunny day, an hour or two before sunset, the ground starts to lose heat and the air near the ground begins to cool rapidly. Due to the absence of wind, a static layer of cold air is produced as the ground cools. This, in turn, induces the condensation of fog. The morning sun cannot initially penetrate this layer of fog. The cold air being dense cannot rise and is trapped by the warm air above. The fog cannot move out of the area due to the surrounding hills. The topographic features resemble a closed chemical reactor in which the pollutants are trapped. This condition often continues through the cool night and reaches its maximum intensity before sunrise. When the morning sun warms the ground, the air near the ground also gets warm and rises within an hour or two. This may be broken up by strong winds. In cold regions, this situation can persist for several days. Such a situation is known as smog (a mix of smoke and fog).

Strong breeze

5.2.4   Effects of Meteorological Conditions on Air Pollution

The velocity of the wind affects the dispersal of pollutants. Strong winds mix the polluted air more rapidly with the surrounding air, diluting the pollutants rapidly. When the wind velocity is low, mixing takes place, but at a slower rate and the concentration of pollutants remains high.

When sulphur dioxide and nitrogen oxides are transported by prevailing winds, they form secondary pollutants such as nitric acid vapour, droplets of sulphuric acid and particles of sulphate and nitrate salts. These chemicals descend on the earth's surface in two forms—wet (acid rain, snow, fog and cloud vapour) and dry (as acidic particles). The resulting mixture is called acid deposition or *acid rain.*

Acid deposition has many harmful effects, especially when the pH falls below 5.1 for terrestrial systems and below 5.5 for aquatic systems. It causes respiratory diseases such as bronchitis and asthma which may lead to premature death. It also damages statues, buildings, metals and the finish of cars. Acid deposition can damage tree foliage directly, but the most serious effect is the weakening of trees so they become more susceptible to other types of damage. The nitric acid and nitrate salts in acid deposition can lead to excessive soil nitrogen levels. This can over-stimulate the growth of some plants and intensify the depletion of other important soil nutrients, such as calcium and magnesium, which in turn can reduce tree growth and vigour.

5.2.5 Effects of Air Pollution on Living Organisms

Our respiratory system has a number of mechanisms that help in protecting us from air pollution. The function of the hair in the nose is to filter out large particles. The sticky mucus in the lining of the upper respiratory tract captures smaller particles and dissolves some gaseous pollutants. When the upper respiratory system is irritated by pollutants, sneezing and coughing expel the contaminated air and mucus. Prolonged smoking or exposure to air pollutants can overload or break down these natural defences, causing or contributing to diseases such as lung cancer, asthma, chronic bronchitis and emphysema. Elderly people, infants, pregnant women and people with heart disease, asthma or other respiratory diseases are especially vulnerable to air pollution.

Cigarette smoking is responsible for the greatest exposure to carbon monoxide. Smoking has been banned in public places in India since 2008. Advertisements against smoking appear on all television programmes warning people of its risk and its carcinogenic properties. Exposure to air containing even 0.001% of carbon monoxide for several hours can cause collapse, coma and death. As carbon monoxide remains attached to the haemoglobin in the blood for a long period, it accumulates and reduces the oxygen carrying capacity of blood. This impairs perception and thinking, slows reflexes and causes headache, drowsiness, dizziness and nausea. Carbon monoxide in heavy traffic causes headache, drowsiness and blurred vision which adds to the risk of road traffic accidents. In large amounts, carbon monoxide can even cause death.

Sulphur dioxide, nitrogen oxides, especially NO, and suspended particles in the air aggravate the respiratory tract, leading to bronchitis and asthma. Prolonged exposure to these particles damages the lung tissue and contributes to the development of chronic respiratory diseases and cancer.

Many volatile organic compounds (such as benzene and formaldehyde) and toxic particulates (such as lead and cadmium) can cause mutations in the foetus, reproductive problems or cancer. The repeated inhalation of ozone, a component of photochemical smog, causes coughing, chest pain, breathlessness and irritation of the eyes, nose and throat (**Fig. 5.1**).

Effects on plants: The leaves of crop plants are damaged when gaseous pollutants enter the pores of the leaf. Chronic exposure of the leaves to air pollutants can break down the waxy coating that helps prevent excessive water loss. This leads to further damage from diseases, pests, droughts and frosts. Such exposure interferes with photosynthesis and plant growth, reduces nutrient uptake and causes the leaves to turn yellow, brown or drop off altogether. At higher concentrations of sulphur dioxide, most of the flower buds become stiff and hard. They eventually fall off from the plants, as they are unable to flower and form seeds.

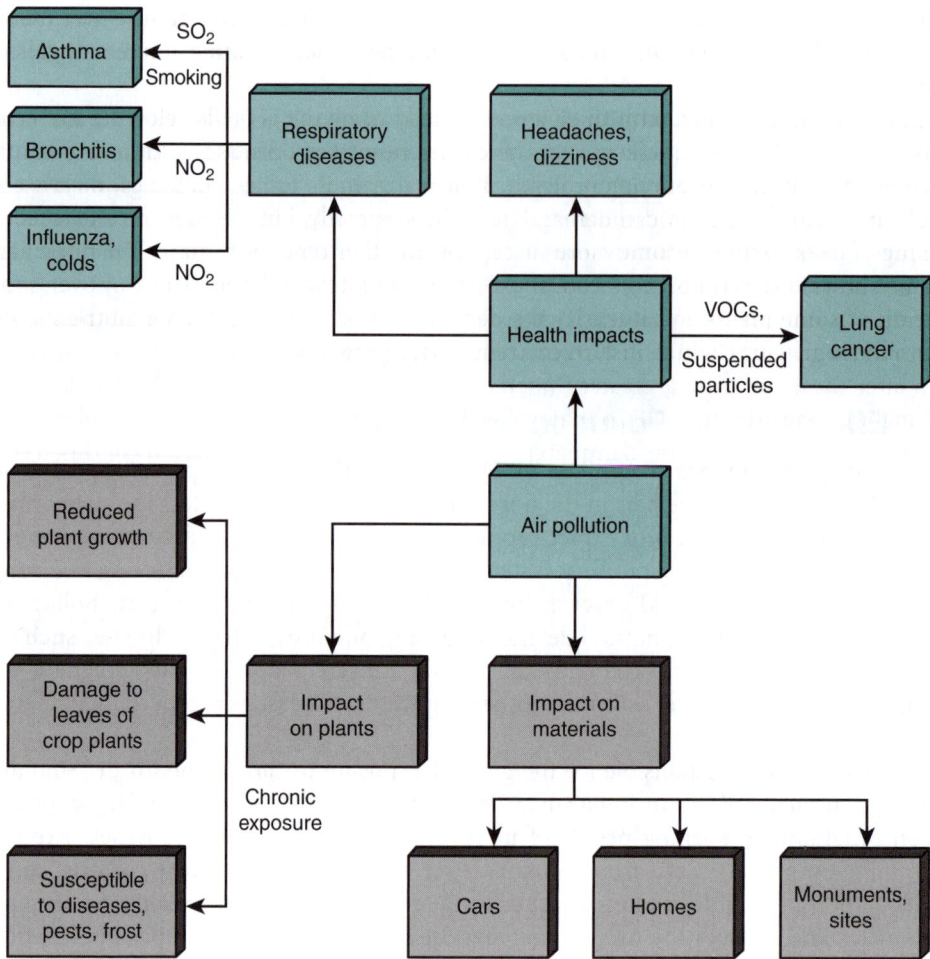

Fig. 5.1   Impact of air pollution

5.2.6   Effects of Air Pollution on Materials

Every year, air pollutants cause damage worth billions of rupees. Air pollutants break down the exterior paint on cars and houses. They cause discolouration of irreplaceable monuments, historic buildings, marble statues, and other heritage sites and sites of natural beauty.

5.2.7   Effects of Air Pollution on the Stratosphere

The upper stratosphere consists of considerable amounts of ozone, which works as an effective screen for UV light. This region, called the *ozone layer*, extends up to 60 km above the surface of the earth. It is, however, most dense in the region between 20 and 25 km from the earth's surface. The ozone layer does not consist solely of ozone, but also has other common atmospheric gases. In the densest portion, there is only 1 ozone molecule in 100,000 gas molecules. Therefore, even small changes to the ozone concentration can produce dramatic effects on the living organisms on earth.

The total amount of ozone in a 'column' of air from the earth's surface up to an altitude of 50 km is the *total column ozone*. This is recorded in *Dobson Units (DU)*, a measure of the thickness of the ozone layer by an equivalent layer of pure ozone gas at normal temperature and pressure at sea level. This means that 100 DU 1 mm of pure ozone gas at normal temperature and pressure at sea level.

Ozone is a form of oxygen with three atoms instead of two. It is produced naturally from the photodissociation of oxygen molecules in the atmosphere. The ozone thus formed is constantly broken down by naturally occurring processes that maintain its balance in the ozone layer. In the absence of pollutants, the creation and breakdown of ozone are purely governed by natural forces, but the presence of certain pollutants can accelerate the breakdown of ozone. It has been known that ozone shows fluctuations in its concentration, sometimes accompanied by ozone depletion. However, it was only in 1985 that the large scale destruction of ozone, also called the *ozone hole*, came into the limelight, when British researchers published measurements about the ozone layer.

Soon after these findings, a greater impetus was given to research on the ozone layer, which convincingly established that CFCs (chlorofluorocarbons) were leading to its depletion. These CFCs are extremely stable, non-flammable, non-toxic and essentially harmless. This has made them ideal for many industrial applications such as aerosols, air conditioners, refrigerators and fire extinguishers. Many perfumes and room fresheners (sprays) used CFCs. CFCs were also used in making foam for mattresses and cushions, disposable styrofoam cups, glasses, packaging material for insulation and cold storage. Their stability gave them a long life-span in the atmosphere. Though they have been banned after the Montreal Protocol in 1989, and less damaging HFCs are used, the effects of ozone depletion is expected to last till 2065.

Halons are similar in structure to the CFCs but contain bromine atoms instead of chlorine. They are more dangerous to the ozone layer than the CFCs. Halons are used as fire-extinguishing agents as they do not harm the people or the equipment exposed to them during a fire.

The CFCs and halons migrate into the upper atmosphere after they are released. As they are heavier than air they have to be carried by air currents up to a point just above the lower atmosphere in which they slowly diffuse into the upper atmosphere. This is a slow process and can take as long as five to fifteen years. In the stratosphere, unfiltered ultraviolet radiation severs the chemical bonds, releasing chlorine from the CFC. This attacks the ozone molecule, resulting in its splitting into an oxygen molecule and an oxygen atom.

Despite the fact that CFCs are evenly distributed over the globe, ozone depletion is especially pronounced over the South Pole, due to the extreme weather conditions in the Antarctic atmosphere. The presence of ice crystals makes the Cl–O bonding easier. The ozone layer over countries like Australia, New Zealand, South Africa and parts of South America is depleted. India signed the Montreal Protocol in 1992, which aimed at controlling the production and consumption of ozone depleting substances. In fact, the Montreal Protocol was set up as an example of a successful international agreement. Its extensive adoption is resulting in a progressive recovery of the ozone layer.

Ozone depletion: Changes in the ozone layer (**Fig. 5.2**) have serious implications for humans.

◈ *Effects on human health*: Sunburn, cataract, ageing of the skin and skin cancer are caused by increased UV radiation as the ozone filter is lost. It weakens the immune system by suppressing the body's resistance to infections such as measles, chickenpox and other viral diseases that elicit rash as well as parasitic diseases introduced through the skin.

◈ *Food production*: UV radiation affects the ability of plants to capture light energy during photosynthesis. This reduces the nutrient content and growth potential of plants. This is especially true in the case of legumes and cabbage.

Phyto- and *zooplankton* are damaged by UV radiation. In zooplankton (microscopic animals), the breeding period is shortened by changes in radiation. As plankton form the basis of the marine

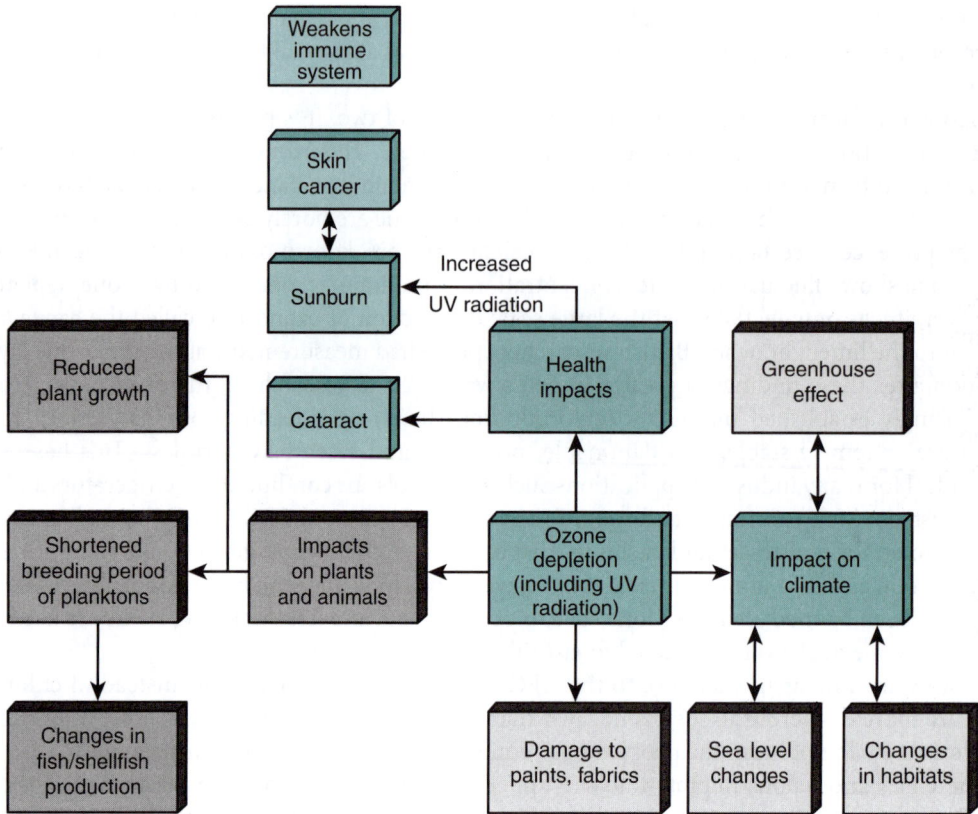

Fig. 5.2 Impact of ozone depletion

food chain, any change in their number and species composition influences fish and shellfish production.

- *Effect on climate:* Atmospheric changes induced by pollution contribute to global warming, a phenomenon caused by the increase in concentration of certain gases such as carbon dioxide, nitrogen oxides, methane and chlorofluorocarbons (CFCs). Observations of the earth have shown beyond doubt that atmospheric constituents such as water vapour, carbon dioxide, methane, nitrogen oxides and CFCs, trap heat in the form of infrared (IR) radiation near the earth's surface. This is known as the greenhouse effect. The phenomenon is similar to what happens in a greenhouse. The glass in a greenhouse allows solar radiation to enter, which is absorbed by the objects inside. These objects radiate heat in the form of terrestrial radiation which is trapped in the greenhouse, increasing the temperature inside and ensuring the luxuriant growth of plants. There are several adverse effects of global warming.
- With a warmer earth, the polar ice caps are melting, causing a rise in ocean levels and flooding of coastal areas.
- In countries like Bangladesh and the Maldives, this would be catastrophic. If the sea level rises by 3 m, the Maldives will disappear completely beneath the waves.
- The rise in temperature will bring about a fall in agricultural produce.
- Changes in the distribution of solar energy can bring about changes in habitat. A previously productive agricultural area will suffer severe droughts, while torrential rains will fall in locations that were once deserts. This could bring about changes in the species of wild plants, agricultural crops, insects, livestock and microorganisms.

◆ In the polar regions, temperature rise caused by global warming has already shown disastrous effects. A vast quantity of methane is trapped beneath the frozen soil of Alaska. When the permafrost melts, methane is released and accelerates the process of global warming.

> **Learning by reflection**
>
> Think about how air pollution affects your life every day.

5.2.8   Control Measures for Air Pollution

Air pollution can be controlled by two fundamental approaches—preventive techniques and effluent control.

One of the effective means of controlling air pollution from industries is to have proper equipment in place. This includes devices for removal of pollutants from the flue gases through scrubbers, closed-collection recovery systems (through which it is possible to collect the pollutants before they escape), the use of dry and wet collectors, filters and electrostatic precipitators.

Building higher smoke stacks (chimneys) facilitates the discharge of pollutants as far away from the ground as possible. Industries should be carefully located in order to minimise the effects of pollution after considering the topography and wind directions. The substitution of raw materials that cause more pollution with those that cause less pollution reduces the level and minimises its effect on human and environmental health.

5.2.9   Air Pollution in India

Vehicle and industrial emissions are the main causes of pollution in the cities of India. The use of cleaner fuels such as Compressed Natural Gas (CNG) in Delhi and other measures such as banning old vehicles in some cities, improving traffic management, implementing the emission standards for vehicles and introducing improved vehicular technology has seen a decline in the levels of NO_2, SO_2 and CO.

On the other hand, the exponential increase in the number of vehicles has meant that the Respirable Suspended Particulate Matter levels exceed the prescribed National Ambient Air Quality Standards (NAAQS) in residential areas in many cities such as Lucknow, Faridabad, Mumbai, Bangalore and Delhi. The Central Pollution Control Board maintains these standards.

Air quality monitoring: The Central Pollution Control Board (CPCB) is executing a nation-wide programme of air quality monitoring known as the National Air Quality Monitoring Programme (NAMP). This programme aims at determining the present air quality status and trends, and controlling and regulating pollution from industries and other sources to meet the ambient air quality standards. The NAMP provides background air quality data that is required for industrial positioning and town planning. The NAMP has an extensive network of 342 operating stations across the country. Under this programme, four air pollutants, sulphur dioxide (SO_2), oxides of nitrogen such as NO_2, suspended particulate matter (SPM) and respirable suspended particulate matter (RSPM/PM10) have been identified for regular monitoring at several locations. The monitoring is carried out for 24 hours, twice a week. The CPCB liaises with other agencies such as the State Pollution Control Boards, Pollution Control Committees and the National Environmental Engineering Research Institute, Nagpur, to ensure robust and consistent air quality data from all the monitoring stations (**Table 5.2**).

SAFAR: A national initiative known as System of Air Quality and Weather Forecasting and Research (SAFAR), by the Ministry of Earth Sciences (MoES), Government of India, under the scheme

Table 5.2 Air quality data

S. No.	Pollutant (µg/m³)	Time weighted average	Conc. in air
			Industrial, residential rural and others
1	SO$_2$	Annual	50
2	NO$_2$	Annual	40
3	PM 10	Annual	60
4	PM 2.5	Annual	40
5	Lead (Pb)	Annual	0.5
6	Carbon monoxide (CO)	Annual	2

National ambient air quality standards http://cpcb.nic.in/air-quality-standard

Metropolitan Advisories for Cities for Sports, Tourism (Metropolitan Air Quality and Weather Services), has been set up for the metropolitan cities of India. It is a research-based system. This was done to provide location-specific information on air quality and its forecast over the next 1–3 days. It has been combined with the warning system on weather parameters. It was developed by the Indian Institute of Tropical Meteorological Development (IMD) and National Centre for Medium Range Weather Forecasting (NCMRWF). An active collaboration of local municipal corporations and various local educational institutions and governmental agencies lead to the implementation of SAFAR.

The objective of SAFAR was to increase awareness among the general public related to the air quality in their city so that mitigation measures and action for improving the air quality and related health issues, could be initiated. The driving forces to realise this objective include educating the public, self-mitigation and development of strategies for policy makers. It provides forecasts on the weather, emissions and UV radiation in the region. Currently, SAFAR is active in two metropolitan regions of India—the National Capital Region of Delhi and the Pune Metropolitan Region.

The SAFAR system integrates several complex components to provide four main products which translates highly scientific air quality information in a simple format for easier understanding.

(i) Air quality forecast, 1–3 days in advance.
(ii) Weather forecast, 1–3 days in advance.
(iii) Location-specific UV index information.
(iv) Emission scenario.

Legal aspects of air pollution control in India: The Air (Prevention and Control of Pollution) Act was legislated in 1981. The Act provided for the prevention, control and abatement of air pollution. In the areas notified under this Act, no industrial pollution-causing activity could be initiated without the permission of the concerned state and CPCB. However, this Act was not strong enough to play a precautionary or corrective role to reduce pollution in the 1980s. After the Bhopal gas disaster, a more comprehensive Environment Protection Act (EPA) was passed in 1986. This Act, for the first time, conferred enforcement agencies with necessary punitive powers to restrict any activity that could harm the environment. To regulate vehicular pollution, the Central Motor Vehicles Act of 1939 was amended in 1989. Following this amendment, the exhaust emission rules for vehicle owners were notified in 1990 and the mass emission standards for vehicle manufacturers were enforced in 1991. The mass emission norms were further revised in 2000. More recently, all transport vehicles require a Pollution Under Control (PUC) certificate, renewable twice a year after the first two years of a new vehicle registration. In 2003, the National Auto Fuel Policy was announced with an implementation schedule of a phased programme for introducing Euro 2 to 4 emission fuel regulations by 2010.

5.3 WATER POLLUTION

'Our liquid planet glows like a soft blue sapphire in the hard-edged darkness of space. There is nothing else like it in the solar system. It is because of water.'
— *John Todd, American speaker*

Water is the essential element that makes life on earth possible. Without water there would be no life on earth. We usually take water for granted. It flows from the taps when they are turned on. Rain replenishes it in the monsoon. Most of us are able to bathe, swim, water our gardens, wash clothes and cook food when we want to.

Learning by reflection
If the world's water supply were only 100 litres, our usable supply of fresh water would be only about 0.003 litres (half a teaspoon!).

Like good health, we take water for granted when we have it. If there is a year of drought or the water is undrinkable, we suffer.

By the middle of this century, almost twice as many people will be trying to share the same amount of fresh water the earth has today. With water wars within and between countries already underway, the access to water resources has become an important factor in sustaining the overall development of any nation's economy.

Water availability on the planet: The water found in streams, rivers, lakes, wetlands and artificial reservoirs is called surface water. Water that percolates into the ground and fills the pores of soil and rock is called ground water. Porous water-saturated layers of sand, gravel or bedrock through which ground water flows are called aquifers. Most aquifers are replenished naturally by rainfall that percolates downwards through the soil and rock. This process is called *natural recharge*. If the withdrawal rate from an aquifer exceeds its natural recharge rate, the water table is lowered. Any pollutant that is discharged on the land and percolates into the aquifer pollutes the ground water. This pollutes well water.

Water availability in India: India receives most of her rainfall during the months of June to September, due to seasonal winds and temperature differences between the land and the sea. These winds blow from opposite directions in the different seasons. They blow into India from the surrounding oceans in the summer and blow out from the subcontinent to the oceans in the winter. The monsoon in India is usually reasonably stable but varies geographically. In some years, the commencement of the rains may be delayed considerably over the entire country or a part of it. The rains may also terminate earlier than usual, or may be heavier than usual over a part of the country. These fluctuations cause local floods or drought. However, in India, even areas that receive adequate rainfall during the monsoon suffer from water shortage in the post-monsoon period due to lack of adequate storage facilities.

5.3.1   Types of Water Pollution

Definition: Water pollution deals with the quality, or composition, of water when it is changed directly or indirectly as a result of human activities and is thus unfit for any purpose.

Biological Oxygen Demand (BOD): The amount of dissolved oxygen that must be present in water in order for microorganisms to decompose the organic matter in the water. It is used as a measure of the degree of pollution.

Chemical Oxygen Demand (COD): The amount of oxygen required to oxidise fully the organic compounds contained in a volume of water. It is used as a measure of water quality. (https://www.lexico.com/en/definition/chemical_oxygen_demand)

Point sources of pollution: When a source of pollution can be readily identified because it has a definite source and place where it enters a lake on river, it is said to come from a point source; for example, municipal and industrial discharge pipes.

When a source of pollution cannot be readily identified, such as an agricultural run-off or acid rain, they are called non-point sources, or diffused sources of pollution.

There are two types of water pollution—organic (sewage) and inorganic (that caused by chemical pollutants from industrial effluents).

5.3.2 Causes of Water Pollution

There are several common causes of water pollution that are caused by human activities. These include:
◆ disease causing agents (pathogens),
◆ oxygen depleting waste,
◆ inorganic plant nutrients,
◆ water soluble inorganic chemicals,
◆ organic chemicals,
◆ sediments of suspended matter,
◆ water soluble radioactive isotopes, and
◆ hot water released by power plants and industries.

5.3.3 Effects of Disease-Causing Agents (Pathogens)

The effects of water pollution include an increase in disease causing agents, depletion of oxygen, eutrophication, loss of potability and pollution of ground water (**Fig. 5.3**).
◆ Biological pollutants include bacteria, viruses, protozoa and parasitic worms that enter the water from domestic sewage and untreated human and animal waste. Human waste contains concentrated populations of coliform bacteria such as *Escherichia coli* and *Streptococcus faecalis*. These bacteria generally grow in the large intestine of the human body, where they are responsible for digestion and for the production of vitamin K. These bacteria are not harmful in low numbers. However, large amounts of human waste in water increases the population of these bacteria, thereby causing gastrointestinal diseases.
◆ Oxygen depleting wastes include organic wastes that can be decomposed by aerobic (oxygen requiring) bacteria. Large populations of bacteria use all the oxygen present in the water. This degrades the water quality and the oxygen available for other aquatic life is depleted. The amount of oxygen required to break down a certain amount of organic matter is called the biological oxygen demand (BOD). The amount of BOD in water is an indicator of the level of pollution. If too much organic matter is added to water, all the available oxygen is used up. This kills fish and other forms of oxygen-dependent aquatic life. Thus, anaerobic bacteria (those that do not require oxygen) begin to break down the waste. Their anaerobic respiration produces chemicals that have a foul odour and an unpleasant taste, which are harmful to human health.
◆ Inorganic plant nutrients include water-soluble nitrates and phosphates such as those found in fertilisers. The quantity of fertilisers used in a field is often many times more than what is actually required by the crops. The surface water run-off carries the excess fertilisers into nearby water bodies, causing high levels of plant nutrients (referred to as eutrophication) that promote the excessive growth of algae and other aquatic plants. This may interfere with the use of the water by clogging the water intake pipes, changing the taste and smell of the water and causing a build up of organic matter in the water. As the organic matter decays, the oxygen levels decrease, and fish and other aquatic species die.

- Water soluble inorganic chemicals include acids, salts and compounds of toxic metals such as mercury and lead. High levels of these chemicals can make the water unfit for drinking, harm fish and other aquatic life, reduce crop yield and accelerate the corrosion of equipment that is in contact with this water.
- Sediments include insoluble particles of soil and other solids that become suspended in water. This occurs when the soil is eroded from the surrounding land. High levels of soil particles suspended in water interfere with the penetration of sunlight into the water. This reduces the photosynthetic activity of aquatic plants and algae, disrupting the ecological balance of the aquatic bodies. When the velocity of water in streams and rivers decreases, the suspended particles settle down at the bottom as sediments. Excessive sediments destroy the feeding and spawning grounds of fish and clog and fill lakes and artificial reservoirs.
- Water soluble radioactive isotopes are damaging pollutants and are contained in radioactive waste. They get concentrated in various human and animal tissues and organs as they pass through food chains and food webs. The ionising radiation emitted by such isotopes can cause birth defects, cancer and genetic damage.
- Hot water is released by power plants and industries that use large volumes of water to cool the plant. This results in a rise in temperature of the local water bodies resulting in thermal pollution. This warm water not only decreases the solubility of oxygen but changes the breeding cycles of various aquatic organisms.
- Organic chemicals that pollute water include oil, gasoline, plastic, pesticides, cleaning solvents, detergents and many others. These are harmful to aquatic life and human health. They enter the water directly through industrial activities—from the improper handling of chemicals, and from the improper and illegal disposal of chemical wastes.

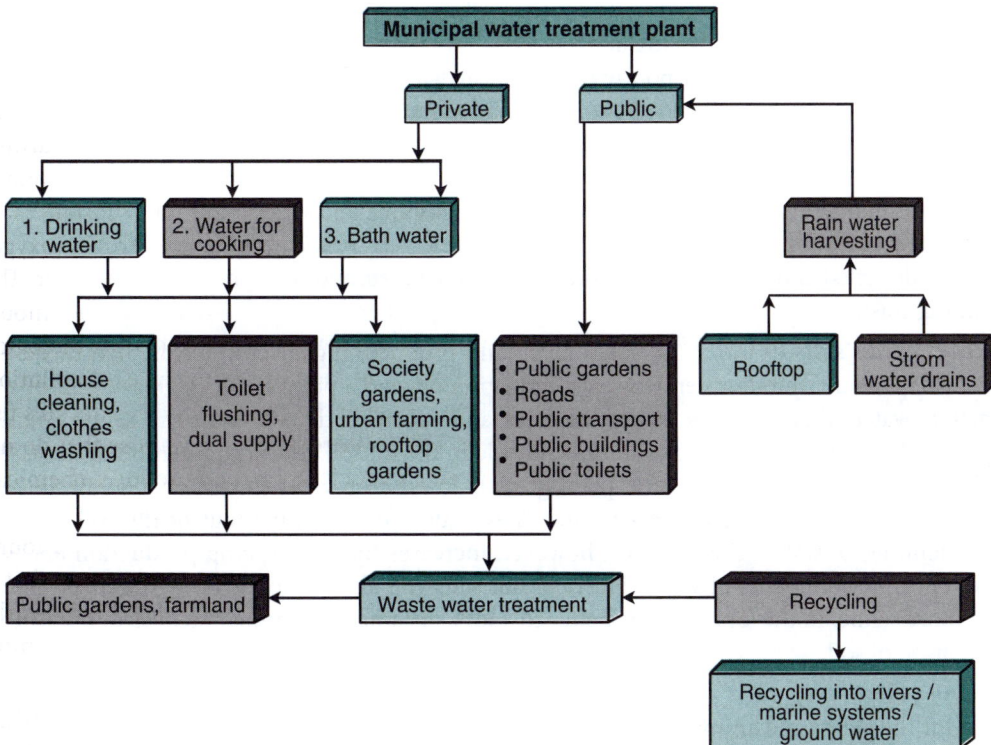

Fig. 5.3   Future management of urban water supply source

Pesticides are an example of organic chemicals that enter the water cycle through agricultural activities. The application of pesticides causes bio-accumulation and bio-magnification of harmful substances in the aquatic food chain. Pesticides enter the water cycle as run-off from agricultural areas. They are absorbed by the phytoplankton and aquatic plants. This microscopic flora and fauna is used as food by herbivorous fish, which are in turn eaten by carnivorous fish, which are eaten by water dependant birds in aquatic systems. At each link in the food chain, these chemicals (which do not pass out of the body) get accumulated (bio-accumulation) and get increasingly concentrated (bio-magnification). One of the effects of accumulation of high levels of pesticides (such as DDT) is that birds that eat the polluted food, lay eggs with shells that are much thinner than normal. This results in the premature breaking of these eggs, killing the immature chicks inside the egg. Birds of prey, like hawks, eagles and other fish-eating birds, are seriously affected by such pollution. DDT has been banned in India for agricultural use and is allowed only for malaria eradication. However its ill effects still persist in our environment.

Oil is another organic chemical that enters surface water bodies through the run-off from roads, parking lots and industries. The leakage from underground oil tanks can also pollute ground water. Accidental oil spills from large transport tankers at sea have been causing significant environmental damage to marine ecosystems.

Effects of pollution on ground water: While oil spills are highly visible and often get a lot of media attention, a much greater threat to human life comes when ground water that is used for drinking and irrigation becomes polluted. While ground water is easy to deplete and pollute, it gets renewed very slowly and hence must be used judiciously. Ground water flows very slowly. Thus, the contaminants are not effectively diluted or dispersed, in contrast to those in surface water. Pumping ground water and treating it for consumption is a slow and expensive process. It is therefore extremely essential to prevent the pollution of ground water in the first place. Some causes of ground water pollution include:

◆ urban run-off of untreated or poorly treated waste water and garbage,
◆ industrial waste storage located above or near aquifers,
◆ agricultural practices such as the application of large amounts of fertilisers, pesticides and animal feed in the rural sector,
◆ leaks from underground storage tanks containing gasoline and other hazardous substances,
◆ leachate from landfills,
◆ poorly designed and inadequately maintained septic tanks, and
◆ mining waste.

Effects of excess salts in water: Irrigated farm land produces higher crop yields than those that are only dependant on rainwater. However, irrigation beyond a point has its own set of ill-effects. Irrigation water dissolves salts in dry climates and much of the water in the saline solution evaporates leaving its salts such as sodium chloride in the topsoil. The accumulation of these salts is called salinisation, which can stunt plant growth, lower crop yield and eventually kill the crop and render the land useless for agriculture. These salts can be flushed out of the soil by using large quantities of water. This practice, however, increases the cost of crop production and also wastes enormous amounts of water. Flushing out salts can also make the downstream irrigation water more saline. In the long run it is better to adopt sustainable farming practices to prevent the degradation of soil.

Effects of pollution of water due to organic wastes: The amount of oxygen dissolved in the water is vital for the plants and animals living in it. Waste, which directly or indirectly affects the oxygen concentration, plays an important role in determining the quality of water. Normally, the greatest volume of waste discharged into water courses, estuaries and the sea is sewage, which is primarily organic in nature and degrades the water quality (**Table 5.3**).

Table 5.3 Water quality criteria

Designated use	Class of water	Parameter			
		pH	DO (mg L^{-1})	BOD (mg L^{-1}, 5 d, 20°C)	Coliforms (MPN / 100 mL)
Drinking water source without conventional treatment but after disinfection	A	6.5–8.5	6	2	50
Drinking water source after conventional treatment and disinfection	B	6 – 9	4	3	5000

Using the oxygen present in the water, organic wastes are broken down into stable inorganic compounds. However, as a result of this bacterial activity, the oxygen concentration in the water is reduced. When the oxygen concentration falls below 1.5 mg L^{-1}, the rate of aerobic oxidation is reduced and replaced by anaerobic bacteria that can oxidise the organic molecules without the use of oxygen. This results in end products such as hydrogen sulphide, ammonia and methane which are toxic to many organisms. This process results in the formation of an anoxic zone which is low in oxygen content from which most life disappears except for anaerobic bacteria, fungi, yeast and some protozoa. The water becomes foul smelling.

5.3.4 The Status of Water Quality and Pollution of India's Rivers

India has always had a tradition of venerating rivers. Most rivers in India are worshipped and named after gods, goddesses or saints. Ironically, these rivers suffer from severe pollution. Urbanisation, industrialisation, excessive withdrawal of water, agricultural run-off, improper agricultural practices and various religious and social practices contribute to river pollution in India. Every river in India, the Ganga, Yamuna, Cauvery or Krishna, has its own share of problems due to pollution. Waters from the Ganga and Yamuna are drawn for irrigation through the network of canals as soon as these rivers reach the plains. This reduces the amount of water that flows downstream. What flows downstream in the river is water from small *nullahs* and streams that carry with them sewage and industrial effluents. The small amount of residual fresh water is unable to dilute the pollutants and the rivers turn into stinking sewers. In spite of data from scientifically competent studies conducted by the Central Pollution Control Board (CPCB), the Government has not been able to tackle this issue. Sewage and municipal effluents account for 75% of the pollution load in rivers while the remaining 25% is from industrial effluents and non-point pollution sources.

> **Observational learning**
> - Visit the local river and look for signs of pollution.
> - Get involved in an NGO programme that studies river water quality through a citizen science approach.

In 1985, India launched the Ganga Action plan (GAP), the largest ever river clean-up operation in the country. GAP Phase-II in 1991 included cleaning operations for the tributaries of the Ganga—the Yamuna, Gomti and Damodar.

In 1995, the National River Conservation Plan (NRCP) was launched. Under this, all the rivers in India were included for clean-up operations. In most of these plans, attempts were made to clean drains and divert sewage to treatment plants, before letting the effluents flow into the rivers. The plan covered 18 rivers in 46 towns in 10 states. Under this plan, the main activities included treating the pollution load from the sewer systems of towns and cities, setting up sewage treatment plants, electric crematoria and low-cost sanitation facilities, riverfront development, afforestation and solid waste management.

The biggest drawback of these river-cleaning programmes was that they failed to allocate responsibility regarding who would pay for running the treatment facilities in the long run. With the power supply being erratic, most of these facilities are under-utilised. The problem of river pollution due to agricultural run-off has not been addressed in this programme.

In order to strengthen the ongoing National River Conservation Plan (NRCP), a new initiative called the National Ganga River Basin Authority (NGRBA) was formed in 2009. The NGRBA has resolved to make the Ganga pollution-free by 2020. (Currently the National Mission for Clean Ganga, NMCG, has launched the Namami Gange programme which is the flagship programme of the GOI).

5.3.5 Control Measures for Water Pollution

The foremost necessity for the management of water pollution is the setting up of effluent treatment plants to treat waste and thereby reduce the pollution load in the recipient water. Where possible, the treated effluent can be reused for gardening or cooling purposes. A few years ago, a new technology called the *Root Zone Process* was developed by Thermax. This system involves running contaminated water through the root zones of specially designed reed beds (**Fig. 5.4**). The reeds, which are essentially wetland plants, have the capacity to absorb oxygen from the surrounding air through their stomata. The oxygen is pushed through the porous stem of the reeds into the hollow roots where it enters the root zone and creates conditions suitable for the growth of numerous bacteria and fungi. These microorganisms oxidise impurities in the waste water so that the water which finally comes out is clean.

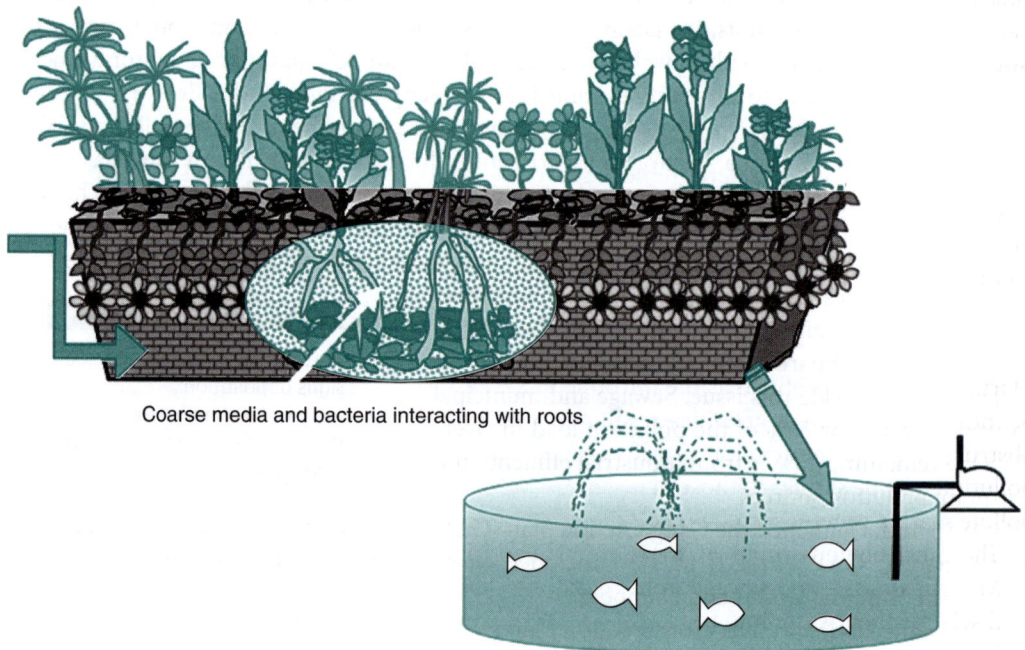

Coarse media and bacteria interacting with roots

Fig. 5.4 Root zone process

Sewage treatment

One way of reducing the pollution load in water is by developing adequate sewage treatment plants. This reduces the biological oxygen demand (BOD) in the water before it is discharged in the receiving waters. Various stages of treatment – primary, secondary or advanced – can be used, depending on the quality of the effluent that is required.

Primary treatment: Primary treatment involves processes such as screening and sedimentation to remove pollutants. This treatment generally removes about 35% of the BOD and 60% of the suspended solids.

Secondary treatment: The main objective of secondary treatment is to substantially reduce the BOD. This includes three commonly used approaches—trickling filters, activated sludge process and oxidation ponds. Secondary treatment can remove at least 85% of the BOD.

5.3.6   Effects of Thermal Pollution on Water Resources

Thermal pollution refers to the degradation of water quality as a result of any process that changes the ambient water temperature. It occurs when an industry removes water from a source (such as a river), uses the water for cooling purposes, and then returns the heated water to its source. Power plants, for example, discharge heated water back into the water body, which is at least 15°C higher than the normal. The warmer temperature decreases the solubility of oxygen and increases the metabolism of fish. This changes the ecological balance of the river. Within certain limits, thermal additions can promote the growth of certain fish and the fish catch may be high in the vicinity of a power plant. However, sudden changes in temperature caused by periodic plant shutdowns, both planned and unintentional, can result in the death of these fish that are acclimatised to living in warmer waters.

Tropical marine animals are generally unable to withstand a temperature increase of 2–3°C and most sponges, molluscs and crustaceans are eliminated at temperatures above 37°C. This results in a change in the diversity of fauna, as only those species that can live in warmer water will survive and proliferate excessively.

Control measures of thermal pollution on water resources: Thermal pollution can be controlled by passing the heated water through: (i) a large shallow cooling pond into which hot water is pumped from one end and cooler water is removed from the other, or (ii) a cooling tower after the hot water leaves the condenser. The heat is dissipated into the air and the water can then be discharged into the river, or pumped back to the plant to be reused for cooling. The disadvantage in both these methods, however, is that large amounts of water are lost to evaporation.

5.3.7   Effects of Pollution on the Marine Environment

Marine pollution is defined as the introduction of substances to the marine environment directly or indirectly by human activities resulting in adverse effects such as hazards to human health, obstruction of marine activities and lowering of the quality of sea water. While the causes of marine pollution may be similar to those of general water pollution, there are some causes that specifically pollute marine waters.

◈   The most obvious input of waste is through pipes directly discharging waste into the sea. Municipal waste and sewage from coastal housing and tourist hotels in coastal towns, directly discharge their waste into the sea.

◈   Pesticides and fertilisers from that are washed off the land by rain, enter water courses and eventually reach the sea.

◈   Petroleum and oil washed off from the roads is expected to enter the sewage system, but storm water overflows carry these materials into rivers and eventually into the seas.

◈   Ship accidents and accidental spillages at sea can be very damaging to the marine environment. Shipping channels in estuaries and at the entrances to ports often require frequent dredging to keep them open. This dredged material contains heavy metals and other contaminants which have been often dumped in the sea.

◇ Offshore oil exploration and extraction pollutes the seawater to a large extent.

Marine pollution due to oil: Oil pollution of the sea attracts great attention because of its visibility. The main source is tanker operations when ships carry seawater as ballast on their return journey. This ballast water is stored in the cargo compartments that previously contained oil. When the cargo is unloaded, a certain amount of oil remains, clinging to the walls of the container. The ballast water thus becomes contaminated with oil. When a fresh cargo of oil is to be loaded, these compartments are cleaned with water, which discharges the dirty ballast along with the oil into the sea. The other sources include dry docking of ships for repairs, offshore oil production and tanker accidents. The 2010 BP oil spill and the 2010 Mumbai oil spill are examples of major disasters that impact our oceans.

Effects of marine pollution on living organisms:

◇ Phytoplankton blooms or 'red tides' causing a whole area of the water to be discoloured as a result of large amounts of organic waste.

◇ Clogging of gills of marine species has an impact on commercially important marine species, reducing the market value of seafood.

◇ Damaging impacts of oil slicks (from liquid oil spilled on the sea) on marine and bird species, salt marshes and mangrove swamps which tend to trap oil, affecting their flowering, fruiting and germination.

◇ 'Tainting' imparts an unpleasant flavour to fish and seafood and is detectable even at extremely low levels of contamination. Thus, the economic loss from oil slicks on fish and shellfish production poses a significant threat to the seafood industry.

◇ Drill cuttings dumped on the sea bed create anoxic conditions and result in the production of toxic sulphides in the bottom sediment, thus eliminating the benthic fauna.

Impacts of oil pollution in the sea on birdlife: If liquid oil coats a bird's plumage, its water-repellent properties are lost. Water then penetrates the plumage and displaces the air trapped between the feathers and skin. This air layer is necessary as it provides buoyancy and thermal insulation. Once affected by oil pollution, the plumage becomes water-logged and the birds may sink and drown. Loss of thermal insulation results in exhaustion of food reserves in an attempt to maintain body temperature, often followed by death. Birds often clean their plumage by preening and in the process consume the oil which, depending on its toxicity, can lead to intestinal, renal or liver failure.

Control measures for oil pollution: Cleaning oil from surface waters and contaminated beaches is a time-consuming and labour-intensive process. The natural process of emulsification of oil in water can be accelerated through the use of chemical dispersants, which can be sprayed on the oil. A variety of slick-lickers in which a continuous belt of absorbent material dips through the oil slick and is passed through rollers to extract the oil have been designed. Rocks and harbour walls can be cleaned with high-pressure steam or dispersants after which the surface must be hosed down.

5.4 SOIL POLLUTION

'We can no more manufacture soil with a tank of chemicals than we can invent a rain forest or produce a single bird. We may enhance the soil by helping its processes along, but we can never recreate what we destroy. The soil is a resource for which there is no substitute.'

—Donald Worster

Donald Worster thus reminds us that fertilisers are not a substitute for fertile soil.

Soil is a thin covering over the land consisting of a mixture of minerals, organic material, living organisms, air and water, that together support the growth of plant life. Several factors contribute to the formation of soil from the parent material. This includes the mechanical weathering of rocks due to temperature changes and abrasion, wind, moving water, glaciers, chemical weathering activities and lichen. Climate and time are also important in the development of soil. In extremely dry or cold climates, soils develop very slowly, while in humid and warm climates, they develop more rapidly. Under ideal climatic conditions, soft parent material may develop into 1 cm of soil within 15 years. Under certain climatic conditions, hard parent material may require thousands of years to develop into new soil.

The top layer, or surface litter layer, is called the *O-horizon*. It consists mostly of freshly fallen and partially decomposed leaves, twigs, animal waste, fungi and other organic materials. Normally, it is brown or black.

The uppermost layer of the soil is called the *A-horizon*. It consists of partially-decomposed organic matter (humus) and some inorganic mineral particles. It is usually darker and looser than the deeper layers. The roots of most plants are found in these two upper layers. As long as these layers are anchored by vegetation, the soil stores water and releases it in a trickle throughout the year instead of rapid run-off with force as seen during a flood. These two top layers also contain a large amount of bacteria, fungi, earthworms and other small soil insects, which form complex food webs in the soil. These soil fauna help recycle soil nutrients and contribute to soil fertility by breeding the decaying detritus into smaller fragments which increases the surface of the particles on which fungi and bacteria can create soil nutrients.

The *B-horizon*, often called the subsoil, contains less organic material and fewer organisms than the A-horizon. The area below the subsoil is called the *C-horizon* and consists of weathered parent material. This parent material does not contain any organic materials. The chemical composition of the C-horizon helps to determine the pH of the soil and also influences the soil's rate of water absorption and retention. Soils vary in their content of clay (very fine particles), silt (fine particles), sand (medium-sized particles) and gravel (coarse to very coarse particles). The relative amounts of the different sizes and types of mineral particles determine the soil texture. Soils with approximately equal mixtures of clay, sand, silt and humus are called loams.

5.4.1 Types of Soil Pollution

Soil pollution occurs by chemical pollution due to the use of fertilisers and pesticides in agricultural landscapes. Chemical effluents when released from industries cause soil pollution. It is also caused due to acid rain as a result of air pollution. Overuse of water in irrigation leads to salinisation of soil.

5.4.2 Causes of Soil Degradation

Erosion: Soil erosion can be defined as the movement of surface litter and topsoil from one place to another. While erosion is a natural process, often caused by wind and flowing water, it is greatly accelerated by various anthropogenic activities such as farming, construction, over-grazing by livestock, burning of grass cover and deforestation. This leads to the loss of soil quality and quantity, which has taken thousands of years to develop.

5.4.3 Effects of Soil Degradation

The loss of topsoil makes soil less fertile and reduces its water-holding capacity. The topsoil, which is washed away, also contributes to water pollution by clogging lakes and increasing the turbidity of the water, ultimately leading to the loss of aquatic life. For one inch of topsoil to be formed, it takes 200–1000 years, depending on the climate and soil type. Thus, if the topsoil erodes faster than it is formed, the soil can become a non-renewable resource during our lifetime and for the future generations.

5.4.4   Control of Soil Degradation

Therefore, it is essential that proper soil conservation measures are used to minimise the loss of topsoil. Several techniques can be used protect the soil from erosion. Both water and soil can be conserved through integrated land management methods. The two types of strategies generally used are—area treatment (**Table 5.4**), which involves treating the land or a drainage-line approach (**Table 5.5**), and managing the natural water courses (nullahs).

Table 5.4   Area treatment

Purpose	Treatment measure	Effect
Reduces the impact of raindrops on the soil	Develop vegetative cover on non-arable land	Minimum disturbance and displacement of soil particles
Infiltration of water where it falls	Apply water infiltration measures on the area	In situ soil and moisture conservation
Minimum surface run-off	Store surplus rainwater by constructing bunds and ponds in the area	Increased soil moisture in the area, facilitates groundwater recharge
Ridge to valley sequencing	Treat the upper catchment first and then proceed towards the outlet	Economically viable, less risk of damage and longer life of structures of the lower catchments

Table 5.5   Drainage-line treatment

Purpose	Treatment measure	Effect
Stop further deepening of gullies and retain sediment run-off	Plug the gullies at formation	Stops erosion, recharges ground water at the upper level
Reduce run-off velocity, pass cleaner water to the downstream side	Create temporary barriers in nullahs	Delayed flow and increased ground water recharge
Low construction cost	Use local material and skills for constructing structures	Structures are locally maintained

Continuous contour trenches can be used to enhance the infiltration of water, reduce the run-off and check soil erosion. These are actually shallow trenches dug across the slope of the land and along the contour lines, for the purpose of soil and water conservation. They are most effective on gentle slopes and in areas of low to medium rainfall. These *bunds* are stabilised ideally by indigenous tree species, shrubs and grasses. In areas with steep slopes where contour *bunds* are not possible, continuous contour benches (CCBs) made of stones are used for the same purpose.

There are several ways in which land can be managed and restored.

◆ Live check-dams, in which barriers are created by planting grass, shrubs and trees across the *gullies*.

◆ A bund constructed out of stones across the stream that can be used for conserving soil and water.

◆ An earthen check-bund constructed out of local soil built across the stream to check soil erosion and the flow of water.

◆ A gabion structure, which is a bund constructed of stone and wrapped in galvanised chain link. A gabion structure has a one-inch-thick, impervious wall of ferrocement at the centre of the structure, which goes below the ground level up to the hard strata. This ferrocement partition, supported by the gabion portion, is able to retain water and withstand the force of the run-off water.

5.5 CHEMICAL POLLUTION

All matter on earth, both abiotic and biotic, consists of chemicals. These are elements in their pure form or combinations that form compounds. An element's properties are based on the structure of its atoms. Molecules depend on the way in which chemical bonding occurs in their atoms to form new substances. A variety of chemical bonds are constantly forming and breaking down in our environment. These chemical reactions affect our lives in many ways.

Definition of chemical pollution: When chemicals are released into our environment they disrupt the balance of our ecosystems, threatening our health, polluting the air we breathe and contaminating our food and water.

Some pollutants are called persistent chemicals as they remain in the ecosystem for a number of years. Residues of chemicals may have adverse effects on soil microorganisms and affect soil fertility. Crops that are grown in contaminated soil contain chemicals and thus become unfit for human and animal consumption. Chemical pollution in soil can be caused by the overuse of fertilisers, pesticides and herbicides. Construction and demolition sites are sources of soil pollution. Mine landfills and foundries are industrial pollution sites where the 'polluter must pay' for wrongs against the environment either when done knowingly or accidentally.

5.5.1 Types of Chemical Pollutants

A chemical substance that enters the environment from consumer goods, industrial, agricultural or other human activities, which poses an immediate or potential hazard to plant, animal or human life is a pollutant. The major chemical pollutants are heavy metals (mercury and lead), aromatic hydrocarbons (benzene and other petrochemicals), organic solvents (toluene and xylene), organohalides [polychlorinated biphenyls (PCBs) and polybrominated biphenyls (PBBs)], dioxins, nitrogen dioxide and sulfur dioxides.

Pollutants can be classified as those that degrade the air we breathe, the water we drink and use, or the noise levels that create problems for society. There are many sources of chemical pollution. Our technological advances have made us largely reliant on chemicals; these disrupt our life and our environment. When human activities of different types alter chemical reactions, their effects on the environment (both abiotic and biotic) can have disastrous consequences. Pollutants (both inorganic and organic) enter the air, water and soil, and affect one or more of our environmental domains. A large number of these pollutants can individually alter our environment and some of them are highly toxic.

Chemical pollutants are classified into those that are broken down in our environment and are thus less harmful in the longer term, while others that are non-degradable remain persistent in the environment over considerable periods of time.

Household chemicals: Aerosols and other regular household cleaning products can act as pollutants.

Chemical fertilisers and pesticides are used in all our modern agricultural processes. They are used to protect crops from insect pests. Excessive use of fertilisers causes water pollution. Chemicals seep into the ground and contaminate soil, water and food.

> **Observational learning**
>
> Look at the products you use on a daily basis and the warnings on their labels.

Several harmful elements find their way into the atmosphere and add to the degradation of the environment. Emissions from vehicles that use fossil fuels lead to air pollution. Cars, planes and other vehicles give off carbon dioxide. The large amount of CO_2 emitted by millions of vehicles in the world today contributes to climate change.

Water: Ships cause chemical pollution, especially those that carry crude oil. There have been several incidents of oil spills which caused serious damage to marine ecosystems. Hazardous waste comes from factories that do not dispose them appropriately. Metals and solvents from industrial processes pollute our water bodies and poison aquatic life.

Industrial and agricultural processes involve the use of many different chemicals that enter water as surface run-off or by percolating into the ground water. This seriously affects the health and well being of all forms of life.

◆ Metals and solvents from industries pollute rivers and lakes. These are poisonous to many forms of aquatic life. They affect their breeding behaviour making them infertile and lead to ill health and death.

◆ Herbicides and pesticides are used in farming to control weeds, insects and fungi. Run-offs of these pesticides cause water pollution and kill aquatic life. Birds, humans and other animals are affected if they eat contaminated fish. As these chemicals travel through the food chain they get concentrated. This process called *biomagnification*, mainly affects species at the apex of the food pyramid leading to destabilisation of the whole ecosystem. Excessive use of pesticides in our homes causes serious health hazards such as allergic, respiratory, and even neurological ill effects.

◆ Petroleum is a chemical pollutant that contaminates water through oil spills when a ship is damaged. Oil spills have serious localised affects on wildlife. These can spread extensively over marine and coastal areas causing the death of hundreds of fish and sticks to the feathers of seabirds which lose their ability to fly. These disasters are expensive to clean up and disrupt fishing and other activities along the sea shore.

Types of chemicals

◆ *Toxic and hazardous chemicals:* Chemicals that have high toxicity levels can cause a disaster due to their physical or chemical properties. They are capable of producing major accidents which are a threat to life and property. These are notified as hazardous chemicals.

◆ *Flammable chemicals:* Gases which at 20°C and at a standard pressure of 101.3 kPa are ignitable when in a mixture of 13% or less by volume with air are notified as being inflammable.

◆ *Explosives:* Explosives are solid/liquid substances which are capable of chemical reactions or producing a gas at a particular temperature and pressure at a speed which causes damage to the surroundings.

5.5.2 Causes of Chemical Pollution

Fluoride, manganese, barium, cadmium, antimony and many other substances are transported into water bodies in various ways. They contaminate both the surface and ground water. Contamination of water occurs as a result of aerial spraying of chemicals on cultivated areas. They are carried from the soil to the water in drainage or from surface run-off. Contaminants are also blown through surface dust from effluents in manufacturing plants. Pesticides and fertilisers that contain nitrates and phosphates are sources of chemicals that cause water pollution. These chemicals percolate into the ground water and mix with run-off into lakes, rivers and the sea.

Nitrates in drinking water are chemical hazards. Certain bacteria in the intestinal canal can convert them into nitrites which after reaching the blood destroy the oxygen carrying capacity of the haemoglobin present in the red blood corpuscles. Infants whose food is contaminated may fall ill or even die. Industrial emissions cause water pollution if the industry does not conform to the norms set out by the government. An example is mercury in waste water from the paper industry. The mercury gets converted to methyl mercury and kills fish.

Fluoride in the water causes fluorosis. It creates stomach ailments and mental disorders. Manganese salts cause blindness. Fluoride and manganese salts are known to affect the growth and development of plants.

A major source of chemical pollution in the air is due to fossil fuels used in industries and motor vehicles. Sulphur dioxide is produced when coal is burnt. It is contained in acid rain and can cause lung damage. Cars, trucks and aeroplanes produce nitrogen oxides (NO_x) as a byproduct which causes acid rain and lung damage. Other chemicals that cause air pollution include ozone, carbon monoxide and lead.

5.5.3   Effects of Chemical Pollution

Life cycle of chemicals: Chemicals that enter the environment usually do so as wastes or as byproducts, and are often mixed with other chemicals. They interact when they are mixed in an additive, synergistic or antagonistic way. They produce breakdown products or byproducts, or react to form new substances in the waste stream, or in the environment.

Effects on animals and plants: Every human being is subjected to contact with dangerous chemicals. These chemicals cause headache, vomiting, blindness, and skin, liver, heart and kidney diseases. Chemicals cause serious problems in plants which affects the whole ecosystem.

Short-term effects: When a chemical pollutant enters a water body, it can impact the surrounding wildlife, watersheds and its residents. For example, if chemicals get into freshwater supply that people and/or animals rely on for drinking, it may no longer be safe for consumption or sanitation purposes. In the short term, toxic releases from industrial plants into the environment and agricultural run-off can threaten water supply.

Long-term effects: Fertilisers or sewage can introduce chemicals containing nitrates or phosphates into water bodies. Nitrates and phosphates are food for algae in the water. An overload of these chemicals causes the algae to bloom. As the excess algae die and decay, dissolved oxygen is used up and the overall quality of the water is degraded. Aquatic life dies from oxygen deprivation. When emissions from industrial plants such as sulfur and nitrogen oxides enter the atmosphere, they can produce acid rain. Acid rain can weaken plant life, stress marine animals and cause leaching of toxic metals into the soil. In some cases, chemical pollution can kill populations of beneficial species such as bees that support ecosystems. When long-term exposure to chemical pollutants causes native species within an ecosystem to die, the area experiences a loss of diversity and becomes more vulnerable to invasive and undesirable species.

5.5.4   Control of Chemical Pollution

Government, regulatory agencies and industrial initiatives have recognised a hierarchy of approaches to prioritise pollutant control measures:

(i) *Replace*: Use of another, more environmentally friendly chemical.

(ii) *Reduce*: Use as little of the priority pollutants as possible.

(iii) *Managing pollutants*: Careful responsible management to minimise accidental or adventitious loss and waste.

Preventing pollution of the environment by chemicals is very complex as chemicals may be released into the environment at any stage from development and testing, through the manufacturing processes, storage, distribution, through their use and finally after disposal.

The release of chemicals into the environment can broadly be categorised into,

◆ point source release, and

◆ diffuse or non-point source release.

Controls have been set in the past, to concentrate on tackling the largest point source and introducing strict standards for the discharge into water or sewers. It is widely recognised that control of pollution is not a single environmental management strategy or single industry issue. The implementation of Integrated Pollution Control (IPC) has played an important role in introducing a more holistic control philosophy to environmental management. IPC applies to releases from the most polluting industrial processes. Direct toxicity assessment is considered a control tool for chemical mixtures for both process and emission control. In order to provide a more integrated view of the state of the environment, ecological monitoring is essential. The ultimate aim of chemical control is the protection of the environment by measuring improvement in ecological quality. The ultimate effectiveness of chemical control has to be measured.

Industrial pollution: types, effects and control

In order to provide for the daily needs of our growing population, various types of industries are setup to manufacture different products. The industries use raw materials, process them and generate finished products. Besides these, several byproducts are generated. Some are in huge quantities and processing them is not cost effective. If the cost of processing the waste is prohibitive, the industrialist throws the waste into the environment in the form of gases, liquids or solids. The gases are released into the atmosphere, the liquids are discharged into aquatic bodies such as canals, rivers or the sea, and solid wastes are either dumped on land or in aquatic bodies. In all these situations, either the air, water or land is polluted due to dumping of wastes.

There are about 17 industries which are declared to be the most polluting. These include factories of caustic soda, cement, distilleries, dyes and dye intermediaries, fertilisers, iron and steel, oil refineries, paper and pulp, pesticides and pharmaceuticals, sugar, textiles, thermal power plants and tanneries. **Table 5.6** lists a few of the industries and their wastes and the type of pollution they create.

Effects of industrial pollution

Effects on human health:

◆ Pollutants differ in their effects on human physiology, wellbeing and health. They cause irritation of the respiratory tract, eyes, nose, throat, and so on.

◆ Pollutants are responsible for increase in mortality and morbidity.

◆ A variety of particulates mainly pollens, initiate asthmatic attacks.

◆ Chronic pulmonary diseases such as bronchitis and asthma are aggravated by high concentration of SO_2, NO_2, particulate matter and photochemical smog.

◆ Certain heavy metals like lead enter the body through the lungs and cause poisoning.

Effects on animal health: The pollutants enter in two steps.

Table 5.6   Type of industry and its ill effects

S. No.	Type of industry	Chemicals	Pollution
1.	Caustic soda	Mercury, chlorine gas	Air, water and land
2.	Cement dust, smoke	Dust, smoke	Particulate matter
3.	Distillery	Organic waste	Land and water
4.	Fertiliser	Ammonia, cyanide, oxides of nitrogen, oxides of sulphur	Air and water
5.	Dye	Inorganic waste pigment	Land and water
6.	Iron and steel	Smoke, gases, coal dust, fly ash, fluorine	Air, water and land
7.	Pesticides	Organic and inorganic waste	Water and land
8.	Oil refineries	Smoke, toxic gases, organic waste	Air and water
9.	Paper and pulp	Smoke, organic waste	Air and water
10.	Sugar	Organic waste, molasses	Land and water
11.	Textiles	Smoke, particulate matter	Land and water
12.	Tanneries	Organic waste	Water
13.	Thermal power	Fly ash, SO_2 gas	Air and water

(i) Accumulation of the airborne contaminants in the vegetation that animals forage for.
(ii) Subsequent poisoning of the animals when they eat the contaminated food. Three pollutants – fluorine, arsenic and lead – are responsible for the loss of livestock.

Effects on plants: Industrial pollution has been shown to have serious adverse effects on plants. In some cases, it is found that vegetation over 150 km away from the source of pollutants has been affected. The major pollutants affecting plants are SO_2, O_3, MO, NO_2, NH_3, HCN, ethylene, herbicides, PAN (peroxyacetyl nitrate), and so on. In the presence of pollutants, healthy plants suffer from a variety of diseases and may die.

Control of industrial pollution

Some important control measures are:
◈ control at source,
◈ selection of industry site,
◈ treatment of industrial waste,
◈ plantation trees,
◈ stringent government action,
◈ assessment of the environmental impact, and
◈ strict implementation of the environmental protection act.

5.5.5   Biological Concentration and Biomagnification

A pollutant present in the environment makes its entry into the food chain through the producers. After entry, these get accumulated in the cells and tissues of the primary consumers when they feed on plants. The pollutants are transferred into the body of the primary consumer in addition to that directly absorbed from the environment. This leads to an increase in the concentration of the pollutant in the body of the primary consumer. The primary consumer is ingested by a secondary consumer and the pollutants are thereby transferred to the latter. Thus, on moving through the food chain, it is seen that the concentration of the pollutants becomes greater in the tissues of the organisms belonging to the higher trophic levels than in the organisms at the entry point in the lower trophic levels. Hence the residual retention of the pollutants is the highest at the upper trophic level.

The pollutants get magnified at the higher levels of the food chain. This type of magnification of the pollutants in a food chain, mediated by biological agents (members of different trophic levels) is known as biological magnification or biomagnification of pollutants in the ecosystem. Since man is omnivorous and has access to different trophic levels for food in the food chain, he receives the pollutants in much larger amounts which get deposited in various tissues. This causes serious health issues depending on the physiological system affected and concentration level of the pollutant.

Six types of chemical industries are: (i) inorganic and organic chemical industries, (ii) fertiliser industries, (iii) refineries and petroleum industry, (iv) pesticide industries, (v) electroplating and heat treatment industries, and (vi) hydro-generated oil and soap industries.

5.6   NOISE POLLUTION

Noise is a non-chemical pollutant that affects humans and animals. Noise leads to deafness if it is sudden and at higher decibels. Our noisy festivals can deter sleep and affect concentration at work. The new types of noise pollution that affect marine animals are a serious concern. Sonar equipment used in shipping affects whales and dolphins leading to beaching and death along long coastlines. Playing back the territorial or breeding calls of wild birds to attract them for photography can confuse them and affect their breeding and nesting behaviour.

Noise may not seem as harmful as the contamination of air or water, but it is a problem that affects human health and can contribute to the general deterioration of environmental quality.

5.6.1   Types of Noise Pollution

Noise is undesirable and unwanted sound. Not all sound is noise. What may be music to one person may be noise to another! It is not a substance that can accumulate in the environment like most other pollutants. Sound is measured in units called decibels (dB).

5.6.2   Causes of Noise Pollution

Several sources of noise contribute to both indoor and outdoor noise pollution. Noise emanating from factories, vehicles and from loudspeakers played during festivals contribute to outdoor noise pollution, while loudly played radios or music systems and other electronic gadgets contribute to indoor noise pollution.

5.6.3   Effects of Noise Pollution

Several harmful effects are caused by exposure to high sound levels. These effects can range in severity from being annoying to being extremely painful and harmful for the eardrum.

Noise is also harmful to wildlife. It scares the animals, especially in Protected Areas. The noise from fire crackers during Deepavali disturbs waterfowl in lakes and rivers. The noise created by propellers of boats leads to serious effects on fish and marine mammals. A specific form of marine noise pollution is from sonar waves used by naval and commercial ships. This disrupts the echo location that marine mammals use for detecting the condition of their habitat and the fish shoals they depend on for food. Many increasing episodes of beaching and death of marine mammals on the shoreline are possibly due to the sonar waves used by ships.

Effects on physical health: The most direct harmful effect of excessive noise is physical damage to the ear causing temporary or permanent hearing loss. Temporary hearing loss is often called a *temporary threshold shift* (TTS). People suffering from this condition are unable to detect low volume sounds of certain frequencies. However, the hearing ability is usually recovered within a month of exposure to a noisy event. Permanent loss, called *noise-induced permanent threshold shift* (NIPTS), represents a loss of hearing ability from which there is no recovery. As an example, in Maharashtra, people living in the close vicinity of Ganesh *pandals* that blare music for the ten days of the Ganesh festival are usually known to suffer from TTS. Below a sound level of 80 dB, hearing loss does not occur. However, temporary effects are noticed at sound levels between 80 and 130 dB. About 50% of the people exposed to 95 dB sound levels at work develop NIPTS and most people exposed to more than 105 dB experience permanent hearing loss to some degree. A sound level of 150 dB or more can physically rupture the human eardrum. The degree of hearing loss depends on the duration as well as the intensity of the noise. For example, a one hour exposure to a 100 dB sound level can produce a TTS that may last for about one day. However, in factories with noisy machinery, workers are subjected to high decibel levels for several hours a day. Workers exposed to 95 dB for 8 hours every day for over 10 years may suffer from NIPTS. In addition to hearing loss, excessive sound levels can harm the circulatory system by raising the blood pressure and altering pulse rates. High noise levels can also create psychological ill effects and disturb the peace and quiet, as well as sleep.

Effects on mental health due to noise: Noise causes emotional or psychological effects such as irritability, anxiety and stress. Lack of concentration and mental fatigue are significant health effects of noise. It has been observed that the performance of school children is poor in comprehension tasks when schools are situated in busy areas of a city and suffer from noise pollution.

As noise interferes with normal auditory communication, it may mask auditory warning signals and hence increase the rate of accidents, especially in industries. It can also lead to lower efficiency and productivity of workers and cause higher accident rates on the job.

Thus, noise is more than just a nuisance. It seriously affects the quality of life. It is therefore important to ensure mitigation or control of noise pollution.

5.6.4   Control Measures for Noise Pollution

Permitted noise levels: A standard safe time limit has been set for exposure to various high noise levels. Beyond this 'safe' time, continuing exposure over a year will lead to hearing loss. There are four fundamental ways in which noise can be controlled: (i) reduce noise at the source, (ii) block the path of noise, (iii) increase the path-length, and (iv) protect the recipient. In general, the best control method is to reduce noise levels at the source.

Source reduction can be done by effectively muffling vehicles and machinery to reduce noise. In industries, noise reduction can be done by using rigid sealed enclosures around machinery, lined with acoustic absorbing material. Isolating machines and their enclosures from the floor, using special spring mounts or absorbent mounts and pads and using flexible couplings for interior pipelines contribute to reducing noise pollution at the source. However, one of the best methods of noise source reduction is the regular and thorough maintenance of operating machinery. Noise levels at construction sites can be controlled through proper construction planning and scheduling techniques. Locating noisy air-compressors and other equipment away from the site boundary, along with creating temporary barriers to physically block the noise, can help in reducing noise pollution.

Most of the vehicular noise comes from the movement of the vehicle tyres on the pavement and wind resistance. However, poorly maintained vehicles can add to the noise levels. Traffic volume and speed significantly affect the overall sound level on a road. For example, doubling the speed increases the sound levels by about 9 dB and doubling the traffic volume (number of vehicles per hour) increases sound levels by about 3 dB. A smooth flow of traffic also causes less noise than a repeated stop-and-go traffic pattern. Proper highway planning and design are essential for controlling traffic noise. Establishing lower speed limits for highways that pass through residential areas, limiting traffic volume and providing alternative routes for truck traffic are effective noise control measures.

The path of traffic noise can be blocked by constructing vertical barriers along the highway. Planting trees around houses is an effective method to reduce noise levels.

In industries, different types of absorptive material can be used to control interior noise. Highly absorptive interior finish material for walls, ceilings and floors can greatly decrease indoor noise levels. Sound levels also drop significantly with increasing distance from the source of noise. *Increasing the path length* between the source and the recipient offers a passive means of control.

Urban industrial and municipal landuse planning, especially pertaining to the location of airports, make use of the attenuating effect of distance on sound levels. The use of earplugs and earmuffs can be very effective in protecting individuals from high noise levels. Specially designed earmuffs can reduce the sound level reaching the eardrum by as much as 40 dB. However, workers tend not to wear them on a regular basis despite company requirements for their use.

5.7 NUCLEAR HAZARDS AND HUMAN HEALTH RISKS

Nuclear energy can be both beneficial and extremely harmful, depending on the way in which it is used. We routinely use X-rays to examine bones for fractures, treat cancer with radiation and diagnose diseases with the help of radioactive isotopes. Approximately 17% of the electrical energy generated in the world comes from nuclear power plants. However, it is impossible to forget the devastation that nuclear bombs caused in the cities of Hiroshima and Nagasaki. The radioactive waste from nuclear energy has caused and continues to cause serious environmental damage in Japan. Disasters in nuclear plants at Chernobyl and in Japan after the tsunami have wreaked havoc in people's lives and that of subsequent generations.

Nuclear fission is the splitting of the nucleus of an atom. The resulting energy is used for a variety of purposes. The first controlled fission of an atom was carried out in Germany in 1938. However, the United States was the first country to develop the atomic bomb, which was subsequently dropped on the Japanese cities of Hiroshima and Nagasaki. The world's first electricity generating reactor was constructed in the United States in 1951. The Soviet Union built its first reactor in 1954.

In December 1953, the US President Dwight D Eisenhower in his 'Atoms for Peace' speech made the following prediction: 'Nuclear reactors will produce electricity so cheaply that it will not be necessary to meter it. The users will pay a fee and use as much electricity as they want. Atoms will provide a safe, clean and dependable source of electricity.'

Although nuclear power is being used today as a reliable source of electricity, the above statement sounds highly optimistic today after its risks have been realised. Several serious accidents have caused worldwide concern about the safety and disposal of radioactive waste.

In a nuclear reactor, low-grade uranium ore which contains 0.2% uranium by weight, is obtained by surface or underground mining. After it is mined, the ore goes through a milling process where it is crushed and treated with a solvent to concentrate the uranium and produces 'yellow cake', a material containing 70%–90% uranium oxide. Naturally occurring uranium contains only 0.7% of fissionable U-235, which is not high enough for most types of reactors. So it is necessary to increase the amount of U-235 by enrichment, although it is a difficult and expensive process. The enrichment process increases the U-235 content from 0.7% to 3%. Fuel fabrication then converts the enriched material into a powder, which is compacted into pellets. These pellets are sealed in metal fuel rods about 4 m in length, which are then loaded into the reactor. As fission occurs, the concentration of U-235 atoms decreases. After about three years, a fuel rod does not have enough radioactive material to sustain a chain reaction and the spent fuel rods must be replaced by new ones. However, these spent rods are still highly radioactive and can affect human health as they contain about 1% U-235 and 1% plutonium. These rods are a major source of radioactive waste material produced by a nuclear reactor.

Initially, it was thought that spent fuel rods could be reprocessed, not only to provide new fuel, but also to reduce the amount of nuclear waste. However, the cost of producing fuel rods by reprocessing was found to be greater than the cost of producing fuel rods from the ore. At each step in the cycle, there is a danger of exposure to harmful radiation and this poses several health and environmental concerns.

5.8   SOLID WASTE MANAGEMENT

In ancient cities, food scraps and other waste were simply thrown onto the streets where they accumulated. Around 320 BC in Athens, the first known law forbidding this practice was established and a system of waste removal began to evolve in several eastern Mediterranean cities. The initial disposal methods were very crude and were often just open pits outside the city walls. As the population increased, efforts were made to transport the waste further away from cities, thus creating city dumps. Until recently, the disposal of municipal solid waste did not attract much public attention. The favoured means of disposal was to dump solid waste outside the city or village limits, and occasionally burn or compact it.

Around most towns and cities in India, the approach roads are littered with multicoloured plastic bags and other garbage. Waste is also burnt to reduce its volume. Modern methods of disposal such as incineration and the development of sanitary landfills are now attempting to solve these problems. The lack of space for dumping solid waste has become a serious problem in several cities and towns all over the world. Dumping and burning waste is not an acceptable practice today, from an environmental or a health perspective. The disposal of solid waste should be part of an integrated waste management plan. The method of collection, transport, processing, resource recovery and final disposal should be synchronised to achieve a common objective.

In recent years, humans have become increasingly wasteful. This is the result of unprecedented economic growth, consumerism and population growth. Thus, solid waste management has become one of our most burning issues. Urban areas which generate huge quantities of waste have caused health concerns such as gastrointestinal diseases, both endemic and epidemic. Municipalities are unable to deal with the enormous mounds of waste that are generated by city dwellers. In rural areas there is very little that panchayats can do for clearing waste. This makes the Swachh Bharat initiative, a key aspect of human well being.

Primarily, solid waste is divided into biodegradable and non-biodegradable waste. Inadequately managed non-biodegradable waste leads to mounds of garbage which persist for decades in the soil and release dangerous leachates into water sources and reduce soil fertility. Industries, especially chemical manufacturers, generate great quantities of toxic material that has to be treated at a high cost. Since plastics are indestructible, they can remain in dumps for generations.

Solid waste consists of household waste, commercial waste, institutional waste, construction and demolition debris, sanitation residue, e-waste, agriculture, horticulture, dairy waste and radioactive waste. Municipal solid waste includes household and commercial wastes which are generated in municipal areas in either solid or semi-solid form. Industrial wastes and hazardous wastes are a part of manufacturing processes. Biomedical wastes are waste products generated in a hospital and are a serious health hazard. Pharmaceutical waste has ill effects, as drugs when discarded carelessly, are harmful to other people.

Municipal Solid Waste (MSW) contains a wide variety of materials. It can contain food waste (like vegetables and meat, left-over food, eggshells) which is classified as wet garbage, and paper, plastic, tetrapacks, plastic cans, newspaper, glass bottles, cardboard boxes, aluminium foil, metal items and wood pieces which are classified as dry garbage.

5.8.1   Problems of Municipal Solid Waste Disposal

The most difficult problem in our cities is that citizens do not segregate their household waste into degradable and non-degradable waste. Major Indian metros such as Mumbai, Delhi, Bengaluru, Kolkata and Chennai generate about 10 million tonnes of garbage every day. Mumbai and Delhi have 3 major landfills each and their biggest/main landfills, Deonar in Mumbai (90 years old) and Ghazipur in Delhi (33 years old) are functioning way beyond their lifespan. To deal with this crisis, the Swachh Bharat Abhiyan was launched in 2014. It has set specific targets for our cities.

If cities continue to dump untreated waste at the present rate, we will need 1240 hectares of land per year and with the projected generation of 165 million tons of waste by 2031, the requirement of setting up of a landfill for 20 years of 10 metres height will require 66,000 hectares of land. Land is a finite entity, which means that it will require conversion of other lands such as natural ecosystems or agricultural land into landfills.

Source segregation is the third basic step of solid waste management. Every household needs to put in place different coloured waste bins so that the waste can be managed appropriately by municipalities and panchayats.

5.8.2   Control Measures for Urban and Industrial Solid Waste

Segregation of waste: As per the new solid waste management rules, all waste generators at an individual level should start segregating their waste into three categories – biodegradables, dry waste (plastic, paper, metal, wood) and domestic hazardous waste (diapers, napkins, mosquito repellents, cleaning agents, packaging material, plastic bags) – before they hand their waste over to the waste collectors (**Fig. 5.5**).

An integrated waste management strategy includes three main components:

◈ source reduction,
◈ recycling, and
◈ disposal.

Source reduction is one of the fundamental ways to reduce waste. This can be done by using less material when making a product, reusing products on site and designing products or

Fig. 5.5   Segregation of waste

packaging to reduce their quantity. On an individual level, we can reduce the use of unnecessary items while shopping, buy items with minimal packaging, avoid buying disposable items and also avoid asking for plastic carry bags.

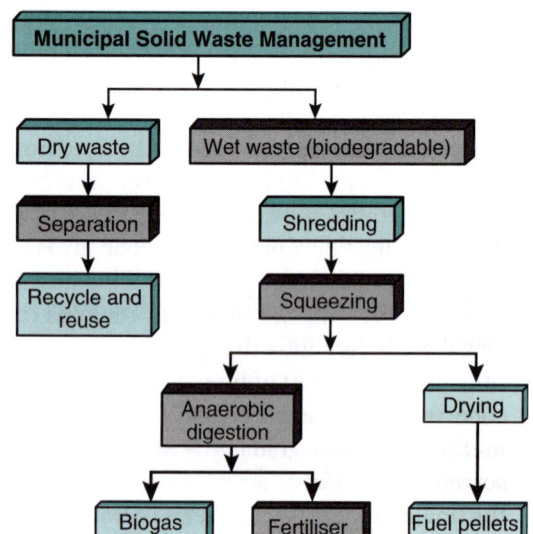

Recycling is reusing some components of the waste that may have some economic value. Recycling has readily visible benefits such as conserving resources, reducing energy used during manufacture and reducing pollution levels. Some materials, such as aluminium and steel, can be recycled many times. Mining of new aluminium is expensive. Hence, recycled aluminium has a strong market value and plays a significant role in the aluminium industry. Metal, paper, glass and plastics are recyclable. Paper recycling helps preserve forests, as it takes about 17 trees to make one ton of paper. Crushed glass (*cullet*) reduces the energy required to manufacture new glass by 50%. Cullet lowers the temperature requirement of the glass making process, thus conserving energy and reducing air pollution. However, even if recycling is a viable alternative, it presents several problems.

The issues associated with recycling are either technical or economic. Plastics are difficult to recycle because of the different types of polymer resins used in their production. Since each type has a distinct chemical composition, different plastics cannot be recycled together. Thus, separation of different plastics before recycling is necessary. Similarly, in recycled paper, the fibres are weakened and it is difficult to control the colour of the recycled product. Recycled paper is banned for use in food containers to prevent the possibility of contamination. It very often costs less to transport raw-paper pulp than scrap paper. Collection, sorting and transport account for about 90% of the cost of paper recycling. The processes of pulping, de-inking and screening wastepaper are generally more expensive than making paper from wood or cellulose fibres. Recycled paper is thus more expensive than virgin paper. However, as technology improves, the cost will come down.

Disposal of solid waste is done most commonly through a sanitary landfill or through incineration. A modern sanitary landfill is a depression in an impermeable soil layer lined with an impermeable membrane.

Highlights of the new solid waste management rules, 2016

Segregation at source: The new rules have mandated the source segregation of waste in order to channelise the waste to wealth by recovery, reuse and recycle. Waste generators would now have to now segregate waste into three streams—biodegradables, dry (plastic, paper, metal, wood, and so on) and domestic hazardous waste (diapers, napkins, mosquito repellants, cleaning agents, and so on) before handing it over to the collector.

5.8.3   Collection and Transport of Solid Waste

Currently in most of the cities, the municipalities are unable to remove urban waste as fast as it is produced. For a more efficient management system, it must be sorted and transported at reasonable frequencies so that it does not accumulate at multiple sites across residential areas.

There are three main collection components—collection containers, collection points and collection frequency.

Transfer station: A transfer station is an intermediate station between the final disposal and collection points and the final disposal dump yard which increases the efficiency of the system. It is useful if the disposal site is far from the collection areas. The transfer station serves as a facility for sorting and recovery of recyclable material. This reduces the cost as well as the space required in the final dumping area. It helps in decreasing the cost of travelling from the collection area to the disposal site as waste is unloaded from smaller collection vehicles and transferred to larger vehicles for transport to the final disposal site.

The main variables which affect the collection system include the collection container types, collection vehicles, management pattern and waste collection routes.

5.8.4 Municipal Sanitary Landfill

The three key characteristics of a municipal sanitary landfill (**Fig. 5.6**) that distinguish it from an open dump are:

◆ Solid waste is placed in a suitably selected and prepared landfill site in a carefully prescribed manner.

◆ The waste material is spread out and compacted with appropriate heavy machinery.

◆ The waste is covered each day with a layer of compacted soil.

Vermicomposting: Nature has perfect solutions for managing the waste that is created, if the natural ecosystem is left undisturbed. The biogeochemical cycles are designed to clear the waste material produced by animals and plants and we can mimic these. All dead and dry leaves and twigs decompose and are broken down by organisms such as worms and insects, and finally by bacteria and fungi, to form a dark rich soil-like material called compost. These organisms in the soil use the organic material as food, which in turn provides them with nutrients for their growth and activities. These nutrients are returned to the soil to be used again by trees and other plants. This process recycles nutrients and mimics nature and can be used in agriculture, horticulture and gardening. This soil is used as manure, thus reducing the need for environmental damaging and unhealthy chemical fertilisers.

Integrated solid waste management system (ISWM): Any current methods of waste disposal cannot deal with all the waste components in an environmentally sustainable way. There is a need to select the most appropriate management options to achieve the desired results (**Fig. 5.7**). Source reduction, waste to energy conversion, combustion, recycling, recovery, composting and landfills,

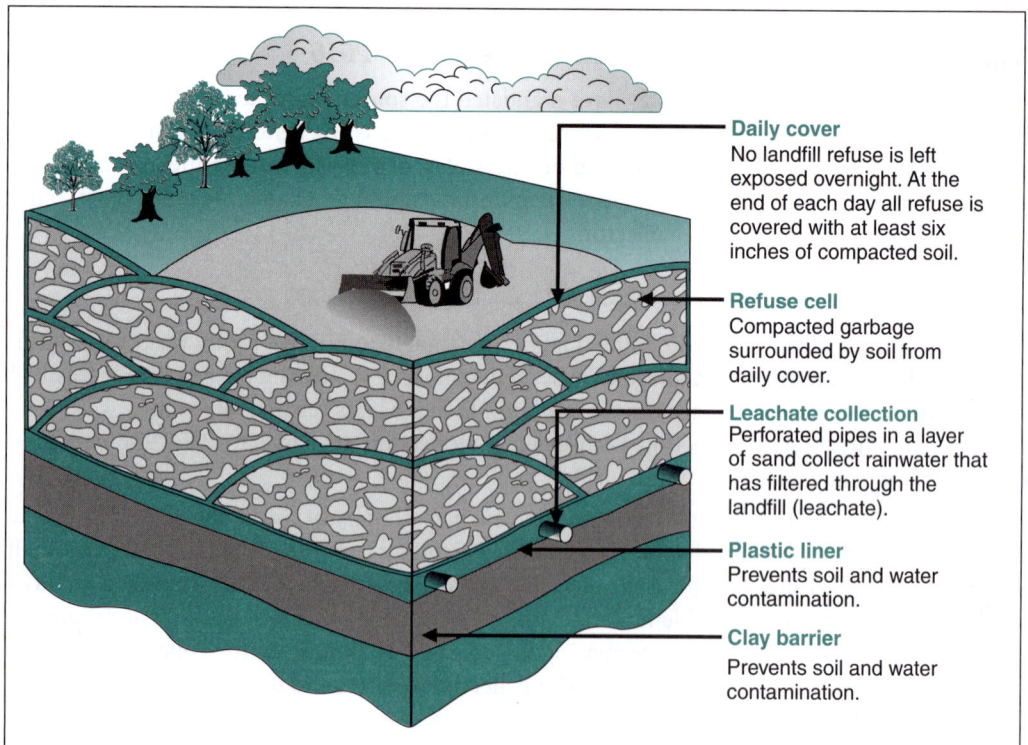

Daily cover
No landfill refuse is left exposed overnight. At the end of each day all refuse is covered with at least six inches of compacted soil.

Refuse cell
Compacted garbage surrounded by soil from daily cover.

Leachate collection
Perforated pipes in a layer of sand collect rainwater that has filtered through the landfill (leachate).

Plastic liner
Prevents soil and water contamination.

Clay barrier
Prevents soil and water contamination.

Fig. 5.6 Cross-section of an active landfill

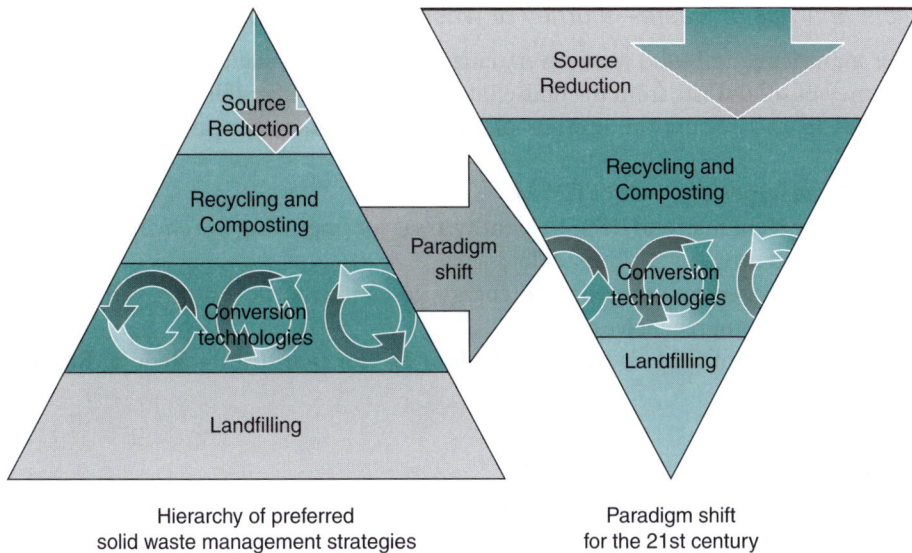

Fig. 5.7 Paradigm shift

all have their own roles to play in comprehensive waste management. Different combinations of these options make modern management of waste environmentally, economically and socially sustainable.

Characteristics of integrated solid waste management systems: An integrated system combines an optimised waste collection system and effective sorting followed by one or more options. It has control over all types and sources of solid waste.

Economic considerations:

◈ Market oriented recycling or production of compost and energy are essential components but depend on the market for these outputs.

◈ Flexibility in design helps to channel the waste according to economic and environmental conditions.

◈ Socially acceptable—objection to waste management sites can be minimised through public consultation and education, appropriate designing and 24/7 management.

◈ Economic considerations and affordability at the urban/rural governance level have to be considered in the planning and designing stage. A long term view must be taken in the selection of the site.

Benefits of the integrated solid waste management system: The essential outcome of the system must lead to:

◈ higher efficiency of resource use,

◈ better business opportunities and economic growth,

◈ higher efficiency, efficacy and safety of the system in relation to residual waste and lecheates,

◈ greater local ownership and responsibilities/participation of citizens,

◈ support for lower underprivileged sectors who work for this socially relevant and important activity,

◈ decreased health hazards for surrounding residents, and

◈ appropriate aesthetic management.

Life cycle analysis of waste—journey of waste products from the cradle to the grave

For better management of solid waste, every citizen should understand the life cycle of waste, that is, the journey of solid waste from the household (cradle), to the landfill site (its grave).

◈ This helps in minimising the impacts of waste on the environment, on people and on the economy.
◈ It is one of the important tools of ISWM.
◈ It provides a system map and helps to identify areas for improving the environment.
◈ It covers a broad range of environmental issues.
◈ It offers the prospect of mapping the energy and material flow along with resources and emissions of the comprehensive system.

The life cycle analysis of waste is based on the four **R**s (for industries)—**Reduce**, **Reuse**, **Recycle** and **Recover**.

Waste systems analysis

◈ Life cycle analysis of the incoming solid waste.
◈ Assessment in the Integrated Solid Waste Management facility.
◈ Information on raw material—analysis of the sources, generation of waste and types of waste.
◈ Overview on the manufacturing units producing waste, regarding segregation, biological/thermal treatment (recovery of heat and energy).
◈ Distribution into use/reuse, 3Rs (**Reduce**, **Reuse**, **Recycle**; for households). Marketing of residual waste for reuse.
◈ Disposal—the management of landfill and its types.
◈ Life cycle analysis to examine every stage of the life cycle of all forms of waste.
◈ Management of every operation/unit process, at each stage and in all stages. For example, the several stages of the overall process of composting.
◈ Assessment and monitoring of the toxicity and/or quantity of products before these reusable products are purchased, used or discarded.

Reducing solid waste

This is the first step to comprehensive solid waste management. We need to reduce the utilisation of commercial goods that eventually find their way to landfills. Thus, using 3Rs is the next step at the household level.

◈ Source reduction is also known as 'waste prevention'. It is mainly done through behaviour changes of all citizens.
◈ Reducing the waste is often related to its reusability.
◈ Two basic routes of waste prevention.
   ▪ For the manufacturers: Change in the design of products, the way of packaging and altering the ways of reducing waste.
   ▪ For consumers: Change in purchasing decisions and the way of using and discarding the end products that are not used.
◈ Purpose of source reduction: Minimise material volume, encourage product reuse, increase product lifetime, change/alter buying practices, support use of eco-friendly products.
◈ Benefits of source reduction of waste:
   ▪ Reducing the waste before it can be generated is the most logical way to save cost and natural resources and to reduce cost of resource utilisation.

- Waste reduction cuts the municipal and commercial costs involved in waste collection and disposal.
- Waste reduction improves productivity by targeting wasteful processes and products.

Recycling: Recycling is the process of converting waste materials into new useful materials and objects.

- Recycling is the most widely recognised form of source reduction.
- It involves the process of segregation, collection, processing of a new product and effective marketing.
- It uses the material that would have otherwise been discarded or dumped.
- It is the fundamental part of a modern waste management plan.
- It can divert a significant portion of a waste stream from disposal in landfill and combustion facilities.

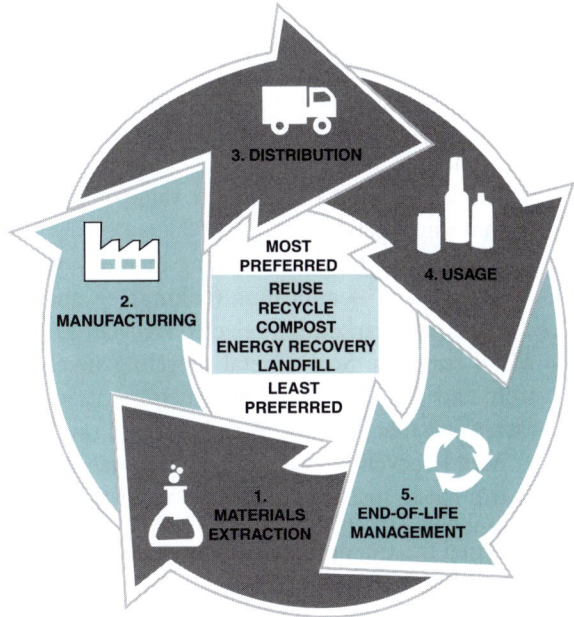

Significance of recycling:

- Recycling helps in reduction of cost in transportation of waste and its disposal.
- It generates economic returns by the sale of recyclables.
- It provides employment by creating additional job opportunities and markets for skilled and unskilled labour.
- Improved health and sanitary conditions in urban areas results from the waste recycling process.
- It reduces the occupational hazards for the workers in the solid waste management system and rag pickers, and it thus reduces the investment in public health programmes.
- Improved recycling increases the economic value of waste.
- It provides an opportunity for people who are engaged in the recycling of waste management to switch to more socially acceptable occupations. It supports rag pickers associations.
- Recycling reduces the volume of waste that has to be finally disposed and thus reduces the pollution at the disposal site.
- Use of recyclables in industrial production relieves the tremendous pressure on precious natural resources used as raw material in the manufacture of thousands of consumer items.

Collection and storage of recyclables:

- Collection of sorted material at the source of solid waste generation (home, service sectors, production units of consumer goods) is a necessary component of a recycling programme.
- Source separation involves segregation of recyclable and reusable material at the point of generation of solid waste.
- Collection of recyclables is carried out with the help of curbside programmes, drop off–buy back programmes and Material Recovery Facility (MRF).

Commonly recycled materials:

- *Paper recycling*: Paper recycling is the process of recovering waste paper and remaking it into new paper products.

- Paper and cardboard form the second largest component of domestic waste after organic waste.
- Paper recycling is practiced extensively.
- It reduces the demand for wood and energy.
- The steps of paper recycling includes, collection and storage → repulping and screening → cleaning and deinking → refining and bleaching → making of recycled paper.

◈ *Glass recycling*: Glass is a commonly recycled material that accounts for 2.5% by weight of solid waste generated. Recycling of broken glass reduces the risk of the occupational hazards of glass manufacture.
- Recycled glass is of economic value only if it can be separated by colour and then crushed to make new glass.
- The process of glass recycling includes, sorting → melting → moulding, blowing, cutting → product → annealing (baking) → quality inspection → finished product → approved (packing, marketing), rejected (recycling).

◈ *Metal recycling*: Scrap metal recycling involves the recovery and processing of scrap metal from end-of-life products or structures, as well as from waste generated during manufacturing (scrap), so that it can be introduced as a raw material in the production of new goods.
- Metal recycling is important because using recycled scrap metal in place of virgin iron ore made insisted or cast iron can yield 75% savings in energy and 90% savings in raw materials used.

◈ *Plastic recycling*: Plastic consists of 8%–9% of the total amount of municipal solid waste. For recycling it has to be segregated by type.
- In India, the total plastic waste generated is 15,342 tonnes per day.
- Processed plastic waste is 9,205 tonnes per day.
- Remaining plastic waste (unprocessed) is 6,137 tonnes per day. (CSE report, Ref: ttps://economictimes.indiatimes.com/articleshow/59301057.cms?utm_source=content ofinterest&utm_medium=text&utm_campaign=cppst)
- The process of plastic recycling involves collecting recyclable plastic waste → categorising → processing the categorised plastic waste into various raw materials → manufacturing new products from recycled plastic. (http://www.sita.com.au/commercial-solutions/resource-recovery-recycling/plastic-polystyrene/)
- Before any plastic waste is recycled, it needs to go through four different stages so that it can be used further for making various types of products.

◈ *Sorting*: It is necessary that every plastic item is separated according to its material and type.

◈ *Washing*: The plastic waste is washed to remove impurities such as labels and adhesives.

◈ *Shredding*: Clean plastic waste is loaded into different conveyer belts that run the waste through several shredders. These shredders tear up the plastic into small pellets, preparing them for recycling into other products.

◈ *Extruding*: This involves melting the shredded plastic so that it can be made into pellets, which are used for making different types of plastic products.

5.9 WASTE TO ENERGY METHODOLOGIES

Recovery of steam and electricity from waste can be done by several processes.

(i) **Incineration**: Solid waste is burnt in a properly designed furnace under suitable temperature and operating conditions.

◈ It is one of the effective methods of reducing the volume of municipal solid waste by 90% and weight by 75%.
◈ The process consists of controlled burning of waste at a high temperature.
◈ It sterilises and stabilises the waste with reduction in the volume.
◈ Waste reduction is immediate. It does not require long term storage or decomposition.
◈ The waste can be incinerated at the site without transport to a distant area.
◈ Air discharge must be effectively controlled for reducing the impact on the environment.
◈ The ash residue is usually non-polluting (sterile).
◈ Technology exists to completely destroy even the most hazardous of materials in a completely effective manner.
◈ It requires a relatively small disposal area as compared to landfill sites.

(ii) *Refuse Derived Fuel (RDF)*
◈ When the solid waste contains large amounts of combustibles, it can be used as a fuel.
◈ In an RDF system, solid waste is pre-processed to remove non-combustible items and to reduce the size of the combustible fractions.
◈ The recovered combustible portion is termed as RDF.

(iii) *Biogasification*: It is an anaerobic treatment in which organic waste is stabilised and produces methane gas as an energy source.
◈ The gas provides income generation, energy and fertiliser substitution and generates macro-economic benefits through decentralised energy generation.

Landfill

A sanitary landfill is an engineered facility for the disposal of municipal solid waste designed and operated to minimise public health hazards and environmental impacts.

Landfill process:
◈ Solid waste is placed in a suitably selected and prepared (lined) landfill site in a carefully prescribed manner.
◈ Waste material is spread out and compacted with appropriate heavy machinery.
◈ Waste is covered each day with a layer of compacted soil. Daily cover usually consists of about 30–60 cm of native soil or alternative material such as compost, sand and tyre shredded fluff.
◈ The purpose of landfilling is to bury/alter the chemical composition of the waste, so that it will not pose any threat to the environment, public health, or impact the lives of the adjacent residents.
◈ Landfills are usually made up of cells in which a discrete volume of waste is kept isolated from adjacent waste cells by a suitable barrier.
◈ The term 'cell' is used to describe the volume of material placed in a landfill during one operating period.
◈ The process of landfilling is carried out in an excavated cell. In the *trench method*, solid waste is placed in cells/trenches excavated in the soil. The *area method* is used when the terrain is unsuitable for excavation of cells, trenches and the ground water table is high.
◈ The most important feature of a modern sanitary landfill design is the technology used to prevent ground water pollution by installation of liners and leachate management systems.
◈ Migration of landfill gas is done by extraction wells and stone-filled vents which are often placed around the periphery of a landfill site.

5.9.1 Types of Waste

Hazardous waste: Modern society produces large quantities of hazardous waste generated by chemical manufacturing companies, petroleum refineries, paper mills, smelters and other industries. Waste is that which can cause harm to humans or the environment. Waste is classified as hazardous when it causes or significantly contributes to an increase in mortality, an increase in serious irreversible or incapacitating reversible illness, and poses a substantial present or potential hazard to human health or the environment when inadequately treated, stored, transported or disposed of.

Characteristics: Waste is classified as hazardous if it exhibits any of the four primary characteristics based on the physical or chemical properties of toxicity, reactivity, ignitability or corrosivity. In addition to this, waste products that are either infectious or radioactive are also classified as hazardous.

Types of hazardous waste and their effects: *Toxic wastes* are substances that are poisonous even in very small or trace amounts. Some may have an acute or immediate effect on humans and animals, causing violent illness or death. Others may have a chronic or long-term effect, slowly causing irreparable harm to the exposed individuals. Acute toxicity is readily apparent as the affected victim has serious repercussions to the toxin shortly after being exposed. Chronic toxicity is much more difficult to determine because the effects may not be seen for years. Certain toxic wastes are known to be carcinogenic (cancer causing) and others may be mutagenic, causing biological changes in the foetuses of exposed people and animals.

Infectious waste includes human tissues removed during surgery, used bandages and hypodermic needles and microbiological materials. This includes hospital and laboratory waste which must be managed separately from all other waste as it can infect people with life threatening bacterial or viral diseases (**Fig. 5.8**, **Table 5.7**).

Radioactive waste is the output from nuclear power plants that can persist in the environment for thousands of years before it decays appreciably. This can cause mutations in new born babies and cancer in adults.

Effects of the environmental problems and health risks caused by hazardous waste: As most hazardous waste is disposed of on land, the most serious environmental impact is contaminated

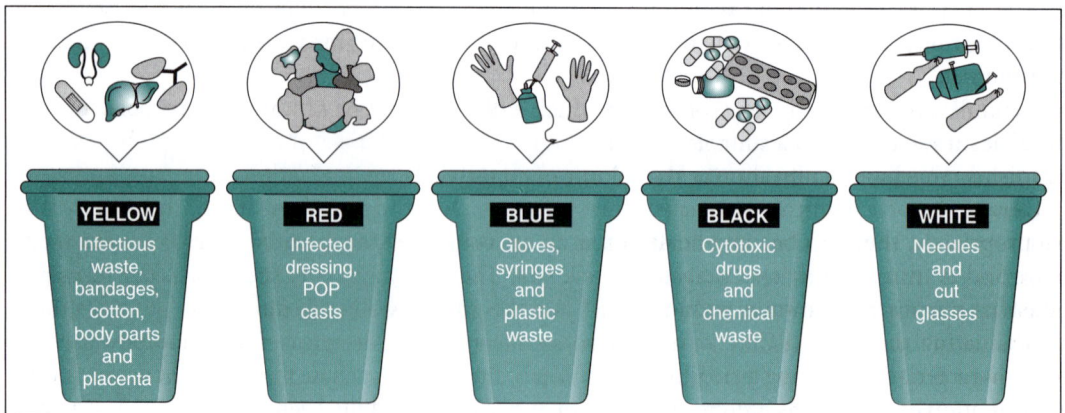

Fig. 5.8 Segregation of bio-medical waste in colour coded bins

Table 5.7 Biomedical wastes categories and their segregation, collection, treatment, processing and disposal options

Category	Type of waste	Type of bag or container to be used	Treatment and disposal options
Yellow	a) Human anatomical waste b) Animal anatomical waste c) Soiled waste d) Expired or discarded medicines e) Chemical waste f) Discarded linen, mattresses, beddings contaminated with blood or body fluid, routine mask and gown	Yellow coloured nonchlorinated plastic bags or containers	Incineration or plasma pyrolysis or deep burial* The discarded medicines shall be either sent back to manufacturer or disposed by incineration
	g) Micro, Bio-t and other clinical lab waste	Autoclave safe plastic bags or containers	
	h) Chemical liquid waste	Separate collection system leading to effluent treatment system	After resource recovery, the chemical liquid waste shall be pretreated before mixing with other wastewater
Red	Contaminated waste (recyclable)	Red coloured nonchlorinated plastic bags or containers	Autoclaving or micro-waving/hydroclaving followed by shredding or mutilation or combination of sterilisation and shredding Treated waste to be sent to recyclers Plastic waste should not be sent to landfill sites.
White (translucent)	Waste sharps including metals	Puncture proof, leak proof, tamper proof containers	Autoclaving or dry heat sterilisation followed by shredding or mutilation or encapsulation in metal container or cement concrete
Blue	Glassware and metallic body implants	Puncture proof and leak proof boxes or containers with blue coloured marking	Disinfection or through autoclaving or microwaving or hydroclaving and then sent for recycling

Source: Bio-Medical Waste Management Rules, 2016 and amended rules 2018

ground water. Once ground water is polluted with hazardous waste, it is frequently not possible to reverse the damage.

Chemicals: Lead, mercury and arsenic are frequently used hazardous substances, often referred to as heavy metals. Lead affects red blood cells by reducing their ability to carry oxygen and shortening their lifespan. Lead may also damage nerve tissues, resulting in neurological (brain) diseases.

Mercury is employed in the production of chlorine, and as a catalyst in the production of some plastics. In the food web, mercury becomes more concentrated as it is taken up by various organisms. This biomagnification is known to cause brain damage.

Thousands of chemicals are used in industries every day. When used incorrectly or inappropriately, they can become health hazards. PCBs (polychlorinated biphenyls) are resistant to fire and do not conduct electricity very well, which makes them excellent materials for several industrial purposes. Rainwater washes out PCBs from disposal areas in dumps and landfills thus contaminating the water. PCBs do not break down rapidly in the environment and retain their toxic characteristics for long periods of time. They cause long-term exposure problems for both humans and wildlife. PCBs are concentrated in the kidneys and liver where they cause damage. They also cause reproductive failure in birds and mammals.

We may not realise it but many household chemicals can be quite toxic to humans as well as to wildlife. Most of the dangerous substances in our homes are found in various kinds of cleaners, solvents and products used in vehicle care.

Today, the most common methods for disposing of hazardous waste are land disposal and incineration. In countries where there is abundant land available for disposal, for example, North America, land disposal is the most widely used method. In countries like Europe and Japan, where land is not readily available and is expensive, incineration is preferred. Despite strong laws, illegal dumping of wastes continues. Hazardous waste management must move beyond burying and burning. Industries need to be encouraged to generate less hazardous waste through their manufacturing processes. Although toxic waste cannot be entirely eliminated, technology is available for minimising, recycling and treating these life threatening wastes. An informed public can also contribute in a big way towards this end. It is essential for us to understand the ill effects of chemicals so that we can make informed decisions about their use. We might decide that the benefits of the use of a toxic substance do not outweigh the risks and choose not to use it at all, or we may decide that it is acceptable to use a substance under specific circumstances where it is adequately controlled and exposure to toxicity is prevented.

5.10 POLLUTION CASE STUDIES

Case Study 5.2 Polluted bottled water

Background: *Ill effects due to bottled water and beverages:* Despite the fact that all bottled water plants are expected to use a range of purification methods, this consumer product is a serious health risk. The fault obviously lies in the treatment methods used.

In bottled water plants, the water is filtered using membranes with ultra-small pores to remove fine suspended solids and is expected to filter out bacteria, protozoa and even larger viruses. Nano-filtration can also remove insecticides and herbicides. However, it is expensive and is thus inadequately used by industries. Most industries also use an activated charcoal adsorption process, which is effective in removing organic pesticides (reverse osmosis and granular activated charcoal) but not heavy metals. Even though most of the manufacturers claim to use these processes, the presence of pesticide residues points to the fact that either the manufacturers do not use the treatment process effectively, or they only treat a part of the raw water. The low concentrations of pesticide residues in bottled water do not cause acute or immediate effects. However, repeated exposure even to extremely miniscule amounts can result in chronic effects like cancer, liver and kidney damage, disorders of the nervous system, damage to the immune system and birth defects.

Issue: Six months after the Centre for Science and Environment (CSE) reported pesticide residues in bottled water, it also found these pesticides in popular cold drink brands sold across the country. This is because the main ingredient in a cold drink or carbonated non-alcoholic beverage is water and there were no standards specified for water to be used in these beverages in India.

There were no standards for bottled water in India till 29 September 2000, when the Union Ministry of Health and Family Welfare issued a notification [no. 759(E)] amending the Prevention of Food Adulteration Rules, 1954. The BIS (Bureau of Indian Standards) certification mark became mandatory for bottled water from 29 March 2001. However, the parameters for pesticide residues remained ambiguous. Following the report published by the Centre for Science and Environment (CSE) in *Down to Earth* (Vol. 11, no. 18), a series of committees were established and eventually on 18 July 2003, amendments were made in the Prevention of Food Adulteration Rules stating that pesticide residues considered individually should not exceed 0.0001 mg/L, that the total pesticide residues should not be more than 0.0005 mg/L and that the analysis should be conducted by using internationally established test methods meeting the residue limits specified. This notification came into force from 1 January 2004.

CASE STUDY 5.3   A waste management model for small towns

Vengurla in Sindhurdurg district, Maharashtra has achieved 95% waste segregation at source. It is one of the only towns in India which generates revenue from its waste. The State Government conferred Vengurla the Vasundhra Award, 2017, for its green initiatives and shortlisted it as a successful model for 100% solid waste management under the Swachh Bharat Abhiyaan.

Vengurla is one of the only towns in India to convert a landfill into a waste management park, called the Swachh Bharat Waste Park. The park now hosts a biogas plant, a briquette-making plant, a segregation yard and a plastic crusher unit. It also has fruit trees and an organic farm. The idea was to make waste management a hygienic and environment friendly place. In 2015, the town with a population of 15,000, adopted waste segregation at source and today has achieved more than 95% segregation. The local administrative body of the town earns ₹1.5 lakhs per month from processing 7 tonnes of its waste generated per day in the township.

The chief officer in 2015 believed that apart from segregation at source, no city can become clean without direct public participation; behavioural change among citizens became a part of the waste management of Vengurla. Under the Solid Waste Management Rules 2016, the usage of plastic bags of less than 50 microns was banned and a fine of ₹500 was imposed for its usage. Littering and non-segregation are fined. Citizens as well as sanitation workers were educated on the need for segregation at source. The landfill was managed by segregating organic content. Compost was made from the organic waste which was used in farms and sold to local residents for gardening. To encourage reuse, unused items were dropped into a box placed under a tree, called the tree of humanity.

CASE STUDY 5.4   Swachh Survekshan 2016

Background: A survey of 73 cities in India for cleanliness was carried out and they were categorised based on the marks scored by each of them in the 'Swachh Survekshan 2016'. Mysuru topped the list and Dhanbad was at the bottom.

Leaders in city urban waste management: Rank City 1. Mysuru, 2. Chandigarh, 3. Tiruchirapalli, 4. New Delhi Municipal council, 5. Visakhapatnam, 6. Surat, 7. Rajkot, 8. Gangtok, 9. Pimpri–Chinchwad, 10. Greater Mumbai, 11. Pune, 12. Navi Mumabi, 13. Vadodara, 14. Ahmedabad, 15. Imphal. (Press Information Bureau Government of India Ministry of Urban Development, 15 February 2016.)

Indore (Madhya Pradesh), Bhopal (Madhya Pradesh) and Vishakhapatnam (Andhra Pradesh) are the top three cleanest cities of the country, as per the results of the Swachh Survekshan, 2017. However, all of them have poor segregation of waste and have adopted a centralised cluster-based approach.

Of the top 50 cities, 31 cities are in three states—Gujarat (12), Madhya Pradesh (11) and Andhra Pradesh (8), all of which are using a cluster-based waste management approach. Cities such as Panjim and Alappuzha, advocate decentralised waste management based on household level segregation, recycle and reuse. (Saturday, 06 May 2017, Down to Earth)

According to a Central Pollution Control Board (CPCB) report, in 2016, India produced some 52 million tonnes of waste each year, or roughly 0.144 million tonnes per day, of which roughly 23% is processed—taken to landfills or disposed of using other technologies.

Alappuzha, Panaji, Mysuru and Bhopal, according to the survey, are India's top four clean cities because they give priority to segregation of waste at the household level, and its reuse.

Alappuzha: In July 2012, Alappuzha was facing a crisis. Known as the Venice of the East for its large network of canals, backwaters, lagoons and beaches, the city looked like a waste dump. Rotten garbage had piled up on roadsides, and canals and drains were clogged with bags of stinking waste. Swarms of mosquitoes and flies had invaded the city, spreading chikungunya and dengue. The municipality had been dumping waste for decades in a six-hectare plot it owned in Sarvodayapuram, a village in the nearby Mararikkulam village panchayat. In June 2012, the residents of Sarvodayapuram rose up in arms against waste dumping in their backyard.

Two-and-a-half-years later, Alappuzha had undergone a dramatic transformation. The dumping spots disappeared. But where did all the waste go? Alappuzha, which has a population of 0.174 million and produces 58 tonnes of solid waste a day, is implementing a project called *Nirmala Bhavanam Nirmala Nagaram* (Clean Home Clean City) since November 2012. The focus of the initiative is segregation and treatment of wet waste at the source and was started in twelve of the most urbanised wards, covering 12,000 households, as a pilot project. The plan was to make households set up portable or fixed biogas plants. Those who did not have enough land were advised to go for pipe composting which is ideal for a small family. It consists of two PVC pipes of 1.25 m length and 20 cm diameter with two caps. The pipes are fixed in a vertical position on the ground with one-fourth of the pipe under the ground. A 30 cm thick layer of gravel is first put into the pipe to absorb the leachates. Waste is put into one pipe for 30–35 days till it is full. Once full, the first pipe is closed with the lid. Then for the next 30–35 days waste is put into the second pipe. By the time the second pipe is full, the waste in the first one would have converted to compost.

In 2013, the clean city drive took a new turn with the entry of Thumburmuzhi, a model aerobic composting unit named after the Thumburmuzhi campus of Kerala Veterinary and Animal Science University, where it was developed to compost carcasses of animals. About two tonnes of waste can be processed into compost in 90 days in this tank. The municipality has set up 12 waste collection centres, with 165 Thumburmuzhi bins at public places and the old waste dumping spots.

Alappuzha, which has a population of 0.174 million and produces 58 tonnes of solid waste per day, has adopted decentralised waste management and is pushing for 100% segregation in all the 23 wards of the city. Moreover, as many as 80% of households now have biogas plants and decentralised composting systems. (Source: Richa Agarwal, Down to Earth, Thursday 30 November 2017)

Surat: The city's garbage disposal system was so poor that there was an outbreak of plague in 1994. Rapid urbanisation during the 1990s in Surat (the economic capital of Gujarat), led to an explosive growth of slums. Waste in the city had become a serious issue. In 1994, the city was hit by plague that killed 52 people and triggered a mass exodus. In just 18 months after the outbreak, Surat was transformed into one of the cleanest cities in the country. Around 1,570 metric tonnes (MT) of waste is collected in the city every day. Citizens pay ₹600 per annum as user charges for waste management.

The city was divided into seven zones to decentralise responsibilities for all civic functions. The entire city was covered under the collection system by November 2004. Waste management in Surat is characterised by 97% collection efficiency and 92% door-to-door collection. Segregation of incoming heterogeneous waste is carried out and waste is converted into compost and RDF pallets. Waste recovery to the tune of 22% was reported during the year 2012–13, which was almost negligible before the implementation of the project. Revenues were obtained from *Carbon Credits*. Land required to dispose of inert waste is less as compared to that required for dumping mixed waste. The change was considered dramatic.

(http://www.nswai.com/DataBank/pdf_mc/Gujarat_mc/City%20Report%20on%20Surat%20SWM%20Project%20under%20JNNURM.pdf)

Mysore: The city of Mysore generates around 400 tonnes of total solid waste a day, of which 60% is wet waste and the rest is dry waste. In terms of solid waste management, the Mysore City Corporation (MCC) has focussed on reduction and segregation of waste at source, door-to-door collection, recycling and reuse of waste, generation of wealth from waste with the involvement of the public and non-profit and educational institutions through public awareness campaigns. Campaigning activities such as daily morning FM radio jingles, pamphlets and door-to-door contact programmes, are also used.

The 65 wards are divided into nine zero-waste management units (ZWM). Each of the nine zero waste management units cover five to ten wards. Each unit has a small composting unit and a resource recovery unit. The biodegradable waste is subjected to composting and within a period of 30–45 days is converted into manure. The dry waste is handled by manual segregation and the saleable material is sold to local recyclers. Inert materials such as dirt and ash are sent to the landfill at Vidyaranyapuram. (*Source:* Waste smart cities, Swati Singh Sambyal @swatisambyal, Down to Earth, Wednesday 15 June 2016)

5.10.1   Dos and Don'ts for Solid Waste Management

◈ Carry your own cloth or jute bag when you go shopping.
◈ Say no to all plastic bags as far as possible. Replace them with paper, cloth and jute bags.
◈ Reuse the soft drinks pet bottles.
◈ Segregate the waste in the house as wet and dry.

◈ Keep two garbage bins and see that the biodegradable and the non-biodegradable material is put into separate bins and disposed of separately.

◈ Dig a compost pit in the garden and put all the biodegradable waste into it to provide you with rich manure for the garden.

◈ See to it that all garbage is thrown into the municipal bin for further disposal of municipal solid waste.

◈ Do not litter on roads or in offices, theatres, market places or any other common public places.

◈ When outside, do not throw paper and other wrappings of leftover food. Make sure it is put into a dustbin.

◈ Do not throw the waste/litter on the streets, drains, open spaces, water bodies, and so on.

◈ Adhere to good citizenship norms of community storage/collection of waste in flats, multi-storied buildings, societies, commercial complexes, and so on.

◈ Manage excreta of pet dogs and cats appropriately.

◈ Provide waste processing/disposal at a community level.

◈ Organise public education and awareness programmes.

◈ Increase awareness in children through interesting educational programmes in schools.

5.11   CONSUMERISM AND WASTE PRODUCTS

Modern societies that are based on using large amounts of consumer goods, especially those that are manufactured for one-time use, are extremely wasteful. The current consumption patterns are depleting non-renewable resources, polluting and degrading ecosystems and altering the natural processes on which life depends.

People in the industrialised countries make up 20% of the world population but consume 80% of the world's resources and produce 80% of waste. This is due to a pattern of economic development which ensures that people continue consuming more than they actually need. India is rapidly moving into this unsustainable pattern of economic growth and development. It is seen that today's consumption patterns are not only depleting natural resources at a rapid rate, but are also widening the inequalities in consumption in different societies. Money is not the only way to measure the cost of an item that we use. When one adds up all the raw material and energy that goes into the manufacture of goods, or the services provided by nature that one uses during a day's activities, the toll on the environment is large. When this cost is multiplied over a lifespan, the amount is staggering. If one considers the over-utilisation in each family, city or country, the impact is incredibly high. For example, two hundred billion cans, bottles, plastic cartons and paper cups are thrown away each year in the developed world. Disposable items greatly increase waste. Rather than compete on quality or reliability, many industrial consumer products are made for one-time use.

How can we reset the level of consumerism and decrease waste in our lives?

◈ Buying quality products that are warranted against failure or wearing out.

◈ By using bioproducts and biofoods.

◈ By learning about the raw materials that things are made of, with an appreciation of their origin from nature's storehouse.

◈ By knowing and being sensitive to the conditions of the workers who make them.

While there may be some new appliances and cars that are more productive and energy efficient, discarding the old ones often leads to an almost total waste of the energy and material already invested in these products. This alone may more than nullify the energy savings of the new product.

Consumerism is related to the constant purchasing of new goods, with little attention to the true needs of their durability, origin, or the environmental consequences of their manufacture and disposal. It is driven by huge sums spent on advertising designed to create both a desire to follow trends and a personal feeling of satisfaction based on acquisition. It interferes with the sustainable use of resources in a society by replacing the normal desire for an adequate supply of life's necessities with an insatiable quest for things that are purchased by growing incomes. In consumerist societies, there is little regard for the true utility of what is bought. An intended consequence of this strategy, which is promoted by those who profit from consumerism, is to accelerate the discarding of the old, either because of lack of durability or a change in fashion.

In developed countries, especially, landfills are being rapidly filled with discarded products that are cheap, or those that fail to work within a short time of purchase and cannot be repaired. In many cases, consumer products are made psychologically obsolete by the advertising industry long before they actually wear out.

The increasing demands of consumption on the finite resources of the planet, increasing level of environmental pollution and the problems of waste disposal must be replaced by the careful utilisation of resources and recovery of used material by waste recycling. Therefore, the reuse of goods and waste utilisation should become a part of the production–consumption cycle.

Current patterns in the industrial sector have led to the disposal of waste in a careless and uneconomical manner. For example, it is estimated that the per capita production of domestic waste is many times higher in a developed country when compared to a developing country. Unfortunately, many developing countries are now working out similar wasteful trends through development, but do not have the same economic potential to handle the waste that this new unsustainable strategy produces.

Large quantities of solid, liquid and gaseous waste are produced by urban industrial communities in the form of plastic, paper, leather, tin cans, bottles, mineral refuse and pathological waste from hospitals. Dead animals, agricultural waste, fertiliser and pesticide overuse and human and animal excreta are essentially rural concerns.

Waste is either discharged into the atmosphere, into water sources or buried underground. This waste is not considered to have any economic value. This attitude towards waste has led to disastrous effects on the environment besides the overexploitation of natural resources.

5.11.1 Role of an Individual in the Prevention of Pollution

Human actions cause a host of environmental problems. If we are to find solutions to these problems, we must recognise that each of us is individually responsible for the quality of the environment we live in. Personal actions can worsen or improve the quality of our environment. Several people may feel that environmental problems can be solved with quick technological solutions. While most individuals want a cleaner environment, not many want to make changes in their lifestyle that would contribute to a cleaner environment. To a large extent, the decisions and actions of individuals determine the quality of life for everyone. This necessitates that individuals should not only be aware of various environmental issues and the consequences of their actions on the environment, but should also make a firm resolve to develop environmentally ethical lifestyles.

With the help of solar energy, natural processes developed over billions of years can indefinitely renew the topsoil, water, air, forests, grasslands and wildlife on which all forms of life depend. However, this can only happen if we do not use these potentially renewable resources faster than they are replenished by nature. All of nature's ecosystems have a carrying capacity beyond which they break down. Some of our waste can be diluted, decomposed and recycled by natural processes indefinitely as long as these processes are not overloaded. Natural processes also provide services of flood prevention and erosion control at no cost at all. We must therefore learn to value these resources and use them sustainably.

Article 51-A(g) which deals with the *Fundamental Duties* of citizens states: 'It shall be the duty of every citizen of India to protect and improve the natural environment including forests, lakes, rivers and wildlife, and to have compassion for living creatures'. Thus, protection and improvement of the natural environment is the duty of the State (Article 48-A) and every citizen [Article 51-A(g)].

Concepts that help individuals to contribute towards a better quality of environment and human life are:

◈ Develop respect for all forms of life. Each individual must try to answer four basic questions: Where do the things that I consume come from? What do I know about the place where I live? How am I connected to the earth and other living things? What is my purpose and responsibility as a human being?

◈ Plant trees wherever you can and more importantly take care of them. They reduce air pollution. Provide a conducive living environment for them.

◈ Reduce the use of wood and paper products wherever possible. Manufacturing paper leads to pollution and loss of forests (which release oxygen and absorb carbon dioxide).

◈ Recycle paper products and use recycled paper wherever possible.

◈ From the mail you receive, reuse as many envelopes as you can. You can reduce the amount of mail you receive by signing up to receive electronic statements/ bills from banks or phone companies.

◈ Do not buy furniture, doors or window frames made from tropical hardwoods such as teak and mahogany. These are made from forest-based resources. Buy furniture made from plantation-based wood such as rubberwood, pine or bamboo.

◈ Help in restoring a degraded area near your home, or join an afforestation programme.

◈ Use pesticides in your home only when absolutely necessary and use them in small amounts. Some insect species help to keep a check on the population of other more devastating pest species.

◈ Advocate organic farming by asking your grocery store to stock organically grown vegetables and fruits. This will automatically help to reduce the use of pesticides.

◈ Reduce the use of fossil fuels by either walking short distances or using a car pool, sharing a bike or using public transport. This reduces air pollution.

◈ Switch off the lights and fans when not needed.

◈ Do not pour pesticides, paints, solvents, oil or other products containing harmful chemicals down the drain or onto the ground.

◈ Buy consumer goods in refillable glass containers instead of cans or throw away bottles that damage the environment.

◈ Use rechargeable batteries.

◈ Avoid plastic carry bags when you buy groceries, vegetables or any other items. Use your own cloth bag instead.

◈ Use sponges and washable cloth napkins, dish towels and handkerchiefs instead of paper ones.

◈ Don't use disposable paper and plastic plates and cups when reusable versions are available.

◈ Recycle all newspaper, glass, aluminium and other items accepted for recycling in your area. You might have to take a little trouble to locate such dealers.

◈ Set up a compost bin in your garden or terrace and use it to produce manure for your plants to reduce the use of fertilisers.

◆ Lobby for setting up garbage segregation and recycling programmes in your locality.

◆ Start individual or community composting in your neighbourhood and motivate people to join.

◆ Do not litter the roads and surroundings just because the sweeper from the municipal corporation will clean it up. Take care to put trash into dustbins or bring it back home with you where it can be appropriately disposed.

◆ You could join any of the several pro-environment NGOs that exist in our country or become a volunteer. Organise small local community meetings to discuss positive approaches to pollution prevention.

◆ Learn about the biodiversity of your own area. Understand the natural and cultural assets provided by flora and fauna. This will help you to develop a sense of pride in your city/town/village and will also help you understand the problems facing the survival of wildlife.

◆ You cannot improve your world by not voting. You have the option to make a choice rather than complain later.

◆ Take care to put into practice what you preach. Remember environment protection begins with YOU.

SUMMARY

■ Pollution is caused by human activity and has a detrimental effect on the environment and human health. Each of the pollution-related concerns requires people's participation.

■ Air pollution is caused by burning of fossil fuels from vehicular pollution, industries and coal fired thermal power plants.

■ Air pollution occurs due to the presence of undesirable chemicals, particulate matter or biological materials in air, in quantities that are harmful to human health and the environment.

■ Air pollution affects the health of living organisms, damages plants, harms buildings and materials worth billions of rupees.

■ There are devices that remove pollutants before they escape into the air. For example, scrubbers, dry and wet collectors, filters and electrostatic precipitators. Effects of air pollution can be minimised by building higher smoke stacks that discharge pollutants further away from the ground.

■ There are several classes of common water pollutants that require our attention.
  ◆ Disease causing agents such as bacteria and viruses.
  ◆ Release of untreated sewage.
  ◆ Inorganic plant nutrients such as nitrates and phosphates from fertilisers and pesticides.
  ◆ Water-soluble inorganic chemicals such as acids, salts and compounds of toxic metals.
  ◆ Organic chemicals such as oil, plastics and cleaning solvents.
  ◆ Sediments of suspended matter.
  ◆ Water soluble radioactive isotopes.
  ◆ Hot water released by thermal power plants.

■ Soil erosion, excessive use of fertilisers and pesticides in farming and soil salinity caused by poor irrigation techniques, contribute to soil pollution.

■ Soil erosion is the process of weathering and movement of solids (sediment, soil and other particles) from one place to another, while erosion is a natural process, often caused by wind or flowing water. It has been greatly increased by human activities such as farming, construction, over-grazing by livestock and deforestation.

- Noise pollution causes various harmful effects on human health and wildlife which are caused by exposure to high sound levels.
- Humans have become increasingly wasteful. This is the result of unprecedented economic growth, consumerism and population growth. The effective management of waste is done through the principle of '3R'. This includes Reduce, Reuse and Recycle, which can solve the problem. This requires participation from every person in society. A sure way of doing this is by refusing to use products that are harmful to the environment. A large proportion of these are consumer products that contain plastic and are not reusable and are non-recyclable.
- A large amount of the waste produced is formed throughout the life cycle of the product—during manufacture and use of consumer goods. Therefore, we have to be wise consumers. This will alter the current behaviour of society and take it from an unsustainable to a sustainable world.
- Recovery from waste to get electricity through various processes, such as, incineration, biogasification and preparing Refuse Derived Fuel (RDF) from waste, are now used.

QUESTIONS

1. Discuss the effects of air pollution on living organisms.
2. Write about some of the control measures for air pollution.
3. What is eutrophication? How is it caused?
4. Describe two control measures for chemical pollution in water.
5. Write a note on the role of 'reduce, reuse and recycle' in solid waste management.
6. Describe various processes that are used to recover energy from waste.
7. Describe three techniques that can protect soil from erosion.
8. What are the effects of noise pollution?
9. What are some of the common problems arising from landfills?
10. What is biomagnification?
11. What are some of the steps that an individual can take to prevent pollution? Mention at least five.

Environmental Policies and Practices

48A: Protection and improvement of environment and safeguarding of forests and wild life: The State shall endeavour to protect and improve the environment and to safeguard the forests and wildlife of the country. —*Indian Constitution, 42nd Amendment*

Article 51-A(g): It shall be duty of every citizen of India to protect and improve the natural environment including forests, lakes, rivers and wildlife and to have compassion for living creatures. —*Fundamental Duties*

Learning Objectives

In this chapter you will learn,

◈ What environmental policies and practices are
◈ What climate change, global warming, ozone layer depletion and acid rain are, and how they impact human communities
◈ About India's environment laws and international agreements
◈ About nature reserves

Purpose

As citizens of our world, our country and the place we live in, we must know about environmental policies, laws, and the effects of climate change and biodiversity management.

Our Role

Awareness and knowledge of these issues makes each of us better citizens. Thus, communication education and public awareness are the most effective way to conserve and protect our environment. Each and every citizen must develop a new ethic towards the world's natural resources and nature's services.

6.1 ENVIRONMENTAL POLICIES AND PRACTICES

Environment policies and practices are critical to good governance based on sustainability. Our country is moving rapidly into a new framework for development. However, this must have a bottom-up approach taking all the different sectors of our society into a massive change process where sustainable development encompasses economics, societal equity and environment management.

We live in a democratic country and our constitution and policies of governance are thus sensitive to the needs of all sectors of society. Our laws have been framed to support the long-term needs of our country. Much foresight has led to strong environmental legislations. Our judiciary has been a strong supporter of sound environment management. The Supreme Court, in its wisdom, has addressed environmental issues related to air, water and noise pollution. It has forced the government to further public awareness and strengthen school and college environmental education. The Supreme Court has saved our forests from being converted to other forms of land use and severely punished offenders against the well-being of forests and their dependent communities. Many of our environmental policies and practices are based on ancient traditional knowledge systems. The Mahabharata and Ramayana provide guidance systems for governance to be followed by rules which are conservation- and biodiversity-sensitive.

Our environmental governance is done through MoEF established in 1985. The Ministry has been given the added responsibility of addressing climate change issues since 2015. The constitution states that every citizen is duty bound to protect and preserve the environment, water bodies and wildlife. Every citizen in India has a right to a clean environment (fundamental right). We also have a right to education so that we as a society are sufficiently empowered to create a better environment. In the recent past the *Swachh Bharat* programme, implemented across the country, deals with clean practices that can provide better wealth and well-being for all citizens.

Rapid industrialisation at unsustainable levels has become a major source of pollution which is leading to climate change. Global climate change is the greatest challenge that human society has experienced during the last several centuries. This change will affect the earth's atmosphere, land, water (fresh and marine) and the biological diversity. It is caused by human activities that have resulted from the misuse of our environment. Our times are referred to as the *anthropocene*, the age when humankind has led to serious impacts on mother earth.

6.2 CLIMATE CHANGE, GLOBAL WARMING, OZONE LAYER DEPLETION, ACID RAIN AND IMPACTS ON HUMAN COMMUNITIES AND AGRICULTURE

6.2.1 Climate Change

Today, climate change is an accepted fact unlike a few decades ago, when scientists and governments were arguing about its existence. There are five key signs of climate change as suggested by data from the National Oceanic and Atmospheric Administration (NOAA), USA.

◈ *Increase in global concentration of carbon dioxide*: Carbon dioxide (CO_2) is an important heat-trapping (greenhouse) gas, released through human activities such as deforestation and burning of fossil fuels, as well as natural processes such as respiration and volcanic eruptions.

◈ *Increase in global surface temperature*: Global surface temperatures in 2005 and 2010 were the warmest on record (Source: NASA/GISS). The average global temperature has increased by 0.8°C since 1880. However after 1975, the rate of increase has been 0.15–0.2°C per decade. This indicates that the rate of change is increasing rapidly. NASA suggests that there has been rapid warming in last few decades and the last decade is the warmest. In India, the highest temperature in the month of April has reached 40–42°C. This shows that summers are becoming warmer than in the past. Ocean temperatures have also shown a significant rise and in 2018 a new high was recorded since 1940.

◈ *Decline of the Arctic sea ice*: The Arctic sea ice is now declining at a rate of 11.5% per decade. Arctic sea ice is at its lowest level in September. In the Antarctic, the decrease in the ice has tripled since 2012 according to studies by NASA and the European Space Agency. This has increased sea levels by 3 mm which is a very significant rise. However, if one looks at the Antarctic sea ice from 1992 to 2017, the loss has raised the global sea levels by 7.6 mm.

◈ *Decrease in land ice*: Data from NASA's satellites shows that the land ice sheets in both Antarctica and Greenland are losing mass. The continent of Antarctica has been losing more than 100 cubic kilometres of ice per year since 2002.

◈ *Sea level rise*: Sea level rise is caused by the expansion of sea water as it warms in response to climate change and the widespread melting of land ice. The above evidence lays to rest all debates about the phenomenon of climate change. Projections of future climate change are derived from a series of experiments made by computer-based global climate models. These are calculated based on factors such as future population growth and energy use. Climatologists of the Intergovernmental Panel on Climate Change (IPCC) have reviewed the results of several experiments in order to estimate changes in climate during the course of this century. These studies have shown that in the near future, the global mean surface temperature will rise by 1.4–5.8°C. This 'warming' will be the greatest over land and at high latitudes. The projected rate of warming is greater than that observed in the last 10,000 years. The frequency of extreme weather events such as storm cyclones and cloud bursts with thunderstorms is likely to increase, leading to floods. Extreme weather events can also be linked to repeated droughts. There will be fewer cold spells but more heat waves. The frequency and intensity of the *El Niño* is likely to increase. The global mean sea level is projected to rise by 9–88 cm by the year 2100. More than half of the world's population now lives within 60 km of the sea. They are likely to be seriously impacted by the ingress of saltwater and by the rising sea level which will inundate the coastal areas and increase the salinity of agricultural land. Some of the most vulnerable regions are the Nile delta in Egypt, the Ganges–Brahmaputra delta in Bangladesh and many small islands including the Marshall Islands and the Maldives (WHO, 2001).

Human societies will be seriously affected by extremes in climate and rainfall leading to droughts and floods. A changing climate would bring about alterations in the frequency and/or intensity of these extreme weather events. This is also a fundamental concern for human health and wellbeing. To a large extent, public health depends on safe drinking water, sufficient food, secure shelter and social equity. All these factors are affected by climate change. Freshwater supplies may be seriously affected, reducing the availability of clean water for drinking and washing during drought as well as floods. Water can be contaminated and sewage systems may be damaged during storm conditions. The risk of spread of infectious diseases such as diarrhoea will increase due to extreme weather events. Food production will be seriously reduced in vulnerable regions directly and also indirectly through an increase in pests and plant or animal diseases. The local reduction in food production would lead to starvation and malnutrition with long-term health consequences, especially for children. Food and water shortages may lead to conflicts in vulnerable regions. Climate change related impacts on human health could lead to displacement of a large number of people, creating environmental refugees and lead to further health issues.

Changes in climate affect the distribution of vector species (such as mosquitoes) which, in turn, will increase the spread of environment-related diseases, such as malaria, dengue, chikungunya and filariasis, to new areas that lack a strong public health infrastructure. The seasonal transmission and distribution of many diseases that are transmitted by mosquitoes (dengue, yellow fever) and by ticks (Lyme disease, tick-borne encephalitis) will spread due to climate change.

A Task Group set up by WHO has warned that climate change will have a serious impact on human health. Climate change will increase existing health problems and may also bring in new unexpected diseases. Strategies aimed at reducing potential health impacts of anticipated climate changes should include the monitoring of infectious diseases and disease vectors to detect early changes in the incidence of diseases and the geographical distribution of vectors, environmental management measures to reduce risk, disaster preparedness for floods or droughts and their health-related consequences. It will be necessary to educate and create early warning systems for epidemic preparedness. Improved water and air pollution control will become increasingly essential for human health. Public health and environment education will have to be directed at changes in personal behaviour. The training of researchers and health professionals is essential for the world to become more responsible for responding to the expected outcome of Global Climate Change (GCC).

6.2.2   Global Warming

About 75% of the solar energy reaching the earth is absorbed by the earth's surface, and leads to an increase in its temperature. The rest of the heat radiates back to the atmosphere. Some of the heat is trapped by greenhouse gases (GHGs), mostly carbon dioxide. As carbon dioxide is released by various human activities, the amount is rapidly increasing. This causes global warming.

The average surface temperature of the earth is about 15°C. Without greenhouse gases, most of the earth's surface would be frozen with a mean air temperature of −18°C. Human activities during the last few decades of industrialisation and population growth have polluted the atmosphere to the extent that it has begun to seriously affect the climate. The carbon dioxide in the atmosphere has increased by 31% since pre-industrial times, causing more heat to be trapped in the lower atmosphere. There is evidence to show that the carbon dioxide levels are still increasing. Many countries have signed a convention to reduce greenhouse gases under the United Nations Framework Convention on Climate Change (UNFCC). However, the current international agreements are not still effective enough to prevent significant changes in climate and a rise in sea levels.

6.2.3   Ozone Layer Depletion

Ozone is formed by the action of sunlight on oxygen. It forms a layer 20–50 km above the surface of the earth. This action takes place naturally in the atmosphere, but is very slow. Ozone is a highly poisonous gas with a strong odour. It is a form of oxygen that has three atoms in each molecule. It is considered a pollutant at ground level and constitutes a health hazard by causing respiratory ailments like asthma and bronchitis. It also causes harm to vegetation and leads to a deterioration of certain materials including plastic and rubber. Ozone in the upper atmosphere however, is vital to all forms of life as it protects the earth from the harmful UV radiations of the sun. The ozone layer in the upper atmosphere absorbs the UV radiation, preventing it from reaching the earth's surface.

In the 1970s, scientists discovered that chemicals called chlorofluorocarbons or CFCs, which were used as refrigerants and aerosol spray propellants, posed a threat to the ozone layer. The CFC molecules are virtually indestructible until they reach the stratosphere, where UV radiation breaks them down to release chlorine atoms. These chlorine atoms react with the ozone molecules to break them down into oxygen molecules. These oxygen molecules do not absorb UV radiation. Since the early 1980s, scientists have detected a thinning of the ozone layer in the atmosphere above Antarctica. This phenomenon is now being detected in other places as well, including Australia.

The destruction of the ozone layer causes an increased incidence of skin cancer and cataracts. It also causes damage to certain crops and plankton, thus affecting natural food chains and food webs. This decrease in vegetation leads to an increase in carbon dioxide. After the signing of the Montreal Protocol in 1987, a treaty for the protection of the ozone layer, the use of CFCs was banned. The ozone layer has been slowly recovering but this process will take over 50 years. Although the use of CFCs has been reduced and is now banned in most countries, other chemicals and industrial compounds such as bromine, halocarbons and nitrous oxides from fertilisers continue to damage the ozone layer.

6.2.4   Acid Rain

When fossil fuels such as coal, oil and natural gas are burned, chemicals like sulphur dioxide and nitrogen oxides are produced. These chemicals react with water and other chemicals in the air to form sulphuric acid, nitric acid and other harmful pollutants like sulphates and nitrates. These acid pollutants spread upwards into the atmosphere and are carried by air currents, to finally return to the ground in the form of acid rain, fog or snow. The corrosive nature of acid rain causes environmental damage in many ways. Acid pollutants occur as dry particles and gases, which when washed from the ground by rain, add to the acids in the rain to form an even more corrosive solution.

Damage from acid rain is widespread in North America, Europe, Japan, China and South-east Asia. In the US, coal-burning power plants contribute to about 70% of sulphur dioxide. In Canada, oil refining, metal melting and other industrial activities account for 61% of the sulphur dioxide pollution. The exhaust fumes of motor vehicles are the main source of nitrogen oxides. The acids in the rain chemically react with many objects which get severely damaged.

Effects of acid rain

◈ Acid rain dissolves and washes away nutrients in the soil. It also dissolves naturally occurring toxic substances like aluminium and mercury, freeing them and thereby polluting water or poisoning plants.

◈ Acid rain indirectly affects plants by removing nutrients from the soil in which they grow. It affects trees more directly by forming holes in the waxy coating of leaves, creating brown dead spots which affect the plant's photosynthesis. Such trees are also more vulnerable to insect infestations, drought and cold. Spruce and fir forests at higher elevations are at greater risk. Farm crops are less affected by acid rain than forests.

◈ Acid rain that falls or flows as ground water to reach rivers, lakes and wetlands, causes the water to become acidic. This affects plant and animal life in aquatic ecosystems.

◈ Acid rain has far reaching effects on wildlife. By adversely affecting one species, the entire food chain can be disrupted, ultimately endangering the entire ecosystem. Different aquatic species can tolerate different levels of acidity. For instance, clams and mayflies have a high mortality when water is acidic. Frogs can tolerate more acidic water, although with the decline in supply of mayflies, frog populations also decline. Several animals that are dependent on aquatic organisms are affected by changes in the water quality due to acid rain.

◈ Acid rain and dry acid deposition damages buildings, automobiles and other structures made of stone or metal. The acid corrodes the materials causing extensive damage and ruins historic and ancient heritage buildings like the Taj Mahal which has been affected by acid rain. In response to a public interest litigation by M C Mehta, the Supreme Court closed down several projects that were creating acid rain.

◈ Although surface water polluted by acid rain may not directly harm people, the toxic substances leached from soil can pollute the water supply. Fish caught in these waters may be harmful for human consumption. Acid, along with other chemicals in the air produces urban smog, which causes respiratory problems.

The best way to prevent the formation of acid rain is to reduce the emissions of sulphur dioxide and nitrogen oxides into the atmosphere. This can be achieved by using less energy from fossil fuels in power plants, vehicles and industries. Switching to cleaner fuels is now urgently essential. For example, using natural gas which is cleaner than coal, or using coal with a lower sulphur content, is a positive step. Developing more efficient vehicles will reduce pollutants from being released into the air. If pollutants are formed by burning fossil fuels, they can be prevented from entering the atmosphere by using scrubbers in the smoke stacks of factories. These spray a mixture of water and lime into the polluting gases to recapture the sulphur.

In catalytic converters, the gases are passed over metal coated beads that convert harmful chemicals into less harmful ones. These are used in cars to reduce the effects of exhaust fumes on the atmosphere. Once acid rain has affected soil, powdered limestone can be added to the soil by a process known as *liming* to neutralise the acidity of the soil.

6.2.5   Impacts on Human Communities and Agriculture

Climate change affects people and constitutes one of the world's greatest challenges that we have ever had to face. It affects human health and well-being through extreme weather events and hurts the poor sectors of society most adversely. As the climate changes, crops and farm practices must be altered. This adaptation process is not easy as it has to be done through creating a heightened level of awareness among the farming community. As climate change cannot be reversed rapidly, mitigating the cause will take decades to be made effective. The people most at risk are coastal communities and island folk who will continue to be affected by the sea level rise as well as by the increasing number and velocity of cyclones and thunderstorms.

6.3   ENVIRONMENTAL LAW

Human beings are by nature hungry for a better means of life. Our lifestyles have altered over the past several decades into thinking that we can go on using the earth's water, air, soil and biological diversity in any way we like. This however will lead to a gradually growing disaster and a crisis which will at some point in time become irreversible. Climate change, desertification, ocean pollution, snow and glacier melt, and polar ice cap disintegration will envelope the globe in a vice, unless we make a change in our lifestyles. We need to change to a more environmentally positive law-abiding way of life. Polluters must pay heavily for damaging the earth. To make this happen, each individual must feel responsible for saving the earth. A series of environmental laws have been made to make potential offenders fall in line with responsible behaviour towards the earth. The law punishes offenders and makes them pay for damages they have caused. There are laws that deal with pollution, wildlife crime, deforestation, forest conservation and animal rights that are all linked to our environment. To force polluters to pay, anyone can register a Public Interest Litigation with the court if they find someone violating an environment law.

Laws to protect the environment, water and air were created very early in India compared to most other countries. The need to preserve the environment was discussed in 1972 during the conference of Human Environment held at Stockholm. India began to see the need for a much more comprehensive environmental act.

6.3.1   Environment Protection Act (1986)

The trigger for formulating this Act was the Bhopal gas tragedy that killed a large number of people and seriously affected the health and well-being of a much larger segment of the population living around the factory site in 1984.

Although there were several existing laws that dealt directly or indirectly with environmental issues, it was necessary to have a general legislation for environmental protection. This was because the existing laws focussed on very specific types of pollution, or specific categories of hazardous substances, or were indirectly related to the environment through laws that control land use, protect our national parks and sanctuaries, and our wildlife. However, there was no overarching legislation and certain aspects of environmental hazards were not covered. There were also gaps in areas that were potential environmental hazards or risks and there were several inadequate measures for handling matters of industrial and environmental safety. This was related to the multiplicity of regulatory agencies. Thus, there was a need for an authority to study, plan and implement the long-term requirements of environmental safety, and to direct and coordinate a system of appropriate responses to emergencies threatening the lives of people and the environmental integrity.

This Act was thus passed to protect the environment, as there was a growing concern over the deteriorating state of the environment. As the impact due to degradation of the environment grew considerably, environmental protection became a national priority in the 1970s. While the wider general legislation to protect our environment is now in place, it has become increasingly evident that our environmental situation continues to deteriorate and we need to implement this Act much more aggressively. The presence of excessive concentrations of harmful chemicals in the atmosphere and aquatic ecosystems leads to the disruption of food chains and a loss of species in several ecologically fragile sites.

Public concern and support are crucial for implementing the EPA. This must be supported by an enlightened media, good administrators, highly aware policy makers, informed judiciary and trained technocrats who can influence and prevent further degradation of our environment. Each of us has a responsibility to make this happen. The Act is implemented through the MOEF and CC with the Central Pollution Control Board which coordinates activities of the State Pollution Control Boards.

The Environment Protection Act provides for the protection and improvement of the environment. It establishes the framework for studying, planning and implementing long-term requirements of environmental safety and laying down a system of speedy and adequate response to situations threatening the environment. It is an umbrella legislation designed to provide a framework for the coordination of central and state authorities established under the Water Act (1974) and the Air Act (1981). The term 'environment' is understood as a very broad term under Section 2(a) of the Environment Act. It includes water, air and land as well as the interrelationship that exists between water, air and land, and human beings, other living creatures, plants, microorganisms and property.

Under the Environment Protection Act, the Central Government is empowered to take necessary measures to protect and improve the quality of the environment by setting standards for emissions and the discharge of pollution into the atmosphere by any person carrying on an industry or activity, regulating the location of industries, management of hazardous wastes, and protection of public health and welfare. From time to time, the Central Government issues notifications under the Environment Act for the protection of ecologically-sensitive areas or issues guidelines for several matters related to the environment.

In the case of any non-compliance or contravention of the Environment Act, or of the rules or directions under the Act, the violator will be punishable with imprisonment up to five years, or a fine of up to ₹1,00,000, or both. In the case of continuation of such a violation, an additional fine of up to ₹5,000 per day during which such failure or contravention continues after the conviction for the first such failure or contravention will be levied. Further, if the violation continues beyond a period of one year after the date of conviction, the offender shall be punishable with imprisonment for a term that may extend to seven years.

CASE STUDY 6.1   Union Carbide Company Vs Union of India, 1989

Context: Cause for enactment of an umbrella legislation, that is, the Environment (Protection) Act, 1986 and the Public Liability Insurance Act, 1991. The case also gave birth to the 'absolute liability' principle.

Situation/Incidence: A massive leak of toxic methyl isocyanate (MIC) gas occurred during the night of December 2-3, 1984, at the Bhopal plant of Union Carbide, India, Ltd. (UCIL). This is a subsidiary of Union Carbide Corporation, a New York based corporation. Union Carbide owned 50.9% of the stock of its Indian subsidiary. The Indian government's reports put the death toll at 2,347, over 1,600 of who were killed as a direct result of the deadly gas leak, while the remaining hundreds died because of its fatal after effects over the next several months. The number of people who were seriously injured was between 30,000 and 40,000 and the Indian Government received 500,000 leak-related claims.

UCIL was the Indian subsidiary of the Union Carbide Corporation (UCC). The Indian Government controlled the banks and the Indian public held 49.1% ownership share. In 1994, the Supreme Court of India allowed UCC to sell its 50.9% share. The Bhopal plant was sold to McLeod Russel (India) Ltd. UCC was purchased by Dow Chemical Company in 2001.

Following the accident there has been a long journey for justice for the victims of the incident. The question was related to the quantification of the liability of the corporations handling the hazardous substances in view of the absence of any established principle. Another grave question was related to the impact of such hazardous substances on the environment and the issue of prevention of such damages in future by the installation of proper safety devices and mechanisms.

Case outcome: Initially the case was filed in the American court as the larger stakes were with the American counterpart of the company. The American courts refused to entertain the case on the grounds of 'forum non-convenient'. The legal battle with the corporate giant was pursued further in India.

However, at the end of 4 years after the accident, the victims remained helpless and their agony was increasing. In its order dated 14 February 1989, the Supreme Court ordered an overall settlement of the claims that arose from the disaster whereby the UCC was to pay an amount of US$ 470 million to the Indian Government as full and final settlement of all the claims, past, present and future, both civil and criminal arising out of the disaster. This amount was immediately paid by UCC. In the year 2010, the Government of India filed a curative petition before the Supreme Court of India wherein additional compensation of more than ₹1000 crores is claimed from the company for the welfare of the victims. The petition is still pending.

CASE STUDY 6.2 Vellore Citizens Welfare Forum Vs Union of India and Others, 1996

Context: Constitution of India, Right to a clean environment, Creation of a Green Bench.

Situation: This petition – public interest litigation – was filed under Article 32 of the Constitution of India by the Vellore Citizens Welfare Forum and is directed against the pollution that was being caused by enormous discharge of untreated effluents by the tanneries and other industries in the State of Tamil Nadu.

Case outcome: The untreated effluents were discharged into the river Palar which is the main source of water supply to the residents of the area. The court directed the Municipal Council to impose a pollution fine of ₹10,000 each on all the tanneries in the districts in the office of the Collector/District Magistrate concerned. Further, the court requested the Chief Justice of the Madras High Court to constitute a special 'Green bench' to deal with this case and other environmental matters.

CASE STUDY 6.3 MC Mehta Vs Union of India, WP 12739/1985 (Oleum gas leak case)

Context: Water (Prevention and Control of Pollution) Act, 1974, Air (Prevention and Control of Pollution) Act, 1981, Section 40(2) of Factories Act, 1948, Section 430(3) Delhi Municipal Corporation Act, 1957, and Section 133(1) of Code of Criminal Procedure, 1973.

Situation: Delhi Cloth Mills Ltd., a public limited company, having its registered office in Delhi runs an enterprise called Shriram Foods and Fertiliser Industries, which has several units engaged in the manufacture of a large number of chemical agents. These various units are all set up in a single complex surrounded by thickly populated colonies. Within a radius of 3 km from this complex, a population of approximately 200,000 people live. The caustic chlorine plant was commissioned in the year 1949 and it had about 263 employees.

On 4 December 1985, a major leakage of oleum gas took place from one of the units of Shriram and this affected a large number of persons, both amongst the workmen and the public. It resulted from the bursting of the tank containing oleum gas.

Case outcome: Questions arose concerning the true scope and ambit of Articles 21 and 32 of the Constitution, the principles and norms for determining the liability of large enterprises engaged in the manufacture and sale of hazardous products, and whether such large enterprises should be allowed to continue to function in thickly populated areas. It was a landmark judgment in which the principle of Absolute Liability was laid down by the Supreme Court of India. The Court held that the permission for carrying out any hazardous industry very close to the human habitation could not be given and the industry must be relocated. The Public Liability Act was passed and the policy for the Abatement of Pollution Control was also established.

CASE STUDY 6.4 Animal Welfare Board of India Vs A Nagaraj and Others, 2014

Context: Prevention of Cruelty to Animals Act, 1960.

Situation: Jallikattu is a sport conducted as a part of Mattu Pongal (the third day of the four-day long harvest, Pongal). Bulls are brought and participants to embrace the bull's hump and try to tame it by bringing the bull to a stop. Bulls are deliberately placed in a terrifying situation in which they are forced to run away from a mob of men. The participants and spectators are at risk. Bulls are often provoked with alcohol, sticks, knives, sickles and even chilli powder in the eyes.

Case outcome: The Supreme Court prohibited Jallikattu and other animal races and fights. The court alluded to Section 3 and Section 11 of the Prevention of Cruelty to Animals Act, 1960, and declared that animal fights incited by humans are illegal, even those carried out under the guise of tradition and culture.

6.3.2 Air (Prevention and Control of Pollution) Act (1981)

The Government passed this Act in 1981 to clean up the air by controlling pollution. It states that the sources of air pollution such as industries, vehicles and power plants are not permitted to release particulate matter, lead, carbon monoxide, sulphur dioxide, nitrogen oxides, volatile organic compounds (VOCs) or other toxic substances beyond a prescribed level. To ensure this, Pollution Control Boards (PCBs) have been setup by the government to measure pollution levels in the atmosphere and at certain sources by testing the air. This is measured in parts per million or in milligrams or micrograms per cubic metre. The particulate matter and gases that are released by industries, cars, buses and two wheelers is measured by using air-sampling equipment. However, the most important aspect is for people themselves to appreciate the dangers of air pollution and reduce their own potential as polluters by seeing that their vehicle or the industry they work in reduces levels of emissions.

This Act is created to take appropriate steps for the preservation of the natural resources of the earth which among other things includes the preservation of high-quality air and ensures controlling the level of air pollution. The main objectives of the Act are:

- to provide for the prevention, control and abatement of air pollution,
- to provide for the establishment of Central and State Boards with a view to implement the Act, and
- to confer the powers to implement the provisions of the Act on the Boards and assign to them functions relating to pollution.

Air pollution is more acute in heavily industrialised and urbanised areas, which are also densely populated. The presence of pollution beyond certain limits due to various pollutants discharged through industrial emission is monitored by the PCBs set up in every state.

Powers and functions of the pollution control boards

Central Board: The main function of the Central Board is to implement the legislations that have been created to improve the quality of air, and prevent and control air pollution in the country.

The Central Pollution Control Board advises the Central Government on matters concerning the improvement of air quality and also coordinates activities, provides technical assistance and guidance to State Boards and lays down standards for the quality of air. It collects and disseminates information with respect to matters relating to air pollution and performs functions as prescribed in the Act.

State Pollution Control Boards: The State Boards have the power to advise the State Government on any matter concerning the prevention and control of air pollution. They have the right to inspect any control equipment, industrial plant, or manufacturing process at any reasonable time and give orders to take the necessary steps to control pollution. They are expected to inspect air pollution control areas at intervals or whenever necessary. They are empowered to provide standards for emissions to be laid down for different industrial plants regarding the quantity and composition of emission of air pollutants into the atmosphere. A State Board may establish or recognise a laboratory to perform this function.

Penalties: An industry that emits air pollutants in excess of the standards laid down by the State Board is liable to pay a penalty. The Board also makes applications to the court for restraining persons causing air pollution.

What can an individual do to control air pollution?
- When you see a polluting vehicle, take down the number and send a letter to the Road Transport Office (RTO) and the PCB.
- If you observe an industry polluting air, inform the PCB in writing and ascertain if action has been taken.
- Use cars only when absolutely necessary; walk or cycle as much as possible instead of using fossil fuel-powered vehicles.
- Use public transport as far as possible, as more people can travel in a single large vehicle rather than using multiple small vehicles which add to pollution.
- Share vehicle space with relatives and friends; car-pooling minimises the use of fossil fuels.
- Do not use air fresheners and other aerosols and sprays that contain CFCs that deplete the ozone layer.
- Do not smoke in a public place. It is illegal and endangers not only your own health but also that of others.
- Coughing can spread bacteria and viruses. Use a handkerchief to prevent droplet infection which is airborne, as it endangers the health of other people.

It is a citizen's duty to report to the local authorities such as the Collector, the PCB and the press about offences made by a polluter so that action can be taken against the offender. It is equally important to prevent and report to the authorities on cutting down of trees, as this reduces nature's ability to maintain the carbon dioxide and oxygen levels. Preventing air pollution and preserving the quality of our air is a responsibility that each individual must support so that we can breathe clean air.

The Air (Prevention and Control of Pollution) Act, 1981 (*Air Act*) is an act to provide for the prevention, control and abatement of air pollution and for the establishment of Boards at the Centre and State levels with a view to implement the act. To counter the problems associated with air pollution, ambient air quality standards were established under the Air Act. The Air Act seeks to combat air pollution by prohibiting the use of polluting fuels and substances, as well as by regulating appliances that give rise to air pollution. The Air Act empowers the State Government, after consultation with the SPCBs, to declare any area/areas within the State as air pollution control area/areas. Under the Act, establishing or operating any industrial plant in the pollution control area requires consent from SPCBs. The Boards are also expected to test the air in air pollution control areas, inspect pollution control equipment, and manufacturing processes.

CASE STUDY 6.5   M C Mehta Vs Union of India 13381 of 1984, decided on 30 December 1996

Context: Air (Prevention and Control of Pollution) Act, 1981 and Environment (Protection) Act, 1986.

Situation: The Taj Mahal is an ivory-white marble mausoleum and is acclaimed to be one of the most priceless national monuments, of unsurpassed beauty and worth. The Taj was threatened with deterioration and damage.

Case outcome: The degradation of the Taj Mahal led M C Mehta, an environmentalist and a public interest attorney, to file a public interest litigation before the Supreme Court in 1984. According to the petitioner, the foundries, chemical/hazardous industries and the refinery at Mathura were the major sources of pollution in the Agra region, as was stated in the report of the Central Pollution Control Board. Industrial emissions, brick kilns, vehicular traffic and generator sets are principally responsible for polluting the ambient air around the Taj Trapezium Zone (TTZ). The petitioner averred that the white marble has yellowed and blackened in places and the decay was more apparent inside.

The Honourable Supreme Court of India passed a series of orders in 1993. The Supreme court directed the UP Pollution Control Board (the Board) to get a survey done of the area and prepare a list of all the industries and foundries that are sources of pollution in the area. It categorised the industries and reported that there were a total of 511 industries in the given area. Notices were issued to all these industries to install anti-pollution mechanisms.

The final judgment was delivered on 30 December 1996 by a Division Bench comprising Justice Kuldip Singh and Justice Faizan Uddin. The Court applied the principle of sustainable development in this case, observing that there needs to be a balance between economic development and environmental protection. The Court indicated that relocation of the industries from TTZ was to be resorted to only if natural gas was not acceptable/available by/to the industries as a substitute for coke/coal.

CASE STUDY 6.6   Arjun Gopal Vs Union of India, 2017

Context: Noise pollution rules covered under the Air (Prevention and Control of Pollution) Act, 1981 and Environment (Protection) Act, 1986.

Situation: Diwali was celebrated on 30 October in 2016. On the next day, it was discovered that PM 2.5 levels in the air had crossed 700 $\mu g/m^3$ being among the highest levels recorded in the world and about 29 times above the standards laid down by the World Health Organisation (WHO). This resulted in many people falling sick and others having to purchase face masks for personal use and install air purifiers in buildings.

Case outcome: The court held that the right to health coupled with the right to breathe clean air leaves no doubt that it is important that air pollution deserves to be eliminated by continuing the suspension of licences for the sale of fireworks and therefore implicitly, prohibiting the bursting of fireworks.

CASE STUDY 6.7   M C Mehta Vs Union of India and Others, 2002

Context: Constitution of India, Air (Prevention and Control of Pollution) Act, 1981 and Environment (Protection) Act, 1986.

Situation: In 1995 and 1996, long before the receipt of the Bhure Lal Committee report, there is a reference to conversion of government vehicles to CNG. The report of the Bhure Lal Committee was accepted by the Ministry, and the orders were passed by the Apex Court on 28 July 1998, fixing the time limit within which the switch over to CNG was to take place.

Case outcome: The order of the court was delivered on the basis of Article 39(e), 47 and 48A and collectively cast a duty on the State to secure the health of the people, improve public health and protect and improve the environment. Lack of concern or effort on the part of various governmental agencies had resulted in spiralling pollution levels. On the basis of this background, the apex court held that the Union of India will give priority to the transport sector including private vehicles all over India with regard to the allocation of CNG. CNG was allocated and made available.

6.3.3   The Water (Prevention and Control of Pollution) Act (1974)

The Government formulated this Act in 1974 to prevent the pollution of water by industrial, agricultural and household waste water that can contaminate water sources. Wastewater with high levels of pollutants that enter wetlands, rivers, lakes and the sea are serious health hazards. Controlling the point sources by monitoring the levels of different pollutants is one way to prevent pollution. The main objectives of the Water Act are to provide for prevention, control and abatement of water pollution and the maintenance or restoration of the quality of water. It is designed to assess pollution levels and punish polluters. The Central Government and State Governments have set up PCBs to monitor water pollution.

Water related functions of the pollution control boards

The Government has given the necessary powers to the PCBs to deal with the problems of water pollution in the country. The Government has also suggested penalties for violation of the provisions of the Act. Central and state water-testing laboratories have been setup to enable the Boards to assess the extent of water pollution and standards have been laid down to establish guilt and default. The Central and State Boards have certain powers and functions which are as follows:

Central Board: It has the power to advise the Central Government on any matter concerning the prevention and control of water pollution. The Board coordinates the activities of the State Boards and also resolves disputes. The Central Board can provide technical assistance and guidelines to State Boards to carry out investigations and research relating to water pollution, and organises training for people involved in the process. The Board organises a comprehensive awareness programme on water pollution through mass media and also publishes data regarding water pollution. The Board lays down or modifies the rules in consultation with the State Boards on standards of disposal of waste water. The main function of the Central Board is to promote the cleanliness of rivers, lakes, streams and wells in the country.

State Boards: They have the power to advise the State Government on any matter concerning water pollution. They plan a comprehensive programme for the prevention of water pollution. They collect and disseminate information on water pollution and participate in research in collaboration with the Central Board in organising training of people involved in the process. The Boards inspect sewage treatment plants, purification plants and the systems of disposal and also evolve economical and reliable methods of treatment of sewage and other effluents. They plan the utilisation of sewage water for agriculture and ensure that if effluents are to be discharged on land, then the waste is diluted. The State Boards advise State Governments with respect to the location of industries. Laboratories have been established to enable the Boards to perform their functions.

The State Boards have the power to obtain information from the officers empowered by it to make surveys, keep records of flow, volume and other characteristics of the water. They are given the power to take samples of effluents and suggest the procedures to be followed in connection with the samples. The concerned board analyst is expected to analyse the sample sent to him and submit a report of the result to the concerned Board. The Board is required to send a copy of the result to the respective industry. The Boards also have the power of inspecting any record, register, document or any material object and conducting a search in any place in which there is reason to believe an offence has been committed under the Act.

Penalties are charged for acts that have caused pollution. This includes failing to furnish information required by the Board, or failing to inform the occurrence of any accident or other unforeseen act. An individual or organisation that fails to comply with the directions given in the subsections of the law can be convicted or punished with imprisonment for a term of three months, with a fine of ₹10,000, or both, and in case the offence continues, an additional fine of ₹5,000 every day. If a person who has already been convicted for any offence is found guilty of the same offence again, after the second and every subsequent conviction, she would be punishable with imprisonment for a term not less than two years but which may extend to seven years with a fine.

What Can Individuals do to Prevent Water Pollution?
- Inform the PCB of any offender who is polluting water and ensure that appropriate action is taken. One can also write to the press.
- Do not dump wastes into a household or industrial drain which can directly enter any water body, such as a stream, river, pond, lake or the sea.
- Do not use toilets for flushing down waste items as they do not disappear but reappear at other places and cause water pollution.
- Use compost instead of chemical fertilisers in gardens.
- Avoid the use of pesticides such as DDT, Malathion and Aldrin at home; use alternative methods like a paste of boric acid mixed with gram flour to kill cockroaches and other insects. Use dried neem leaves or other natural solutions as cleaning agents and to help keep away insects.

CASE STUDY 6.8   M C Mehta vs Union of India and Others, 1988

Context: Water (Prevention and Control of Pollution) Act, 1972.

Situation: In 1987, the Supreme Court had issued certain directions with regard to the tanning near Kanpur on the banks of the Ganga. The Court had directed that the case in respect of the municipal bodies and the industries which were responsible for the pollution of the water in the river Ganga would be taken up. Kanpur was one of the biggest cities on the banks of the Ganga.

Case outcome: The Court directed that the Mahapalika should submit its proposals to the State Board within six months (from the date of this judgment). The court ordered that one hour in a week lessons on the protection and improvement of the natural environment including forests, lakes, rivers and wildlife are taken in the first ten classes.

CASE STUDY 6.9   Paryavaran Suraksha Samiti and ANR vs Union of India and Others, 2017

Context: Water (Prevention and Control of Pollution) Act, 1972.

Situation: For the establishment of any industry and to bring it into operation, prior consent of the relevant State Pollution Control Board is mandatory. It is therefore apparent, that all running industrial units, which require 'consent to operate' from the concerned Pollution Control Board, have a functional primary effluent treatment plant (PETP), in place. But the problem is that the PETPs are not maintained due to the exorbitant operational cost after the grant of permission.

Case outcome: The Supreme Court directed that the industrial units may resume their function only after they make their effluent treatments plants functional. Local civic authorities may formulate norms to levy cess from users if they face a financial crunch in the setting up of and running the CETPs, it added. In case it is not done, the respective State governments would have to bear the cost of running the CETPs. The Bench, however, left the issue of setting up of zero liquid discharge (ZLD) plants to the authorities concerned after they complete the first round with regard to the CETPs.

6.3.4   The Wildlife Protection Act (1972)

This Act, passed in 1972, deals with the declaration of National Parks and Wildlife Sanctuaries and their notification. It establishes the structure of the State's wildlife management and the appointment of posts designated for wildlife management. It provides for the setting up Wildlife Advisory Boards. It prohibits hunting of all animals specified in Schedules I to IV of the Act. These are notified in the order of their danger of extinction. Plants that are protected are included in Schedule VI.

The Wildlife (Protection) Act, 1972, was enacted with the objective of effectively protecting the wildlife of our country and to control poaching, smuggling and illegal trade in wildlife and its derivatives. The Act helps in the establishment of protected areas and lays down norms for the regulation of these areas. The Act was amended several times to incorporate changing concepts and conventions in the field of wildlife conservation. The Ministry of Environment Forest and Climate Change has proposed further amendments in the law by introducing more rigid measures to strengthen the Act. The objective is to provide protection to endangered flora and fauna and ecologically important areas rich in biological diversity.

The amendment to the Wildlife Protection Act in 2002 is more stringent and prevents the commercial use of resources by local people. It has brought in new concepts such as the creation of Conservation Reserves and Community Reserves. It has also altered several definitions. For instance, fish are now included under animals. Forest produce has been redefined to ensure the protection of ecosystems.

The Act still has serious issues concerning its implementation. Laws are only as good as the ones that can be enforced. The Act is expected to deter people from breaking the law. However, there are serious problems due to poaching. One cannot expect to use the Act to reduce this without increasing forest staff, providing weapons, jeeps and radio equipment to establish a strong deterrent force.

Penalties: A person who breaks any of the conditions of any license or permit granted under this Act shall be guilty of an offence against this Act. The offence is punishable with imprisonment for a term which may extend to three years, a fine of ₹25,000 or both. An offence committed in relation to any animal specified in Schedule I or Part II of Schedule II, such as the use of meat of any such animal, or animal articles such as trophies, shall be punishable with imprisonment for a term of not less than one year and may extend to six years and a fine of ₹25,000. In the case of a second or subsequent offence of the same nature mentioned in this sub-section, the term of imprisonment may extend to six years and not less than two years with a penalty of ₹10,000.

What can an individual do to protect nature?

- If you observe an act of poaching, or see a poached animal, inform the local Forest Department Official at the highest possible level. One can also report the event through the press. Follow up to check that action is taken by the concerned authority. If no action is taken, one must take it up to the Chief Wildlife Warden of the State.
- Say 'no' to the use of wildlife products and also try to convince other people not to buy them.
- Reduce the use of wood and its products wherever possible.
- Avoid the misuse of paper because it is made from bamboo and wood, the cutting of trees destroys wildlife habitats. Paper and envelopes can always be reused.
- Create a pressure group and ask the government to ensure that the biodiversity of our country is conserved.
- Do not harm animals and dissuade others from cruelty to animals.
- Do not disturb birds' nests and fledglings.
- When you visit the zoo, do not tease the animals by throwing stones or feeding them. Prevent others from doing so.
- If you come across an injured animal, do what you can to help it. There are registered organisations that have facilities to do this, such as Blue Cross.

- If the animal needs medical care and expert attention contact the Society for the Prevention of Cruelty to Animals (SPCA) or Blue Cross in your city.
- Create awareness about biodiversity conservation in your own way to family and friends.
- Join organisations that are concerned with protection of biodiversity, such as the Worldwide Fund for Nature–India (WWF–I), Bombay Natural History Society (BNHS) or a local conservation NGO.

CASE STUDY 6.10   Sansar Chand vs State of Rajasthan, 20 October 2010

Context: Wildlife (Protection) Act, 1974.

Situation: Sansar Chand, his associates and relatives were considered responsible for the killing of all tigers in Sariska. He was also accused of smuggling body parts of the wild animals. Accused of killing over 200 tigers besides thousands of other wildlife species, Chand argued against being charged under the Maharashtra Control of Organised Crime Act (MCOCA) on technical grounds.

Case outcome: In order to highlight the extent of the organised nature of wildlife crimes being committed by the appellant, the court took into consideration that it was not just Sansar Chand, but other members of his family and associates, who were also involved in the illegal trade in wildlife. It was alleged that the appellant's younger brother Narayan Chand was mentioned in FIR No. 82/2005, Kamla Market Police Station, New Delhi, involving the seizure of 2 tiger skins, 38 leopard skins and 1 snow leopard skin and was named as an accused in the complaint filed under Section 55 of the Wild Life (Protection) Act, 1972 in this case. Sansar Chand's wife Rani and son Akash were also accused in the case, involving the seizure of leopard paws and claws. CBI, in the year 2005, invoked the MCOCA against Sansar Chand and his family members and associates and the case is pending trial in a Delhi court.

State (WLI) vs Shahid, 2013

Context: Illegal trade in wild bird species under the Wildlife (Protection) Act, 1972.

Situation: On 27 August 2006, WLI V B Dasan received information that one person was dealing with live wild birds at Wildlife vs Shahid 1 of 9 Pigeon Market, Jama Masjid, Delhi. A raid was conducted along with wildlife officials and wildlife guards who recovered 100 black headed munias in one cage, 34 rose ringed parakeets in one cage and 10 Alexandrine parakeets in another cage from the possession of accused without any permit or licence to possess/keep or to deal with the wild birds. These wild birds are covered in Schedule IV and are protected under the provisions of Wildlife (Protection) Act, 1972.

Case outcome: The court held that the prosecution has proved its case beyond reasonable doubt. Accordingly, the accused were found guilty and were convicted for the offence u/s 9,44 and 49 of the Wildlife (Protection) Act, 1972.

The Indian Forest Act, 1927

The Indian Forest Act, 1927, was largely based on the previous Indian Forest Acts implemented under the British rule. The objective of the Indian Forest Act of 1878 was to consolidate the law relating to forests, the transit of forest produce and the duty leviable on timber and other forest produce. The Act substantiated the procedure for notification of Reserve Forest, Protected Forest and village forests. Consolidation of offences, punishments and regulation of forest produce is done even today under this Act.

   According to the Indian forest Act there are three forest management categories,
   (i) Reserve forests,
  (ii) Protected forests, and
 (iii) Village forests.

After the Forest Act was enacted in 1865 and amended in 1878, the Act divided forests into three categories, as above. The villagers could not take anything from these forests as the British wanted to exploit the forest resources for themselves.

Reserved forests

Land rights to forests declared to be reserved forests are typically acquired (if not already owned) and are owned by the Government of India. Unlike the National Parks or wildlife sanctuaries of India, reserved forests are declared by the respective State Governments. In reserved forests, all activities such as hunting, grazing, and so on, are banned unless specific orders are issued otherwise. (Ref: Legislations on Environment and Forest and Wildlife, MOEF.)

Protected forests

A protected forest is an area notified under the provisions of the Indian Forest Act or the State Forests Act having a limited degree of protection. In protected forests, all activities are permitted unless it is prohibited by a State Government under the provisions of Section 29 of the Indian Forest Act 1927. In protected areas, rights to activities like hunting and grazing are sometimes given to communities living on the fringes of the forest, who sustain their livelihood partially or wholly from forest resources or products. (Ref: Legislations on Environment and Forest and Wildlife, MOEF.)

Village forests

Some villages were allowed to stay on in the reserved forests on the condition that they worked free for the forest department in cutting and transporting trees and protecting the forest from fires. These forests came to be known as *forest villages*. (Ref.: India's Forest Conservation Legislation, MoEF).

SWOT analysis

Strengths: The term 'Forest' as defined by the Supreme Court must be understood according to the dictionary meaning. The Court clarified that this description covers all statutorily recognised forests, whether designated as reserved, protected or otherwise for the purpose of Section 2(i) of the Forest Conservation Act.

Weakness: The terms can be used only when it comes to diversion of forest land but not when it comes to conservation of forest lands.

Opportunities: The judgement gave an opportunity to the government and civil society to keep control over diversion of forests.

Threats: It gives power to the government to decide whether the forest is in accordance with the 'dictionary meaning'.

6.3.5   Forest Conservation Act (1980)

To appreciate the importance of the Forest Conservation Act of 1980, which was amended in 1988, it is essential to understand its historical background. The Indian Forest Act of 1927 consolidated all the previous laws regarding forests that were passed before the 1920s. The Act gave the Government and the Forest Department the power to create Reserved Forests, and the right to use Reserved Forests for Government use alone. It also created Protected Forests, in which the use of resources by local people was controlled. Some forests were to be controlled by the village community, and these were called Village Forests.

The Act remained in force till the 1980s when it was realised that protecting forests for timber production alone was not acceptable. Protecting the services that forests provide and its valuable assets such as biodiversity began to overshadow the importance of the revenue earnings from timber. Thus, a new Act was essential. This led to the Forest Conservation Act of 1980 and its amendment in 1988.

India's first Forest Policy was enunciated in 1952. Between 1952 and 1988, the extent of deforestation was so great that it became essential to formulate a new policy on forests and their utilisation. The earlier forest policies had focussed only on revenue generation. In the 1980s it became clear that forests must be protected for other functions such as the maintenance of soil and water regimes which are prime ecological concerns. It also provided for the use of goods and services of the forest for its local inhabitants.

The new policy framework made conversion of forests into land for other uses much less possible. Conservation of the forests as a natural heritage finds a place in the new policy, which includes the preservation of its biological diversity and genetic resources. It also values meeting the needs of local people for food, fuelwood, fodder and NTFPs. It gives priority to maintaining environmental stability and ecological balance. It expressly states that the network of Protected Areas should be strengthened and extended.

In 1992, the 73rd and 74th Amendments to the constitution furthered governance through *panchayats*. The amendments allow the states to invest the local *panchayats* with the authority to manage local forest resources.

The Forest Conservation Act, 1980, was enacted to help conserve the country's diminishing forests. It restricts and regulates the de-reservation of forests or use of forest land for non-forest purposes without the prior approval of the Central Government. To this end, the Act lays down the pre-requisites for the diversion of forest land for non-forest purposes. Numerous guidelines have been issued by the Central Empowered Committee constituted by the Supreme Court of India with respect to diversion of forest lands and the net present value to be recovered for the afforestation on the equivalent land to be given by the project proponents.

The Supreme Court, in its order dated December 12, 1996 [W.P 202/1995], stated that the word **forest** *'must be understood according to its dictionary meaning. This description covers all statutorily recognised forests, whether designated as reserved, protected or otherwise for the purpose of Section 2(i) of the Forest Conservation Act (1980)'. A 'reserve forest is the forest land wherein everything is prohibited except otherwise permitted by the notification to that effect'.*

All the provisions of forest Conservation Act relating to reserved forests shall (so far as they are not inconsistent with the rules so made) apply to village forests.

A *'protected forest is the forest land wherein everything (pro-forest) is permitted except otherwise prohibited by the notification to that effect'.*

CASE STUDY 6.12 Rural Litigation and Entitlement Kendra vs State of UP [1988] INSC 254 (30 August 1988)

Context: Forest (Conservation) Act, 1980.

Situation: The case arose from haphazard and dangerous limestone quarrying practices in the Mussoorie hill range of the Himalayas. Miners blasted out the hills with dynamite, extracting limestone from thousands of acres. The mines are dug deep into the hillsides, an illegal practice that resulted in the cave-ins and slumping. As a result, the hillsides were stripped of vegetation. Landslides killed villagers and destroyed their homes, cattle and agricultural lands.

Case outcome: The Supreme Court held that the Forest (Conservation) Act, 1980, does not permit mining in the forest areas. This tourist zone in its natural setting would certainly be at its best if its serenity is restored in the fullest way. It was stated by the Court that the permanent assets of mankind are not to be exhausted in one generation. The natural resources should be used with requisite attention and care so that ecology and environment may not be affected in any serious way.

Case Study 6.13   T N Godavarman Thirumulpad vs Union of India Writ Petition (Civil) No. 202, 1995

Context: Forest (Conservation) Act, 1980.

Situation: In September 1995, Godavarman Thirumulpad observed the destruction of pristine wooded areas in Gudalur in the Nilgiris, Tamil Nadu. These wooded areas, Janmam Lands (absolute proprietary lands), of the Nilambur Kovilakam, had been taken over by the State of Kerala following the enactment of the Gudalur Janmam Estates (Abolition and Conversion into Ryotwari) Act of 1969. However, the State was unable to protect the natural assets in the areas. Trees were being felled and logs rolled down the mountain slopes and stacked along the highway for kilometres on end. Godavarman Thirumulpad filed a writ petition in the Supreme Court.

Case outcome: The Supreme Court said that forests would be defined by their 'dictionary meaning'. The court assumed responsibility for implementing the Forest Conservation Act. The court ordered all non-forest activity like sawmills and mining to be suspended in forest areas and stopped felling of trees. It kept the Godavarman case open using the device of a 'continuing mandamus' and has heard hundreds of matters related to the implementation of the Forest Conservation Act over several decades. The Supreme Court is the sole administrator of the law when it involves forest matters. This was followed till the creation of the National Green Tribunal in 2010 to 'dispose of cases relating to environmental protection and conservation of forests and other natural resources', as the court found it impossible to attend to the number of petitions pouring into the courts over the years. It thus set up a Central Empowered Committee.

Central Empowered Committee (CEC)

The Supreme Court constituted a National Level Committee as well as State Level Committees under the Environment (Protection) Act, 1986. The National Level Committee would serve in the nature of a supervisory or appellate authority over the State authorities. In 2002, the Court constituted an Authority at the National Level called the Central Empowered Committee (CEC). The task assigned to it included the monitoring of the implementation of the orders of the Court, encroachment removal, implementation of working plan, compensatory afforestation, plantations and other conservation issues.

In terms of the scope of CEC's intervention, the Notification did not limit the CEC only to the Forest (Conservation) Act, 1980, but also to the implementation of the Indian Forest Act, 1927, Wildlife (Protection) Act, 1972, and the National Forest Policy, 1988 including the rules, regulations and guidelines framed under these laws.

The Forest Conservation Act of 1980 was thus enacted to control deforestation. It ensured that forestlands could not be de-reserved without prior approval of the Central Government. This was created as some states had begun to de-reserve the Reserved Forests for non-forest use. These states had regularised encroachments and resettled 'project affected people' from development projects such as dams in these de-reserved areas. The need for a new legislation became urgent. The Act made it possible to retain a greater control over the serious level of deforestation in the country and specified penalties for offenders.

Penalties for offences in Reserved Forests: No person is allowed to make clearings or set fire to a Reserved Forest. Cattle are not permitted to trespass into the Reserved Forest. Felling, collecting of timber, bark or leaves, quarrying or collecting any forest product is punishable with imprisonment for a term of six months or a fine which may extend to ₹500, or both.

Penalties for offences in Protected Forests: A person who commits any of the offences such as felling trees, stripping the bark or leaves of trees, setting fire to such forests, kindling a fire without taking precautions to prevent its spreading, dragging timber, or permitting cattle to damage any tree, shall be punishable with imprisonment for a term which may extend to six months or with a fine which may extend to ₹500, or both.

When there is a reason to believe that a forest offence has been committed pertaining to forest produce, the produce together with all tools used in committing such offences may be seized by a forest officer or police officer. Every officer seizing any property under this section shall put on the property a mark indicating the seizure and report the seizure to the magistrate who has the jurisdiction to try the offence. Any forest officer, even without an order from the magistrate or a warrant, can arrest any person against whom a reasonable suspicion exists.

What can an individual do to support the Forest Conservation Act?
- Report destructive activities in your local green areas such as Reserved Forests and Protected Forests, and in Protected Areas (National Parks and Wildlife Sanctuaries). Provide a report to the Forest Department as well as to the press. Reports of violations can be made to the Conservator of Forests, District Forest Officer, Range Forest Officer, and Forest Guard, the District Commissioner or local civic body.
- Acquaint yourself with the laws, detailed rules and orders issued by the Government.
- Be in touch with concerned local NGOs and associations. Set up an NGO with other like-minded people if it does not exist in your area.
- Create awareness about the existence and value of national parks and sanctuaries and build up a public opinion against illegal activities in the forest or disturbance to wildlife.
- Pressurise the authorities to implement forest and wildlife laws and rules to protect green areas.
- Take legal action if necessary and if possible through a Public Interest Litigation (PIL) against the offending party. Use the help of NGOs who can undertake legal action.
- Help to create public pressure to change rules, laws and procedures when necessary.
- Use better, ecologically sensitive public transport and bicycle tracks. Do not litter in a forest area.
- Participate in preservation of greenery by planting, watering and caring for plants.

To whom should you report forest offences?

If you as a citizen come across anyone felling trees, encroaching on forest land, dumping garbage, cutting greenwood, lighting a fire, or creating a clearing in a Reserved Forest, Protected Forest, National Park, Sanctuary or other forest areas, you must report it to the forest/wildlife officers concerned. For urgent action, one can contact the police. In fact, you should file an FIR in any case because it serves as an important proof that you have made the report.

Hazardous Waste Management Regulations

Hazardous waste means, any waste which by reason of any of its physical, chemical, reactive, toxic, flammable, explosive or corrosive characteristics, causes danger or is likely to cause danger to health or environment, whether alone or when in contact with other wastes or substances. There are several legislations that directly or indirectly deal with hazardous waste management. The relevant legislations are the Factories Act, 1948, the Public Liability Insurance Act, 1991, the National Environment Tribunal Act, 1995 and rules and notifications under the Environmental Act. Some of the rules dealing with hazardous waste management are discussed below:

- **Hazardous Wastes (Management, Handling and Transboundary) Rules, 2008:** The rules brought out a guide for manufacture, storage and import of hazardous chemicals and for the management of hazardous wastes.
- **Biomedical Waste (Management and Handling) Rules, 1998:** They were formulated along parallel lines, for proper disposal, segregation, transport, and so on, of infectious wastes.
- **Municipal Solid Wastes (Management and Handling) Rules, 2000:** These aim at enabling municipalities to dispose municipal solid waste in a scientific manner.

In view of the short-comings and overlaps of some categories causing inconvenience in implementation of the Biomedical Waste (Management and Handling) Rules, 1998, as well as the Municipal Solid Wastes (Management and Handling) Rules, 2000, the Ministry of Environment, Forest and Climate Change has formulated the Biomedical Waste (Management and Handling) Rules, 2015, and the Solid Waste Management Rules, 2016.

The Draft BMW Rules are to replace the Biomedical Waste (Management and Handling) Rules, 1998, and the Draft SWM Rules are to replace the Municipal Solid Waste (Management and Handling) Rules, 2000. The objective of the Draft BMW Rules is to enable the prescribed authorities to implement the rules more effectively, thereby, reducing the generation of biomedical waste, its proper treatment and disposal, and to ensure environmentally sound management of these wastes. The Draft SWM Rules aim at dealing with the management of solid waste including its segregation at source, transportation of waste, treatment and final disposal.

◈ *e-Waste (Management and Handling) Rules, 2011* were notified in May 2011 and came into effect from 1 May 2012, with the primary objective of reducing the use of hazardous substances in electrical and electronic equipment by specifying the threshold for use of hazardous material and to channelise the e-waste generated in the country for environmentally sound recycling. The Rules apply to every producer, consumer or bulk consumer, collection centre, dismantler and recycler of e-waste involved in the manufacture, sale, purchase and processing of electrical and electronic equipment or components as detailed in the Rules.

◈ *Batteries (Management and Handling) Rules, 2001* deal with the proper and effective management and handling of used lead acid batteries. The Act requires all manufacturers, assemblers, re-conditioners, importers, dealers, auctioneers, bulk consumers, consumers, those involved in manufacture, processing, sale, purchase and use of batteries or components thereof, to comply with the provisions of Batteries (Management and Handling) Rules, 2001.

Coastal Regulation Zone Notification

The Ministry of Environment and Forests issued the Coastal Regulation Zone Notification *vide* Notification no. S O 19(E), dated January 06, 2011 with the objective to ensure livelihood security to the fishing communities and other local communities living in the coastal areas, to conserve and protect coastal stretches and to promote development in a sustainable manner based on scientific principles, taking into account the dangers of natural hazards in the coastal areas and sea level rise due to climate change.

The Biological Diversity Act, 2002

The Biological Diversity Act, 2002 was born out of India's attempt to realise the objectives enshrined in the United Nations Convention on Biological Diversity (CBD), 1992, which recognises the sovereign rights of States over their own biological resources. The Act aims at the conservation of biological resources and associated knowledge as well as facilitating access to them in a sustainable manner. The three chief principles accepted by the Act are conservation of biological resources, access with sustainable use of biological resources and equitable sharing of benefits arising out of the use of these resources. The Act established a three tier system. The National Biodiversity Authority as a Central Nodal Agency, State Biodiversity Boards and local level Biodiversity Management Committees. The National Biodiversity Authority in Chennai has been established for the purposes of implementing the objects of the Act.

The National Green Tribunal Act, 2010

The National Green Tribunal Act, 2010 (No. 19 of 2010, NGT Act) has been enacted with the objectives to provide for the establishment of a National Green Tribunal (NGT) for the effective and expeditious disposal of cases relating to environment protection and conservation of forests and other natural resources including enforcement of any legal right relating to the environment and giving relief and compensation for damages to persons and property and for other connected matters.

The Act received the assent of the President of India on 2 June 2010, and was enforced by the Central Government vide Notification no. SO 2569(E) dated 18 October 2010, with effect from

18 October 2010. The Act envisages the establishment of a Natural Green Tribunal (NGT) in order to deal with all environmental legal cases relating to air and water pollution, the Environment Protection Act, the Forest Conservation Act and the Biodiversity Act which have been set out in Schedule I of the NGT Act.

Issues involved in the enforcement of environmental legislation

Environmental practices are linked to administrative functions and their legal aspects. Environmental legislation has evolved to protect our environment as a whole, our health and the earth's resources. The presence of a legislation to protect air, water and soil does not necessarily mean that the problem has been addressed. Once a legislation is made at the global, national or state level, it has to be implemented. For an environmental legislation to be successfully implemented, there has to be an effective agency to collect relevant data, process it and pass it on to a law enforcement agency. If a law or rule is broken by an individual or institution, they have to be punished through the legal process. Information to law enforcement officials must also come from concerned individuals. In most situations, if no cognizance is taken, the interested and concerned individual must file a Public Interest Litigation (PIL) for the protection of the environment. There are several NGOs in the country which take these matters to court in the interest of conservation. Anyone can request them to help in such matters. There are also legal experts such as M C Mehta and Ritvick Dutta who have successfully fought cases in the courts to support environmental causes. The general public must act as a watchdog not only to inform the concerned authorities, but also to see that action is taken against offenders, as mentioned as a citizen's duty in the constitution.

6.3.6   International Agreements

International law is based on conventions and treaties signed by multiple countries. Conventions and treaties bind nation states and private actors to the terms of that agreement through the domestic application of the treaty provisions. Every nation has its own domestic policy defining how to adopt international conventions in the state. These agreements may be bilateral (between two countries/states) or multilateral (between three or more countries/states). India has adopted dualism, meaning, all the conventions ratified by India needs its translation into domestic legislation to be enforced.

There is an 'official' version of every agreement/convention and only that version has the essential binding force. Many times, republished versions are available on Google, but this is not always binding.

6.3.7   Vienna Convention and Montreal Protocol

The Montreal Protocol, part of the Vienna Convention, was entered into by multiple nations in 1987. The convention is to protect the stratospheric ozone layer by phasing out the production and consumption of ozone-depleting substances (ODS). The stratospheric ozone layer filters out harmful ultraviolet (UV) radiation, which is associated with an increased prevalence of skin cancer and cataracts, reduced agricultural productivity and disruption of marine ecosystems. The Montreal Protocol has proven to be innovative and successful, and is the first treaty to achieve universal ratification by all countries in the world. It has led to the phasing out of production and consumption of several major Ozone Depleting Substances (ODSs) such as chlorofluorocabons (CFCs), carbontetrachloride (CTC) and halons globally from 1 January 2010. The Montreal Protocol has not only contributed to protect the ozone layer but also has reduced greenhouse gas (GHG) emissions.

6.3.8   Kyoto Protocol

The Kyoto Protocol gave rise to the concept of Clean Development Mechanisms (CDM). They were defined in different forms such as joint-implementation of projects in furtherance to reducing carbon emissions in developed countries, financial assistance by developed countries to developing countries for sustainable development, technology transfer to have efficient processes, and to develop a procedure for carbon trading.

In furtherance to the principles accepted in the United Nations Framework Convention on Climate Change, 1992, the Kyoto protocol was proposed in December 1997 and came into effect in 2005. Article 3 of the UNFCCC promoted the benefits of the present and future generations, equity and common but differential responsibilities as principles to be followed. The Kyoto Protocol was framed for the implementation of these principles with respect to carbon emissions. Various pressure groups having vested interests in oil, automobile and coal businesses were working at the background to reduce the emission restrictions to be proposed by the Kyoto protocol. The North–South divide was even more prominent at the time of the Kyoto protocol.

The primary objective of the protocol was 'to achieve stabilisation of Green House Gas (GHG) concentrations in the atmosphere that would prevent anthropogenic interference with the climatic system…'. The protocol called for emission reduction of at least 5% below the 1990s levels within the commitment period of 2008–12 by the industrialised countries. A concession was given that the industrialised countries could achieve this target either jointly or individually. It was also expected that by 2005, the countries should make demonstrable progress by creating an effective greenhouse gas accounting system. The protocol insisted that developed countries should assist developing countries to become aware to the impacts of climate change.

The first period of implementation of the Kyoto Protocol (2008–2012) was full of controversies on different issues, especially the stand of the USA who signed the protocol but refused to ratify the same. The second period for implementation of the principles was from 2013–2020 and till 2017, only 65 countries ratified this second period of commission; India is one of them. Meanwhile, the Paris Agreement was adopted in 2015 to keep global warming below 2°C, which would be applicable to all countries.

India was amongst non-annexed countries who were supposed to act immediately on the GHG emissions but in a different manner than developed countries. India has already started taking actions to reduce its emissions by promoting renewable energy, adopting cleaner fuel use, and has also started implementing CDM projects wherever viable.

6.3.9   Convention on Biological Diversity (CBD)

In the 1970s, scientists from various disciplines began to appreciate the value of all forms of life as an integrated earth system. They referred to this as *biodiversity* found in ecosystems; all the different species of flora and fauna and the unique genetic makeup of individuals in all species. They also began to appreciate that several species were being lost forever to extinction processes as a result of human activities in the anthropocene. This era has been dominated by human activities over the last 200 years, of our modern consumerist society whose exploding population is destroying biological diversity. The genetic makeup of our crops and livestock breeds that were developed and used by our farmers for hundreds of years is also at threat and this is seen as a major loss for developing future important strains of new disease-resistant livestock and crops by using modern biotechnology.

All this thinking culminated in debates at the Rio conference in 1992, which lead to a document called *Caring for the Earth* drafted by the United Nations. The debates were led by developing nations whose biodiversity, both wild and domesticated, was been exploited by western industrialised nations. The initiators of the Convention of Biological Diversity (CBD) recognised the sovereign rights over biodiversity for every nation in which unique features had been preserved by local stakeholders. CBD also ensured that the nation's states preserve their own biodiversity and share its resources equitably among all its citizens. CBD therefore prevents economically deprived nations from being exploited by the rich and powerful nations of the world.

In response to this challenge, India drafted her own Biological Diversity Act in 2002. This brings the legal concerns upfront, which suggest that people have to conserve biodiversity, and also helps to use bioresources equitably among all the people of India. The act prevents other countries from exploiting our resources for their economic benefits.

The mechanism to do this through was through a post step process at the ground level which is closely linked to our *Panchayati Raj* system. The BD act makes it clear to the Government and people at large, and then certain legally binding mechanisms have been put into place. This includes the following.

- Every village, town and city must ensure that the local governance includes a Biodiversity Management Committee (BMC).
- Every local community must document its own Peoples Biodiversity Register (PBR) which includes all its wild and domesticated floral and faunal biological resources, along with the Traditional Knowledge Systems (TKS) for the use of bioresources. This prevents illegal exploitation by other nations or other national commercial ventures.
- The economic benefit from such resources must be shared equitably with the local community from which the resources have been used. Products manufactured for food, medicines, cosmetics or other consumer products developed for commercial use, must give 2%–3% of their revenue to the people who have preserved the resources over long periods of time. This is referred to in the Biodiversity act as *Access and Benefit Sharing* (ABS) which prevents unauthorised exploitation of any biological resource. For implementing these important measures, India has established the National Biodiversity Authority in Chennai. Every state in India has a State Biodiversity Board which Acts as an independent regulatory body for protecting the biodiversity resources of the state.
- The Local Biodiversity Management Committees (which are elected members of gram Panchayat/ Municipal Corporation) form the backbone of protecting our biodiversity from external pressures from outside India as well as at the national level. Thus, access to resources by any industry must have prior consent of the Panchayat/ municipality.
- The economic importance of biological resources from which all our pharmaceutical, agricultural and livestock, genetic material is used in modern technology, is enormous. In India, bioresources and wildlife has always had ethical and spiritual value which is impossible to factor into purely economic terms. Thus, the Biodiversity Act 2002 not only protects the biodiversity but ensures a sustainable level of equitable, well managed, resource use for long term security of sources of food, drug, bio-pesticides, bioenergy and other industrial goods.
- The act stipulated that any local biodiversity management committee can create a local biodiversity heritage site to preserve ecosystems, species and genes by banning or controlling the resource use from the area.
- *Peoples participation in biodiversity management:* The Biodiversity Act ensures that people cooperate with the government to conserve and sustainably use our vital bioresources, our wildlife within and outside the protected areas and the thousands of varieties of indigenous crops and many rare livestock breeds. The key concern is to make people from all walks of life realise the enormous importance preserving biological diversity. Thus, India has complied with the requirements of the CBD.

6.3.10   The Chemical Weapons Convention

Chemical weapon warfare is not new to civilisation. Many references of poisoning of arrows, water channels and food are found in ancient war literature. The references of chemical wars and rules and regulations associated with it are present in the Ramayana, Mahabharata, Manusmriti as well as in Greek literature.

In the present era, the convention on such chemical weapons was entered into in the year 1993. The formal name of the Chemical Weapons Convention (CWC) is, the *Convention on the Prohibition of the Development, Production, Stockpiling and Use of Chemical Weapons and on their Destruction.* The countries which sign and ratify this convention are obliged to agree to prohibition of use and production of chemical weapons, as well as the destruction of all chemical weapons. India became party to the convention in the year 1993 and ratified it in 1997. India enacted a Chemical Weapons Convention Act, 2000.

As of November 2011, around 71% of the (declared) stockpile of chemical weapons has been destroyed. The convention also has provisions for systematic evaluation of chemical and military plants, as well as for investigations of allegations of use and production of chemical weapons based on intelligence of other state parties.

Environment Impact Assessment (EIA)

This is one of the most important aspects related to India's environmental practices. For all development projects, whether government or private, the Ministry of Environment, Forests and Climate Change (MoEF and CC) requires an impact assessment to be carried out by a competent organisation. The EIA must look into physical, biological and social parameters. EIAs are expected to indicate the likely impact on the project site if it is passed and implemented. The MoEF and CC has identified a large number of projects that need clearance on environmental grounds before they are setup.

What must a project's EIA address?

◈ An EIA must define the impact of the project on water, soil and air.
◈ It must list the flora and fauna, including any endangered species in the region and specify if any species whose habitat or life could be adversely affected.
◈ It must list the impact on the lives of local people if any.

After the Environmental Protection Act of 1986 was passed, an EIA to get an environmental clearance for a project became mandatory. The impact created by each type of industry differs and the biotic pressure on the proposed sites also varies in its sensitivity. Some areas are more fragile than others, some have unique ecosystems, while others are the wildlife habitats or the home of endangered species of plants or animals. All these aspects require evaluation before a development project or a site for an industry is cleared.

New projects where no development has been done are called *green field projects.* Projects that already exist but require expansion must also apply for clearance. These are called *brown field projects.*

Project proponents are expected to select a competent agency to undertake an EIA. Projects can be classified into those with a mild impact, a moderate impact or a serious impact. Some may have a temporary major impact, during the construction phase, which could later become less damaging, or be mitigated by a variety of measures. In other situations, the impact may continue and even increase, for example, where toxic solid waste is constantly generated. Some projects could thus cause temporary reversible damage while others can have irreversible or even permanent impact.

Environmental clearance—what is involved?

◆ The proposer of the project is expected to apply to the State Pollution Control Board (PCB).

◆ The PCB checks and confirms that the EIA can be initiated.

◆ The agency that does the assessment submits a report to the proposer. This may take several months.

◆ The report of the Environmental Statement is forwarded to the MoEF, which is the impact assessment authority.

After 1997, the MoEF has stipulated that a public hearing should be done at the local level. The PCB publishes an advertisement about the hearing in the local vernacular press. An Environmental Impact Statement (EIS), which is an executive summary of the EIA, is kept for the public to read. The venue and time of the public hearing is declared.

Once the hearing is held and opinions have been expressed, both for and against the project, the minutes of the meeting are sent to the MoEF and CC.

Drawbacks

Experience shows that a large number of EIAs are inadequately researched and are frequently biased as they are funded by the proposer of the project.

While most EIAs are adequate for studies on the possibilities of air, water and soil pollution, they generally deal inadequately with issues such as the preservation of biodiversity and the social issues that may arise from future environmental impact. Frequently, biodiversity concerns are sketchily dealt with and consist mostly of a list of species without population assessments, census figures of wildlife, or a study of the effects on the ecosystem as a whole. Changes in land use patterns affect whole communities of living organisms. This is rarely taken into account, as such issues are difficult to assess in quantifiable terms. Impacts on environmental services are rarely addressed in economic or social parameters.

Issues related to equity of resources that are inevitably altered by development related projects are not fully addressed. These cryptic concerns must be dealt with more seriously in environmental assessments and the public at large should know and appreciate these inadequacies.

It is crucial that environmental awareness becomes a part of public thinking so that public hearings can be made to a more environmentally conscious audience.

It is not sufficient to say that an EIA has been done. It is the quality and sincerity of the EIA that is of importance. It is also important to look at the cumulative impact of several adjacent projects on an area. An EIA is not intended to stop all types of development. The site for an industry can be selected carefully and if it is likely to damage a fragile area, an alternate, less sensitive area must be selected. In some cases, it is essential to drop projects altogether if the anticipated impacts are likely to be very severe. In other cases, it is necessary for the project to counter balance its effects by mitigating the ill effects on the environment. This means compensating for the environmental damage by compensatory afforestation or creating a Protected Area in the neighbourhood at the cost of the project. Rehabilitation and resettlement of the project-affected people is a key concern, which should be given adequate funds and should be done only after consent is clearly obtained from the people living in the area. In most cases it is advisable to avoid resettlement altogether. If an area's vegetation is being affected, the project costs must include the cost of compensatory afforestation and other protective measures.

Citizens actions and action groups

Some of the most relevant practices for providing conservation measures is linked to citizens actions and action groups. This can only be made possible only through Communication Education and public awareness (CEPA). Currently citizen science groups can provide reasonably accurate data which can lead to conservation. Citizens must learn to act as watch dogs to protect their own environment from the consequences of unsustainable projects around them. Well-informed citizens not only have rights but also have a duty to perform in this regard. They can join action groups to develop a lobby to strengthen the environmental movements in the country, their state, town or village. Individuals can take one or several possible actions when they observe offenders who for their own self-interest damage the environment. An individual has the right to bring an environmental offence or nuisance to the attention of concerned authorities. This ranges from Government-line agencies such as the police, the forest department, the Collector or Commissioner of the area as the case maybe. At times the concerned officials may not be able to easily appreciate complex environmental concerns and the individual may have to learn how to communicate these issues in a way in which it becomes essential for the concerned officer to act in a pro-environmental fashion. If this does not work, a citizen can seek legal redress under the relevant statutes of law. The EPA and the WPA are the most frequently used legal instruments for these purposes. It is possible to move courts by a PIL, and take this up to the apex court – the Supreme Court of India – which in the recent past has given several highly enlightened pro-conservation judgements.

Individuals can also elicit public support through the press and electronic media. Among the many environmental battles that have been fought in this country, some have been won while many others have been lost. These projects have led to serious environmental degradation in spite of the laws intended to control such damage.

6.4 NATURE RESERVES

The great variety of life on earth has provided for the needs of human beings over thousands of years. This diversity of living creatures forms a support system that has been used by each civilisation for its growth and development. Those that used this bounty of nature carefully and sustainably survived, while those that overused or misused it disintegrated.

Science has attempted to classify and categorise variations in nature for over two centuries, resulting in an understanding of its organisation into communities of plants and animals and their behavioural patterns. This information has helped us utilise the earth's biological wealth for the benefit of humanity. It has been integral to the process of development, including better health care, better crops and the use of plants and animals as raw materials for food, medicines and industrial products. This has led to a higher standard of living for the developed world. However, this has also produced the modern consumerist society, which adversely affects the diversity of biological resources upon which it is based. The diversity of life on earth is so great that if we use it sustainably, we can go on developing new useful products for many generations to come. This can only happen if we manage biodiversity as a precious resource and prevent the extinction of species.

Our constitution is sensitive to the importance of preserving all living creatures and the protection of nature and natural habitats. It stresses that this is the duty of each and every citizen of India. This has been stressed by the amendment to the constitution in 1976.

> In the Constitution of India, Part IV (Directive principles of State Policy) Article no. 48A states that, 'the state shall endeavour to protect and improve the environment and to safeguard the forests and wildlife of the country'.
>
> In the Constitution of India, Part IV A (Fundamental duties), Article no. 51A states that 'every citizen of India must protect and improve the natural environment including forests, lakes, rivers and wildlife and have compassion for living creatures'.

In view of this great and far reaching thinking, a Protected Area network of National Parks and Wildlife Sanctuaries has been created in India to which conservation reserves and community reserves have been added recently (**Fig. 6.1**).

This concept is not new. In ancient India, there were several *abhyaranya* to protect areas where elephants could breed freely and multiply in the wild. These were then caught and trained by the rulers to act as mounts used in battle and also for other activities such as logging. Animals have always been venerated from the ancient vedic period to the present times. Gods and goddesses have specific animal '*vahans*' to travel through the universe. They are given great importance in religious and cultural thinking. Many tribal gods are depicted as dangerous animals such as tigers and cobras. In the early vedic period and in the Mahabharata and Ramayana, plants and animals were to be protected, venerated as gods and have been retained over generations as part of this great mythological storehouse of local knowledge.

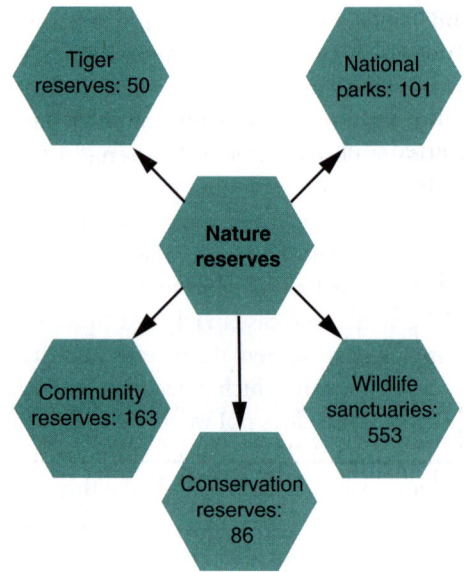

Fig. 6.1 Nature reserves

Community conserved areas under local panchayats
• Sacred groves: Western Ghats, North-east
• Sacred mountains: Himalayan peaks
• Sacred rivers: Ganga, Narmada
• Temples with ancient trees
• Temple ponds

There are references to many species of plants and animals which show that there was a strong feeling about the need to protect plants and animals in ancient India. Ashoka, in 268–231 BC, stopped killing animals for food and proclaimed an edict to make sure that certain species would be protected by law.

Buddhism and Jainism gave great importance to protecting all forms of life. It has taken thousands of years for the world, for the Government of India and the International Union for the Conservation of Nature and Natural resources (IUCN), to develop a list of the Schedule Species of plants and animals that have to be protected and to realise the importance of biological diversity. Since the Rio conference in 1992 and the CBD in 1992, there have been international agreements between nations, to protect species and sites rich in biodiversity for future generations. Thus, most nations of the world signed the Convention on Biological Diversity after the Rio conference in 1972. In India it led to the Parliament passing the Biological Diversity Act in 2002.

During the last few decades, the modern science of Conservation Biology has appreciated that preservation of biodiversity is critically important for the wellbeing of humanity. Biodiversity is now considered a great storehouse of wealth that provides us with food, medicines, industrial products and tourism opportunities. The Protected Areas are islands of nature in a matrix of other types of land use such as agriculture, urbanisation and industry. These Protected Areas provide several environmental services. They support water sources that emerge from their forests. Insects, birds and bats that live in the forests, pollinate our food crops. Birds spread seeds and thereby support plant life in and around Protected Areas as well as in agricultural and horticultural areas. Protected Areas in grasslands, wetlands and the sea are rich in their own species diversity. Marine ecosystems provide fish and crustaceans as food for millions of people.

It is part of human nature for people to be fascinated by wild animals, birds and plants. This is referred to as biophilia. Ethical values are important aspects that make us love and protect all forms of life.

Learning by reflection

Biophilia means an innate and genetically determined affinity of human beings with the natural world. E O Wilson
    Reflect on your own level of sensitivity to nature's wonders.

Scientists have identified, named and studied over 1.9 million plants and animals in the world. However, there are likely to be twenty to a hundred times more species in nature that have still not been identified. The largest number of species occurs in the insect world, followed by flowering plants. Together they form about three fourths of the world's species.

As a part of policy for nature conservation, India is a leader among nations. India's Wildlife Protection Act predates such initiatives done by other nations by several decades. India's efforts at developing its network of PAs in the form of an integrated PA system in 1988 was also an attempt at advanced planning with a scientific basis on size, shape, people issues, corresponding postulates brought in new concepts of genetic, species and ecosystem inputs through a biogeographic zone assessment. In 2003, MOeF and CC developed two new types of protected areas—Conservation Reserves and Community Reserves (**Fig. 6.2**). Several new initiatives were taken by the National Biodiversity Authority to create Community Conserved Areas under the Biodiversity Act, 2002. This act also is meant to identify Biodiversity Heritage Sites. India now has several BHSs managed by local people.

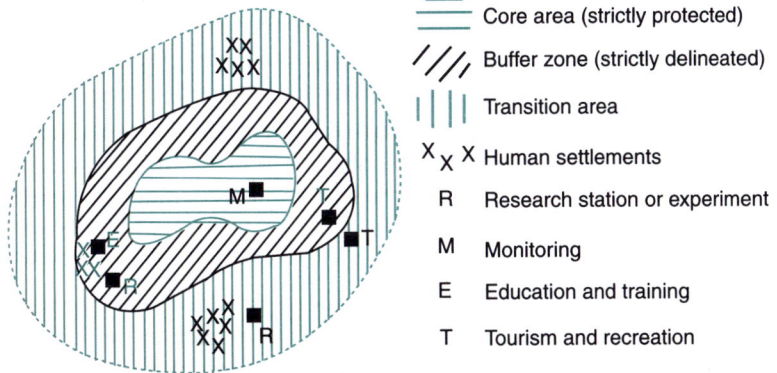

Fig. 6.2   Formally protected areas under the Wildlife Protection Act

6.4.1   Tribal Populations and Rights

India has several tribal communities that have retained their Traditional Knowledge Systems, values and folklore of great importance to society. Their cultures are a part of India's great heritage which includes biodiversity resources, wild food sources and tribal medicines. The process of homogenisation is likely to wipe out their folklore, as well as their traditional knowledge and their various complex systems of land management.

Tribal cultures have led sequestered lives in many parts of India due to which development processes, health care, education and poverty alleviation has not penetrated into their lives and livelihoods. Their forests have been overexploited by non-tribal people often leaving them without the resource base required by traditional indigenous forager communities. To redress their right to life and livelihood, the Government of India has notified and implemented the Forest Right Act, 2006. We need to acknowledge their special rights to their resources, traditionally grown crop varieties and local medicines that they have used for generations.

6.4.2   Human–Wildlife Conflict in the Indian Context

Crop damage by wild herbivores and cattle-lifting by wild carnivores around Protected Areas is a complex issue. It is dealt with by providing compensation. The spread of human population into the periphery of National Parks and Sanctuaries is now a major concern and causes serious conflicts between wildlife and people.

A major cause of damage to crops is due to herbivores such as nilgai and wildboar. Cattle-lifting by tigers and leopards is a growing concern. As PA management improves through modern tools such as satellite imaging, geoinformatics, radio tagging of tigers and other species, special software such as m-stripes to identify and count individual tigers, and drone technology to study habitats, if wildlife populations exceed the carrying capacity of our PAs. This has begun to increase the people–wildlife conflict outside the PAs. Increasing wildlife tourism and tourism complexes around PAs is a serious conflict of interest between wildlife conservation and tourism interests.

SUMMARY

- Environment policies and practices are critical for good governance. Our country is moving rapidly into a new framework for development. For this, it is a must to have a sound approach taking all the different sectors of our society through a massive change process. The country needs support from every citizen. Sustainable development encompasses economics, societal equity and environment management for long term intergenerational equity.

- Climate change affects people and constitutes one of the world's greatest challenges ever faced. Climate change is used to describe any process that causes changes in the earth's climate. Scientists have estimated that the world will be significantly warmer at the end of this decade. The relatively small amount of warming has led to world-wide impacts on human society, flora and fauna.

- Several human activities have translated into various environmental problems besides climate change, such as acid rain and ozone layer depletion, impacting human health and natural ecosystems. Mitigating and adapting to the causes of these problems is possible only if countries across the world cooperate with each other.

- Human wildlife conflicts are increasing around protected areas in India due to successful scientific management of Tiger Reserves, National Parks and Wildlife Sanctuaries. The population of several wild species has increased in the recent past resulting in overpopulation beyond the carrying capacity of protected areas. This results in their dispersal into surrounding human dominated landscapes. Crop damage by wild herbivores and cattle-lifting by wild carnivores around Protected Areas is a complex issue. It is dealt with by providing compensation by forest department to local people.

- The Environment (Protection) Act, 1986 is an overarching legislation that aims to protect and improve all aspects of the environment. The Air Act, Water Act, Wildlife Protection Act, Forest Conservation Act and Biodiversity Act aim to respectively protect and improve the specific conservation aspects of the environment. As protected areas limit resource use of local people, the Forest Rights Act and measures such as ecodevelopment have been created to mitigate problems between protected area management and local people.

QUESTIONS

1. What do you understand by climate change? Discuss measures that can be adopted at a personal/ community level to combat climate change.
2. Which are the five most important environment related Acts?
3. How does the Biological Diversity Act conserve nature and ensure sustainable use of the resources?
4. What do you understand by human–wildlife conflicts? What are the causes?
5. How is agriculture in India affected by climate change?
6. Write a note on the causes of acid rain.
7. What is ozone hole? What are the causes? What can an individual do to mitigate this?

Human Communities and the Environment

मां निषाद प्रतिष्ठां त्वमगमः शाश्वतीः समाः।
यत्क्रौंचमिथुनादेकम् अवधीः काममोहितम्॥

mā niṣāda pratiṣṭhāṁ tvamagamaḥ śāśvatīḥ samāḥ
yat krauñcamithunādekam avadhīḥ kāmamohitam

The verse roughly translates to, "Oh hunter, may you repent for life and suffer, find no rest or fame, for you have killed one of the unsuspecting, devoted and loving krauñcha couple."

Learning Objectives

In this chapter you will learn,

◆ About the human population and its growth
◆ What carbon footprint is
◆ How to rehabilitate people affected due to new projects
◆ What disaster management is and the mitigation strategies involved
◆ About environmental movements, ethics, communication and public awareness

Purpose

There are discussions of contemporary environmental concerns by experts from different fields who see these issues from their own perceptive. As knowledgeable and environmentally conscious individuals we need appreciate and take carefully considered balanced views on these concerns.

Our Role

Once we have a considered objective and scientific rationale for these issues, we can act as positive influencers on good environmental governance based on good ethical values. We then become the torch bearers for appropriate communication and awareness (which is our duty according to the Indian constitution).

7.1 HUMAN POPULATION AND GROWTH

Human communities are intimately connected to their environment. The landscape they live in has natural ecosystem elements that they adapt to and which the community modifies over time to maximise the resources that they need. These processes of change in the environment from natural undisturbed ecosystems to their cultural landscape elements have taken hundreds of years. Humans in the Stone Age began to initiate these changes by hunting and gathering food locally from forests, grasslands, wetlands and the sea. Their miniscule population living in isolated pockets must have had a very small impact on nature.

The level and patterns of use of their environmental resources began to alter when hunter–gatherers settled into the life of a basic farmer who also herded livestock. Ancient urban–rural landscapes began making immense alterations in their ecosystems. The pace of landscape change became rapid when the traditional agrarian and agro-pastoralist farmer and livestock herder started the industrial revolution around 200 years ago. The pressures on the environment escalated. Forests were destroyed. Wetlands were drained or polluted. River systems were modified by dams.

Cart tracks grew into roads and expressways. Industry and thermal power plants spewed smoke into the atmosphere initiating what we recognise today as global climate change.

Village life and livelihoods became increasingly urbanised with the development of commercial establishments and industrial settlements. Townships grew into cities attracting a large migrant population. This resulted in expanding informal housing, requiring increasing resources. These natural resources came from the forests surrounding the rural settings. (see Changing Landscapes, E Bharucha, Harper Collins)

Better health facilities and nutritional status accompanied by the rapid and unprecedented population growth led to the overexploitation of natural resources. This led to a rapid loss and the degradation of the earth's natural resource base. As wealth among urban communities grew, people began to live an increasingly consumerist lifestyle. Rich countries exploited the resources of poor nations. India's resources were plundered by the earlier industrialised capitalist European nations. The British used our forest resources for building their ships. Our cotton and other farm products supported two great European World Wars.

When India became an independent country in 1947, its natural resources had been depleted. It took over five decades to move the economy forward into what we now recognise as one of the world's fastest growing economies. We now label the country as an emerging economy. We base this on a new knowledge-based society. Our industrial growth and per capita incomes are growing. But we are also becoming increasingly resource hungry. We use more and more of our environmental resources and waste increasing amounts of consumer products. The result has been an ever-increasing footprint of human society on nature's resources. These pressures have become an unsurmountable threat to nature, which in turn adds to the pressure on human society. We have extended resource use beyond the carrying capacity of the earth.

Deforestation has led to degraded landscapes. River pollution has led to health problems. Overuse of water for irrigated farmland, for industry and urban household use, has led to severe water shortages. The phenomenal increase in energy use for industry and an exploding population has become a part of an unsustainable modern way of life. Air pollution has placed a stress on the healthcare systems to prevent and treat a number of respiratory diseases. Finally, the climate change that industrialisation has produced has become a dangling sword over human communities. It has triggered extreme weather events, a rise in sea level due to melting ice caps, and a loss of biodiversity that is the tip of the iceberg of changing landscapes. It is a threat that can only be countered by altering our lifestyle towards more responsible behaviour. If we want a better life for the future generations, we need to bring about more sustainable living patterns that are environmentally friendly.

The rate of population growth worldwide is decreasing gradually. However, India's population is still increasing and contributes a large part to the world's population (**Fig. 7.1**). Our country is currently the second largest population on earth, second only to China. India adds over two lakh people per year to the earth's human population (**Fig. 7.2**). In the post-independence era, we realised that population control measures had to be brought into our governance systems. Infant mortality from gastrointestinal disease was brought down and anti-tubercular measures reduced mortality from chronic respiratory diseases. Better healthcare brought about increased longevity and contributed to the population growth.

The impacts on India's environmental assets have grown into threats to human life and wellbeing. This is eroding our natural capital. There is an unprecedented loss of all our resources—air, water, soil, energy, minerals and biodiversity. The effects are poverty, ill health, social unrest and inadequate landuse, all of which are unsustainable.

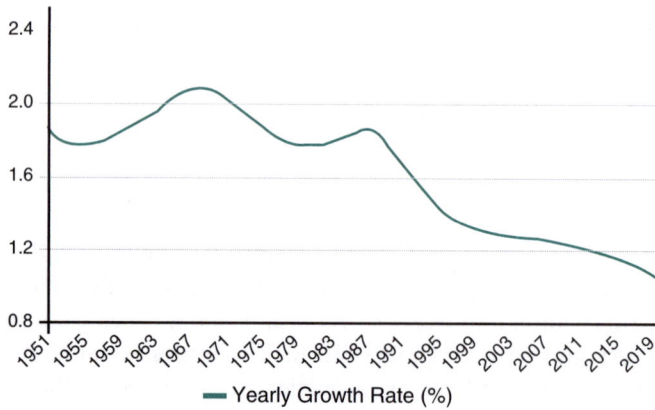

Fig. 7.1 The world population (according to the United States Census Bureau) is 7,477,529,965

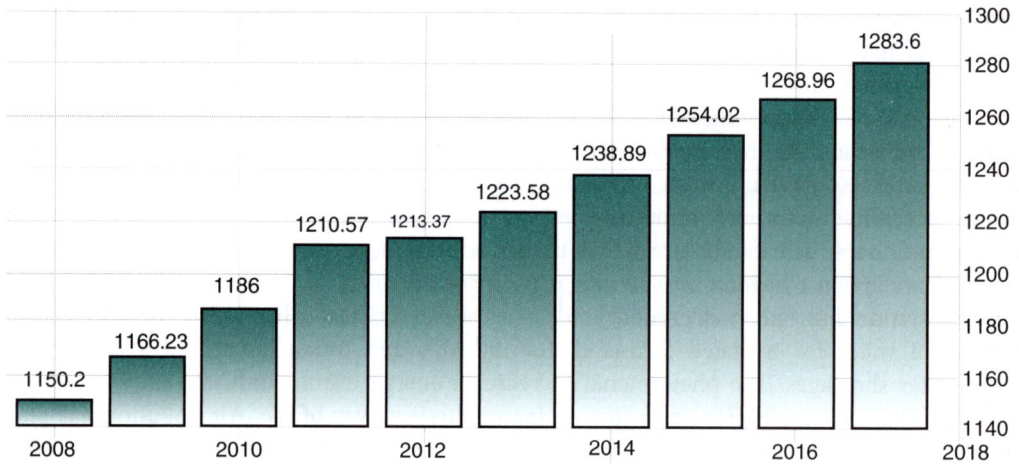

Fig. 7.2 Total population of India: 1283.6 million people in 2017

India's population is well above the carrying capacity of the resources that the country can provide. Nature's physical, chemical and biological resources are constantly being depleted. Energy sources are inadequately distributed. All of nature's services are stretched beyond limits at the current rate at which we are exploiting and misusing these essential services. To balance the depletion of goods and services and maintain our natural capital, we need to stabilise population growth, improve waste management, establish recycling of waste, use biologically efficient pest control, prevent diseases and curtail biodiversity loss. Enhancing the management of climatic change through mitigation and adaptation requires urgent attention.

The anthropocene

Should the anthropocene frighten us? Or, is it a new challenge to mankind's ingenuity? If so, how do we meet these challenges by reducing our individual and societal footprint on earth? Why is this era known as the anthropocene? It is the age when human beings have set the stage for controlling and using resources in a way that have begun to make rapid alterations on our living space.

We can use the earth's resources and services logically with care and sensitivity. Or we can become increasingly careless and damage the earth. These conflicting ways of managing resources, either conservatively or unsustainably, is why the present era is called anthropocene.

The population of the hunters and gatherers who lived three million years ago was small and scattered across the world. As the population grew, agriculture provided more food for subsistence. By AD 100 the population grew to 300 million and continued to grow at a moderate rate. The 18th century marked the start of the Industrial Revolution. Human population expanded as the standard of living rose and famines and epidemics decreased in several regions. From 760 million in 1750, the world's population reached 1 billion by 1800. After World War II, the growth of population accelerated mainly in the less developed countries. A billion people were added between 1960 and 1970 and another billion between 1975 and 1987. The 20th century started with 1.6 billion people and ended with 6.1 billion! It is not the census figures alone that need to be stressed, but an appreciation of its impact on natural resources. The extent of this depletion is further increased by affluent societies that consume more energy and resources per capita, than less fortunate people.

Figure 7.2 shows the explosive population growth in the last 50 years, from 2.5 billion people in 1950 to 6.5 billion in 2005. By 2050, this number could rise to more than 9 billion. This explosive growth in the last 200 years has had its impact on (i) the disparity in living standards between the rich and poor nations and within nations between different sectors of society, (ii) the level of resource use, and (iii) the changes in our environment.

Our landscapes have changed drastically and will continue to change with further growth in human population. The increasing pressures on resources place great demands on the in-built buffering action of nature that has a limited ability to maintain a balance in the environment. However, current development strategies have led to a breakdown of the earth's ability to replenish the resources on which we all depend. There are cultural, economic, political and demographic reasons that explain the differences in the rate of population growth in different countries. This also varies in different parts of certain countries and is linked with community and/or religious thinking.

Global population growth

While the rate at which the world's population is increasing has slowed, the overall population continues to grow. The rate of the world's annual population change can be expressed as a percentage:

$$\text{Annual rate of natural population change (\%)} = \frac{\text{Birth rate} - \text{Death rate}}{1,000 \text{ persons}} \times 100$$

Birth rates and death rates are decreasing worldwide. However, better access to medical care, widespread immunisation and improved sanitation, have made the death rates fall more steeply than birth rates. So, while the exponential growth rate has declined to a slower rate – from 2.2% to 1.2% – between 1963 and 2005, the same period has seen the population base being almost doubled—from 3.2 billion to 6.5 billion (Miller, 2006). Furthermore, the exponential growth rate has been drastically different in developed countries (0.1% annually) compared with developing nations (1.5% annually). This means that the population in developed countries, which is currently 1.2 billion, is expected to change very little in the next 50 years. On the other hand, developing countries will see a steady rise in their population from 5.3 billion in 2005 to over 8 billion in 2050.

Another measure of population growth is *doubling time*. This is the time taken for a population to double in size. At the present rate, the world's population will be doubled in approximately 58 years.

Population explosion

In response to our phenomenal population growth, India seriously took up the challenge to 'reduce birth rates to the extent necessary to stabilise the population at a level which is consistent with the requirement of the national economy'.

Informing the public about the various contraceptive measures available, is of primary importance. This must be done actively by government agencies such as Health and Family Welfare, as well as through all other education and extension strategies. It is of great importance for policy makers and elected representatives of the people – ministers, MPs, MLAs at the central and state levels – to understand the great and urgent need to support measures to control population growth. The media must keep people informed about the need to limit family size and the ill effects of a growing population on the world's resources.

The importance of limiting population growth is crucial for India's sustainable development. The first green revolution in the 1960s produced a large amount of food. However, the strategies that advocated the use of chemical fertilisers and pesticides have led to several environmental problems. Now, a new green revolution is needed, to provide enough food for our growing population without ill effects on the environment due to chemicals and pesticides. Agricultural practices should not damage land, change river courses by building large dams, or change landuse at the cost of critically important forests, grasslands and wetlands. Growing human populations will inevitably expand from farmlands into the remaining adjacent forests. Many such encroachments in India have been regularised over the last few decades. Forest loss has long term negative effects on water and air quality. Unfortunately, the loss of the great resource of biodiversity is still not seen as a major deterrent to human wellbeing. Energy use is growing, both due to the increasing population and a lifestyle that increasingly uses consumer goods which require large amounts of energy for their production, packaging and transport. Our growing population also adds to the enormous amount of waste.

With all these links between population growth and the environment, population stabilisation has become critical to human existence.

Urbanisation

In 1975, only 27% of the people in the developing world lived in urban areas. By 2000, this had grown to 40% and by 2030, well informed estimates are that this will grow to 56%. The developed world is already highly urbanised with 75% of its population living in urban areas. The urban population growth is both due to the migration of people to towns and cities from rural areas in search of better job opportunities, as well as population growth within the cities.

As a town grows into a city, it not only spreads outwards into the surrounding agricultural lands or natural areas such as forests, grasslands and wetlands, but also grows skywards with high-rise buildings. The town often loses its open spaces and green cover unless it is consciously preserved. This leads to the destruction of the quality of life in urban areas.

Good urban planning is essential for rational landuse, for upgrading slum areas, improving water supply and drainage systems, providing adequate sanitation, solid waste recycling, developing effective waste water treatment plants and efficient public transport systems.

While all these issues appear to be under the purview of the local Municipal Corporations, better living conditions can become a reality only if every citizen plays an active role in managing the environment. This includes a variety of 'Dos and Don'ts' that should become an integral part of our personal lives. Unplanned and haphazard growth of urban complexes has serious environmental impacts. Increasing solid waste, improper garbage disposal and air and water pollution are frequent side effects of urban expansion that affect human health and wellbeing.

Apart from undertaking actions that support the environment, every urban individual has the ability to influence a city's environmental management. Citizens must see that the city's natural green spaces, parks and gardens are maintained, river and water fronts are managed appropriately, road side tree cover is preserved, hill slopes are afforested, and architectural and heritage sites are protected. Failure to do this leads to increasing problems which eventually destroy the city's ability to maintain a healthy and happy lifestyle for its dwellers. All these aspects are closely linked to the population growth in the urban sector. In many cities, population growth often outstrips the planner's ability to respond in time.

Pull and push factors

In India small urban centres will grow rapidly during the next decade and several rural areas will require reclassification as urban centres. These urban areas will grow by several million residents. People move to the cities from rural areas in the hope of earning better incomes. This is the 'pull' factor. Poor opportunities in the rural sector stimulates migration to cities. The loss of agricultural land to urbanisation and industry, the inability of governments to sustainably develop the rural sector and a lack of supporting infrastructure in rural areas are all factors that 'push' people from the agricultural and natural wilderness ecosystems into the urban sector.

As populations in urban centres grow, they draw on the resources of more and more distant areas. The ecological footprint corresponds to the land necessary to supply natural resources to an urban community and for the disposal of its waste. At present, the average ecological footprint of an individual at the global level is said to be 2.3 ha of land per capita. It is estimated that the world has only 1.7 ha of land per individual to manage these needs, thus leading to an unsustainable use of land.

The pull factor of the urban centres is not only due to better job opportunities, but also better education, healthcare and relatively higher living standards. During the last few decades in India, improvements in the supply of clean water, sanitation, waste management, education and health care have all been urban centric, even though the stated policy has been to support rural development. In reality, development has lagged behind in the rural sector where the population is rapidly expanding. For people living in our forests and mountain regions, development has been most neglected. It is not appropriate to use the same development methods used for other rural communities for tribal people dependent on collecting natural resources from the forests. A different pattern of development that is based on the sustainable extraction of resources from their own surroundings would be ideal. In general, the growing human populations in the rural sector will opt to live where they are, only if they are given an equally satisfying lifestyle as what they would get by migrating into the urban sector.

Urban poverty and the environment

The number of poor people living in urban areas is rapidly increasing. A third of all the poor people in the world live in urban centres. These people live in the unorganised sector of urban slums and suffer from water shortages and unsanitary conditions. In most cases, while the unorganised housing sector invariably has unhygienic surroundings, the dwellings themselves are kept relatively clean. It is the common areas used by the community that lacks the infrastructure to maintain a hygienic environment. This is now being addressed by the Swachch Bharat Programme of the GOI that is already making a difference.

One billion urban people in the world live in inadequate housing, mostly in slums. The majority of buildings are temporary structures. However, low-income groups that live in high-rise buildings may also be living in poor unhygienic conditions in certain areas of cities. Illegal slums often develop on government land, along railway tracks, on hill-slopes, riverbanks and marshes that are unsuitable for formal urban development. On the riverbanks, floods can render these poor people homeless. Adequate organised planned low-income legal housing for the urban poor remains a serious environmental concern.

Urban poverty is even more serious than rural poverty. Unlike in the rural sector, the urban poor have no direct access to natural resources such as relatively clean river water, fuelwood and Non-Timber Forest Products (NTFPs). The urban poor can only depend on cash to buy the goods they need, while in the rural sector they can grow a substantial part of their own food. Living conditions for the urban poor are frequently worse than that for rural poor. Both outdoor and indoor air pollution due to high levels of particulate matter and sulphur dioxide from industrial and vehicle emissions lead to high death rates from respiratory diseases in cities. Most environmental efforts are targeted at reducing outdoor air pollution whereas indoor air pollution, due to the use of fuelwood, waste material and coal in chulhas, is an equally major health issue. This can be addressed by using better designed 'smokeless' chulhas, hoods and chimneys to remove indoor smoke.

With the growing urban population, a new crisis of unimaginable proportions will develop in the next few years. Crime rates, terrorism, unemployment and serious environmental health-related issues can be expected to escalate. This can only be altered by stabilising population growth on a war footing, supporting education and preventive health measures.

7.1.1   Impacts on Environment, Human Health and Welfare

In many situations, valuable ecological assets are turned into serious environmental problems. This is because we, as a society, do not strongly resist forces that bring about ecological degradation. The societies that use a get-rich-quick approach are what cause the damage. While ecological degradation has frequently been blamed on the need of fuelwood and fodder for growing numbers of rural people, the rich, urbanised, industrial sector is responsible for even greater ecological damage. The changes in landuse from natural ecosystems to more intensive utilisation (such as turning forests into monoculture forestry plantations, or tea and coffee estates) creates serious pressures on biological diversity. The use of marginal lands as intensive agricultural areas (such as sugarcane fields) or changing the land to suit urban or industrial purposes carry an enormous ecological price. For example, wetlands provide usable resources and a variety of services not easily valued in economic terms. In many cases when these are destroyed to provide additional farmland, they produce poor returns. A natural forest provides valuable NTFPs, which outweighs the long-term economic returns provided by felling the forest for timber. These values must form a part of a new conservation ethic. We cannot permit unsustainable development to run onwards at a pace in which our lives will be overtaken by a development strategy that must eventually fail. Unsustainable use of the earth's resources and ecosystem services is frequently irreparable in the short-term. Preventing ecological damage is more cost effective than repairing the damage.

Human health and welfare

Pneumonia: Acute respiratory infection (ARI), most frequently pneumonia, is a major cause of death in children under five years, killing over two million children annually (WHO, 2009).

Measles: Measles is a rash with fever and body ache in children and is caused by a virus. It infects over 200,000 children and kills over 150,000 children under the age of five (WHO, 2009). Its prevention includes greater immunisation coverage, rapid referral of serious cases, prompt recognition of conditions that occur in association with measles and improved nutrition, including breastfeeding and vitamin A supplements. Measles can be prevented by a vaccine. Effective prevention and treatment could save at least 150,000 lives per year.

Poverty–environment–malnutrition: There is a close association between poverty, a degraded environment and malnutrition. This is further aggravated by a lack of awareness on what causes malnourishment of children.

Malnutrition: Although malnutrition is rarely listed as the direct cause of death, it contributes to about half of all childhood deaths. Lack of access to food, poor feeding practices (inadequate breastfeeding, providing the wrong type of food or insufficient food) and associated infections, or a combination of the two, are the major causes of mortality.

Changing family habits and the kinds of food offered to children are important preventive measures. There are strong connections between the status of the environment and the welfare of women and children in India. Women, especially in lower-income groups, both in the rural and urban sectors, work longer hours than men. Their work patterns differ and these activities are more likely to cause health hazards. In urban centres, a number of women eke out a living by garbage picking. They separate plastics, metal and other recyclable material from the waste. During this process, they get several infections.

Environment-related issues that affect human health have been one of the most important triggers in the increasing awareness of the need for better environmental management. The changes in our environment induced by human activities in nearly every sphere of life have had an influence on our health patterns. The assumption that the only indicator of human progress is economic growth is incorrect. We expect urbanisation and industrialisation to bring in prosperity. However, it leads to diseases related to overcrowding and poor quality drinking water, resulting in an increase in water-borne diseases like infective diarrhea and air-borne bacterial diseases such as tuberculosis. High-density city traffic leads to an increase in respiratory diseases such as asthma and bronchitis. Agricultural pesticides that enhanced food supplies during the green revolution have affected both the farm worker and all those who consume the produce. Modern medicine promised to solve many health problems, especially those associated with infectious diseases by the use of antibiotics. However, bacteria have developed many resistant strains, frequently even changing their behaviour in the process, and making it necessary to keep on creating newer antibiotics. Many drugs have been found to have serious side-effects. Sometimes, the cure is as damaging as the disease itself.

Thus, development has created several long-term health problems. Improving the health status of the society will bring about a better way of life only if it is coupled with stabilising the population growth.

Environmental health

Environmental health, as defined by WHO, comprises those aspects of human health, including the quality of life, that are determined by physical, chemical, biological, social and psychosocial factors in the environment. It also refers to the theory and practice of assessing, correcting, controlling and preventing those factors in the environment that adversely affect the health of present and future generations.

Climate and weather affect human health. Public health depends on sufficient amounts of good quality food, safe drinking water and adequate shelter. Natural disasters such as storms, hurricanes and floods still kill thousands of people every year. Unprecedented rainfall and stagnant pools of water trigger epidemics of malaria and water-borne diseases.

Global climate change has serious health implications. Many countries will have to adapt to uncertain climatic conditions due to global warming. As our climate is changing, we may no longer know what diseases to expect. There are increasing storms in some countries, drought in others and a temperature rise throughout the world.

Development strategies that do not incorporate ecological safeguards often lead to ill health, while strategies that can promote health invariably also protect the environment (Fig. 7.3). Thus, environmental health and human health are closely interlinked. An improvement in health is central to sound environmental management. However, this is rarely given sufficient importance in planning development strategies.

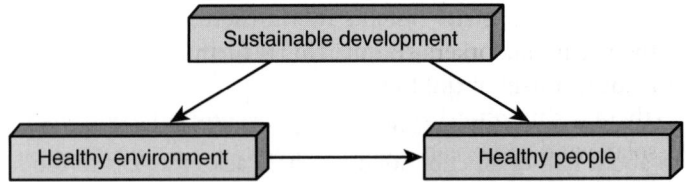

Fig. 7.3   Importance of sustainable development

Examples of the links between the environment and health

◈ Millions of children die every year due to diarrhea from consuming contaminated water or food. An estimated 2000 million people are affected by these diseases and more than 3 million children die each year from water-borne diseases across the world. In India, it is estimated that every fifth child under the age of 5 dies due to diarrhea (UNICEF—Water, Sanitation and Hygiene 2006). This is a result of poor environmental management and is mainly due to inadequate purification of drinking water. Waste water and/or sewage entering water sources without being treated leads to gastrointestinal diseases in the community and even severe sporadic epidemics.

◈ Millions of people, mainly children, have poor health due to parasitic infections such as amebiasis and worms. This occurs from eating infected food or using poor quality water for cooking food. It is estimated that 36% of children in low-income countries and 12% in middle-income countries are malnourished. In India, about half the children under the age of four are malnourished and 30% of new-borns are significantly underweight (World Bank, 2009).

◈ Hundreds of millions of people suffer serious respiratory diseases, including lung cancer and tuberculosis resulting from ill-ventilated homes and public places. Motor vehicle exhaust fumes, industrial gases, tobacco smoke and cooking food on improper chulhas, contribute to respiratory diseases.

◈ Millions of people are exposed to hazardous chemicals from industrial products when controls are not adhered to in their work place or homes that lead to poor health.

◈ Tens of thousands of people in the world die due to traffic accidents owing to inadequate management of traffic conditions. Ineffective first aid at the accident site and the frequent inability to reach a hospital within an hour causes a large number of deaths, especially from head injuries.

Important strategic concerns

◈ Strategies to provide clean potable water and nutrition to people are an important part of a healthy living environment.

◈ Providing clean energy sources that do not affect health is a key to reducing respiratory diseases.

◈ Reducing the environmental consequences of industrial and other pollutants such as transport emissions, can improve public health.

◈ Changing agricultural patterns to reduce the use of harmful pesticides, herbicides and insecticides which are injurious to the health of farmers and consumers by using alternatives, such as Integrated Pest Management (IPM) and non-toxic biopesticides, can improve the health of agricultural communities as well as consumers.

◈ Changing industrial systems into those that do not use or release toxic chemicals that affect the health of workers and people living in the vicinity of industries leads to preventive health measures and a more conducive environment.

◈ There is a need to change from using conventional energy to cleaner and safer sources such as solar, wind and ocean power, that do not affect human health.

◈ Poverty is closely related to health and is itself a consequence of improper environmental management. An inequitable sharing of natural resources and environmental goods and services is linked to poor health.

Definition of Health Impact Assessment (HIA): The WHO defines HIA as a combination of procedures, methods and tools by which a policy, programme or project may be judged on the basis of its potential effects on the health of a population and the distribution of those effects within the population. HIA helps decision makers to make choices that promote health by preventing disease/injuries. For example, transport is a major contributor to traffic-related injuries, noise and air pollution. An effective HIA can promote a healthy transport policy/plan that reduces such risks by encouraging the use of walking/cycle tracks.

Climate and health

Centuries of human civilisation have helped us adapt to living in a wide range of climatic conditions—from the hot tropics to the cold arctic, in deserts, marshlands and in the high mountains. Both climate and weather have a powerful impact on human life and health issues. Natural disasters (heavy rains, floods, hurricanes) can severely affect the health of a community. Poor people are more vulnerable to the health impacts of climate variability than the rich. Of approximately 80,000 deaths that occur world-wide each year as a result of natural disasters, about 95% are in poor countries. In weather-triggered disasters, hundreds of people and animals die, homes are destroyed, crops and other resources are lost. Public health infrastructure, such as sewage disposal systems, waste management, hospitals and roads are damaged. The cyclone in Odisha in 1999 caused 10,000 deaths. The total number of people affected was estimated at 10–15 million!

Within certain limits, human physiology can adapt to changes in weather. However, marked short-term fluctuations in weather lead to serious health issues. Heat waves cause heat-related illness (heatstroke) and death. The elderly and persons with existing heart or respiratory diseases are more vulnerable.

Why have infectious diseases that are related to our environment that were under control suddenly made a comeback? Diseases like tuberculosis have been effectively treated with anti-tubercular drugs for decades. These antibiotics kill the bacteria that cause the disease. However, nature's evolutionary processes permitted the bacteria to mutate by creating new strains. These drug resistant strains are not affected by the routinely used antibiotics and spread rapidly. This leads to a re-emergence of the disease. In the case of tuberculosis, this has led to multi-drug resistant tuberculosis. This is frequently related to HIV, which reduces an individual's immunity to bacteria such as *Mycobacterium tuberculosis* that causes tuberculosis.

The newer broad-spectrum antibiotics, antiseptics, disinfectants and vaccines once thought of as the complete answer to infectious diseases have thus failed to eradicate infectious diseases. In fact, experts now feel that these diseases will cause a higher mortality rate in the future than diseases such as cancer and heart disease. While antibiotic resistance is a well-known phenomenon, there are other reasons for the re-emergence of diseases:

◈ Overcrowding in slums in urban areas leads to several health hazards, including an easier spread of respiratory diseases.

◈ Poor quality of drinking water and poor disposal of human waste due to the absence of a closed sewage system and poor garbage management are serious urban health issues. This has led to the reappearance of diseases such as cholera and an increased incidence of diarrhea and dysentery as well as infectious hepatitis (jaundice).

◈ Impacts of climate change are responsible for a change in disease patterns and their distribution. For example, warmer climate will change the distribution of dengue, malaria and yellow fever spreading them further away from the equator. Warmer, wetter climates can cause serious epidemics of diseases such as cholera. The El Niño, which causes periodic warming in the marine environment affects coastal people. Changes in temperature affect rodent populations which could bring back diseases such as plague.

Globalisation and infectious diseases

Several diseases have been known to spread very rapidly across the world during the last few decades. In 2020 disease spread by Corona virus (Covid-19) has spread rapidly from China across the world.

Globalisation is a world-wide process which includes the internationalisation of communication, trade and economic changes. It involves parallel effects due to the need for rapid social, economic and political adjustments. Whilst globalisation has the potential to enhance the lives and living standards of certain population groups, it increases economic inequalities. It affects the poor and marginalised populations in both the non-formal as well as formal economic sectors of developing countries.

Tuberculosis: TB kills approximately 2 million people each year, including persons infected with HIV. In India, the disease has re-emerged and is now more difficult to treat as it is contagious and airborne. It is a disease that affects mostly young adults in their most productive years.

It is estimated that left untreated, each patient with active tuberculosis will infect on an average, 10–15 people every year. When an individual's immune system is weakened, the chances of getting active TB are greater.

Malaria: This is a life-threatening parasitic disease transmitted by mosquitoes. The cause of malaria, a single-celled parasite called plasmodium, was discovered in 1880. Later, it was found that the parasite is transmitted from person to person through the bite of the female Anopheles mosquito, which requires blood for the growth of her eggs.

At present, approximately 40% of the world's population, mostly those living in the world's poorest countries, risk getting malaria. The disease was once widespread, but was successfully eliminated in the temperate zone during the mid-20th century. Malaria has returned with a vengeance and is found throughout the tropical and sub-tropical regions of the world. It causes more than 200 million acute cases and at least one million deaths annually (WHO, 2009).

Malarial parasites are developing high levels of resistance to drugs. Besides this, many insecticides are no longer useful against the mosquitoes which transmit the disease.

Measures to prevent malaria:

◈ eliminate pools of stagnant water during the monsoons,

◈ use mosquito nets that have been treated with pesticides,

◈ be prompt with treatment with effective up-to-date medicines, and

◈ use mosquito repellants in risk-prone areas.

Water-related diseases

Diseases that are linked to the aquatic ecosystem are of two types. First, those that are due to bad water quality. These are also called water-borne diseases (such as diarrhea). Second, those that are spread by vectors such as mosquitoes that breed in water and cause diseases such as malaria, dengue and chikungunya.

Water supply, sanitation and hygiene development

Globally, about 2.4 billion people live in highly unsanitary conditions. Poor hygiene increases the risk of incidence and spread of infectious diseases. Improperly stored water in homes is frequently contaminated by inadequate management at the household level. This can easily be reduced through education and awareness about how water-borne diseases are transmitted.

Providing access to sufficient quantities of safe water, facilities for the sanitary disposal of excreta, and introducing sound hygiene-related behaviour can reduce morbidity and mortality caused in high risk situations.

Health and water resources development

An important aspect of water-related diseases, in particular, water-related vector-borne diseases, is the way water resources are developed and managed. In many parts of the world, the adverse health impacts of dam construction, irrigation development and flood control are related to an increased incidence of malaria, Japanese encephalitis, schistosomiasis, lymphatic filariasis and other conditions. Other health issues that are indirectly associated with water resources development include nutritional status, exposure to agricultural pesticides and their residues.

Arid areas with rapidly expanding populations are already facing the crisis of insufficient water. Conservation of water and better management of existing resources is an urgent need. The demand and supply balance is a vital part of developing the sustainable use of water. This is being termed as the *blue revolution* and needs governments, NGOs and people to work together towards a better water policy at the international, national, state, regional and local levels. The present patterns of development are water-hungry and wasteful. Current strategies do not adequately address pollution, overuse and misuse the precious resource. The links between managing water resources and health issues have not been prioritised as a major source of environmental problems that require policy change, administrative capacity building and increased financial support.

There are four major types of water-related diseases linked to the management of water resources.

Water-borne diseases: These are caused by dirty water contaminated by human and animal wastes especially from urban sewage, or by chemical wastes from industry and agriculture. Some of these diseases, such as cholera and typhoid, cause serious epidemics. Diarrhea, dysentery, polio, meningitis and hepatitis A and E, are caused due to improper drinking water, excessive levels of nitrates cause blood disorders when they pollute water sources. Pesticides entering drinking water in rural areas cause cancer, neurological diseases and infertility. Improving sanitation and providing drinking water that has been treated reduces the incidence of these diseases.

Water-based diseases: These are caused by aquatic organisms that live part of their life cycle in water and another part as a parasite in man. In India, the guinea worm affects the feet of people. Round worms live in the small intestine, especially of children.

Water-related vector diseases: These diseases are caused by insects such as mosquitoes that breed in stagnant water and spread diseases such as malaria and filariasis. Changes in climate are leading to the formation of new breeding sites. Other vector-borne diseases in India include dengue, chikungunya and filariasis. Dengue fever carries a high mortality. Filariasis leads to fever and chronic swelling of the legs. Eliminating mosquito-breeding sites where pooling of water occurs in the monsoon and using fish to control mosquito larval populations, are two ways to reduce these diseases without using toxic insecticides that have ill-effects on human health.

Water-scarcity diseases: In areas where water and sanitation is poor, there is a high incidence of diseases, such as tuberculosis, leprosy and tetanus which occur when one's hands are not properly washed. In other words, the lack of water leads to poor hygiene, and related diseases.

Diarrhea: There are several types of diarrhea, which give rise to loose stools and dehydration. About 4,500 children, mostly in developing countries, die each day from unsafe water and lack of basic sanitation facilities (UNICEF, 2006).

Repeated diarrhea and poor nutritional status forms a vicious cycle in children. Chemical or non-infectious intestinal conditions can also result in diarrhea. There are several causes for the onset of diarrhea:

◆ It can be caused by several bacterial, viral or parasitic organisms.
◆ It spreads through contaminated water due to poor sanitation, personal and domestic hygiene. Sources include human feces surrounding a rural water source, or from municipal sewage septic tanks and toilets in urban centres.
◆ The feces of domestic animals also contain microorganisms that can cause diarrhea through water.
◆ Contaminated food is a major cause of diarrhea when meals are prepared in unhygienic conditions.
◆ Fish and seafood from polluted water can cause severe diarrhea.
◆ Polluted water can also contaminate vegetables in farmland during irrigation.

Interventions: Key measures to reduce the number of cases of diarrhea include:
◆ access to safe drinking water,
◆ improved sanitation,
◆ good personal and food hygiene, and
◆ health education about how these infections spread.

Key measures to treat diarrhea include:
◆ giving more fluids than usual (oral rehydration) with salt and sugar, to prevent dehydration, and
◆ consulting a medical practitioner or health worker if there are signs of dehydration or other problems.

In rural India, during the last decade, public education through posters and other types of communication strategies has decreased the infant mortality rate due to diarrhea in several states. Posters depicting a child with diarrhea being given a salt and sugar solution orally to reduce death from dehydration have gone a long way in reducing serious conditions requiring hospitalisation and avoiding the use of intravenous fluids, thus reducing mortality.

Risks due to chemicals in food

Food contaminated by chemicals is an important public health concern. This contamination may occur through environmental pollution of the air, water or soil. Toxic metals, PCBs and dioxins, or the intentional use of various chemicals, such as pesticides, drugs and other agrochemicals have serious consequences on human health. Food additives and contaminants used during food production and processing also adversely affect human health.

Diseases spread by food: Some food-borne diseases, though well recognised, have recently become more common. For example, outbreaks of salmonellosis which have been reported for decades, have increased within the last 25 years.

While cholera has devastated much of Asia and Africa in the past, its resurgence for the first time in almost a century on the South American continent in 1991, is an example of a well-recognised infectious disease re-emerging in a region after decades. While cholera is often water-borne, many foods also transmit the infection. In America, iced and raw or under-processed seafood are important causes for cholera transmission.

Cancer and the environment

Cancer is caused by the uncontrolled growth and spread of abnormal cells that may affect almost any tissue of the body. Lung, colon, rectal and stomach cancer are among the five most common cancers in the world for both men and women. Among men, lung and stomach cancer are the most common cancers worldwide. For women, the most common cancers are breast and cervical cancers. In India, oral and pharyngeal cancers are the most common type of cancers, and are related to tobacco chewing. Cigarette and beedi smoking causes lung cancer

More than 10 million people are diagnosed with cancer in the world every year. It is estimated that there will be 15 million new cases every year by 2020. Cancer causes 6 million deaths every year or 12% of deaths worldwide.

The causes of some cancers are known. Thus, the prevention of at least one-third of all cancers is possible. Cancer is preventable by stopping smoking, ingesting healthy food and avoiding exposure to cancer-causing agents (carcinogens). Early detection and effective treatment is possible for a further one-third of cases. Most of the common cancers are curable by a combination of surgery, chemotherapy (drugs) or radiotherapy (X-ray technology). The chance of the cure increases if the cancer is detected at an early stage.

Cancer control is based on the prevention and control of cancer by:

◈ promoting and strengthening comprehensive national cancer-control programmes,
◈ building international networks and partnerships for cancer control,
◈ promoting organised, evidence-based interventions for the early detection of cervical and breast cancers in women,
◈ developing guidelines on disease and programme management,
◈ advocating a rational early approach to effective treatments for potentially curable tumours, and
◈ supporting low-cost approaches to respond to global needs for pain relief and palliative care.

Prevention of cancer: Tobacco smoking is the single largest preventable cause of cancer in the world. It causes 80% to 90% of all lung cancer deaths. Another 30% of all cancer deaths, especially in developing countries, include deaths from cancer of the oral cavity, larynx, esophagus and stomach, which are related to tobacco chewing. Preventive measures include bans on tobacco advertising and sponsorship, increased tax on tobacco products and educational programmes to reduce tobacco consumption.

7.2   CARBON FOOTPRINT

Human impacts on our environment are much larger than our footprint. This shows that human beings create a much larger impact on earth than is generally believed.

> **Learning by reflection**
>
> Think about your footprint on earth and find ways to reduce it.

Environmental footprint and handprints for change

Learning about the environment through case studies is an exciting way to understand and act in a more pro-environmental context in day to day life. The case studies demonstrate how our environment has been managed – sustainably or mismanaged – leading to environmental degradation, loss of species, pollution, ill health, inequality and poverty. A practical way to learn through a case study approach is to carry out a SWOT (Strength, Weakness, Opportunities and

Threat assessment) of a case based on the environment. Many case studies demonstrate how human society has been proactive and prudent. This may be by following laws that prohibit misadventures and illegal activities. Environmental crimes lead to legal action being taken and offenders are punished which encourages responsible citizens' actions. In India, citizens can file a PIL against an environmental offender and even take the Government to court for irresponsible environmental behaviour.

If you review important case studies, your own behaviour can be suitably modified to become more environmentally proactive by making appropriate observations of your environment. This leads to a heightened concern for the environment and sustainable actions.

7.3 RESETTLEMENT AND REHABILITATION OF PROJECT-AFFECTED PERSONS

Any major project such as a dam, mine, expressway or the notification of a National Park, disrupts the lives of the people who live in that area and often requires relocating them to an alternative site. Displacing people is a serious issue. It reduces their ability to subsist on their traditional natural resource base and also creates great psychological pressures. This is especially true of tribal people, whose lives are closely woven around their own natural resources and find it hard to adapt to a new way of life in an alternate ecosystem. Thus, no major project that is likely to displace people can be carried out without the consent of the local people. In India, lakhs of people have been arbitrarily displaced by the thousands of dams built since independence to drive the green revolution. The dams have been built virtually at the cost of these poor local people who have been powerless to resist these giant projects. The Government is expected to find 'good' arable land to resettle these displaced persons and provide them with an adequate rehabilitation package to recover from the disruption. This has rarely occurred to the satisfaction of the individuals affected by the project. In many cases across the country, this has not been implemented satisfactorily for decades.

Resettlement not only puts pressure on the project-affected people, but also on the people who have been living in the area that has been selected for resettlement. Thus, both the communities suffer and conflict over resources is likely to occur in the future. There are, however, situations where the communities themselves request that they be relocated to a new site. This is often observed where people live inside or on the periphery of a National Park or wildlife sanctuary. In these situations, such as the Gir in Gujarat, the local people have asked to be relocated to another site where they can live peacefully away from lions that kill their cattle. However, for decades the Government has been unable to find suitable areas where they can be shifted.

CASE STUDY 7.1   Narmada bachao andolan

Creation of the Narmada dam displaced thousands of local tribal farmers who depended on the waters of the Narmada. Their resettlement took several decades. Ms Medha Patkar fought for the rights of the tribal folk to improve their resettlement and rehabilitation.

CASE STUDY 7.2   Rehabilitation of people from tiger reserves

Major rehabilitation and resettlement projects have been successfully carried out to move people out of the core areas of the tiger reserves so that they can live a better life. This reduces the conflict between humans and wildlife. Crop damage by wild herbivores and cattle lifting by carnivores can be significantly redued. The current compensation to each family includes provision of ₹10 lakhs and agriculture land. This enables them to improve their livelihood, health care and education.

7.4 DISASTER MANAGEMENT

The Indian subcontinent is very vulnerable to droughts, floods, cyclones, earthquakes, landslides, avalanches and forest fires. Among the 36 states and union territories in the country, 22 are prone to disasters.

Defining disaster

'A serious disruption of the functioning of a community or a society involving widespread human, material, economic or environmental losses and impacts which exceed the ability of the affected community or society to cope using its own resources. (UNISDR 2009)

> **Definition of disaster**
> 'Disaster' is a catastrophe, mishap, calamity or grave occurrence in any area, arising from natural or manmade causes, or by accident or negligence which results in substantial loss of life, or human suffering, or damage due to and destruction of property, or damage to or degradation of environment, and is of such a nature or magnitude as to be beyond the coping capacity of the community of the affected area. (DM Act 2005)

Floods are the most frequently occurring natural disasters, due to the irregularities of the Indian monsoon. About 75% of the annual rainfall in India is concentrated in the three to four months of the monsoon season. As a result, there is very heavy discharge from the rivers during this period, causing widespread floods. Approximately 40 mHa of land in the country has been identified as being prone to floods. The major floods are caused in the Ganga–Brahmaputra basin, which carries 60% of the total river flow of our country. India has a long coastline of 5700 km, which is exposed to tropical cyclones arising in the Bay of Bengal and the Arabian Sea. The Indian Ocean is one of the six major cyclone prone regions of the world. In India, cyclones occur usually between April and May, and also between October and December. The eastern coastline is more prone to cyclones and is hit by about 80% of the total cyclones generated in the region.

Floods occur due to torrential monsoon rains. Cyclones that hit the eastern coastal areas repeatedly bring about immense damage to life and property. Monsoon rains hit coastal areas causing floods that kill people, damage their homes and croplands. This is followed by environment related gastrointestinal conditions such as diarrhea and dysentry, and vector borne diseases due to pooling of water which forms the habitat for malaria and dengue causing mosquitoes to breed.

There is a huge loss of life and property to floods in India. Floods are also the most frequent cause of disasters. They not only lead to immediate effects but have long lasting impacts on life, livelihood and the environment itself. The floods in Kashmir in 2014 submerged Srinagar. Mumbai was paralysed by incessant rain and flooding in 2005. The frequency of floods in river valleys in the plains and in Himalayan rivers cause the largest number of disasters which are clustered during the monsoon months. Annually India's economic loss from floods amounts to an average of ₹550 crores.

Earthquakes are considered one of the most destructive natural hazards. The impact of this phenomenon occurs with so little warning that it is almost impossible to make preparations against damages and the collapse of buildings. About 50%–60% of India is vulnerable to seismic activity of varying intensity. Most of the vulnerable areas are located in the Himalayan and sub-Himalayan regions.

Earthquakes are sporadic in nature and suddenness by which earthquakes occur make them difficult to predict. While their severity can be easily estimated, the impact is also related to the site and human density of the affected region. In 2001, Gujarat suffered an earthquake that took about ten thousand lives. The boundaries between tectonic plates create a seismic prone zone around the

Himalayas. These relatively young mountains have loose boulders which lead to landslides. Snow covered peaks give rise to avalanches. India also has earthquake prone areas along the western coasts. The development of settlements and the growth of villages along flood and earthquake prone Himalayan areas worsens these natural calamities.

The term **tsunami** comes from the Japanese language, meaning harbour (*tsu*) and wave (*nami*). A tsunami can be generated by any disturbance that rapidly displaces a large mass of water, like an undersea earthquake, volcanic eruption or submarine landslide. The giant wave produced travels across the ocean at speeds of 500–1000 km/h. As the wave approaches the land, it 'compresses' – sometimes up to a height of 30 metres – and the sheer weight of the water is enough to crush objects in its path, often reducing buildings to their foundations and damaging the exposed sea shore down to the bedrock. The tsunami on 26 December 2004 killed 310,000 people, making it the deadliest tsunami in recorded history. It hit the Andaman Nicobar Islands and the east coast of India causing an enormous amount of destruction.

Cyclones occur frequently along the coast of India. Thousands of people suffer from large and small cyclonic storms. These tropical storms from neighbouring seas sweep over our coasts hitting villages, towns and cities. The floods in the plains recede slowly over days making the land saline and cause increasing health management concerns. In 1999, Orissa suffered the worst ever super cyclone affecting the lives of lakhs of people. Cyclonic winds at 250 kmph raced over a 250 km stretch of coastal land and killed an estimated ten thousand.

Landslides hit the hilly areas of the Himalayas and the Western Ghats. These catastrophes affect people living in vulnerable hill slopes and mountains and leave large numbers of people dead and many more homeless.

Other disasters such as droughts have longer term ill effects due to water shortages and crop failure, leading to starvation over several months or even years. **Droughts** are a perennial feature in some states of India; 16% of the country's total area is drought prone. Drought is a significant environmental problem as it is caused by a lower than average rainfall over a long period of time. Most of the drought prone areas identified by the government lie in the arid and semi-arid areas of the country.

The erratic monsoonal climate is a major cause of drought. Droughts can occur year after year leading to death by starvation. Droughts are frequently followed by diseases due to contamination of water. Water is an important life-supporting resource that we take for granted till a drought occurs. Drinking water shortages can cause acute problems. Loss of the annual crop leads to non-payment of the loans that farmers take to grow crops. This has led to several farmers committing suicide. India has initiated Watershed Development Programmes so that soil and water conservation measures can lessen the ill effects of droughts by permitting rain water to penetrate into the subsoil and be held in natural underground aquifers. Deforestation enhances the ill effects of droughts.

Droughts have impacted human life in India for thousands of years. In spite of the development strategies, our country is still hit repeatedly by droughts. Monsoon climates are inherently variable and not easily predictable. Forecasting a drought period for an area remains highly speculative. Thus, providing resilience to drought conditions alone can minimise the immensity of these calamitous conditions. Drought hit areas bring about a disruption in agricultural produce as well as high livestock mortality. Human mortality results from starvation, malnutrition, disease and lack of potable water.

The National Agricultural Drought Assessment and Monitoring System (NADAMS) is a key mission that helps by responding to a drought situation. Several types of satellite data help in predicting and suggesting the cumulative damage that could occur due to drought conditions.

Approximately 70% of sown areas in the country are considered drought prone and are affected every 3–4 years. This affects the most disadvantaged farmer community who is dependent on rain-fed traditional forms of agriculture. Most of these lands are still not under irrigation. Monsoon rainfall distribution shifts sporadically and is relatively unpredictable.

Management of drought and drought prone areas: As scientific evidence has shown where the drought prone belts in India lie, it is possible to have a management system based on preparedness and early warning. The immediate short-term measures include monitoring of early warning systems and providing assistance to the drought prone areas. Under long-term management, such drought prone areas must have a mitigation strategy through irrigation, soil and water regime management, afforestation and development projects. Altering the crop pattern can help limit the agony of farmers suffering from drought.

Drought relief: State Governments monitor the monsoon and document drought conditions. The Centre provides drought relief measures. Relief measures include providing food grains from neighbouring areas that have sufficient grain. A substitute economic drive comes from employment generation schemes. In the past, traditional agricultural practices of planting several crops rather than a single crop, was a means to mitigate and adapt to drought conditions. Even if one of the crops failed, the drought resistant varieties of crops would tide over the crisis. Alternative food was gathered from forests where wild relatives of food crops were collected. Food substitutes were collected from wild vegetables, yams and tubers found in the forest. A comprehensive drought management programme must thus include a preparedness phase, a prevention phase, and a relief plan.

Human-induced disasters occur from industrial accidents, plane, train and bus accidents, fires in cityscapes, house collapses and accidents at sea from shipping due to oil spills, explosions and fires at industrial sites. While each of these accidents individually contributes to a relatively smaller number of deaths than floods or drought, the total number across India is huge. Traffic accidents though not generally thought of as disasters, lead to a more regular loss of life across India in comparison to the natural disasters. A stampede at a pilgrimage site or during a fire can lead to serious loss of life and property. Flouting environmental laws by an industry can lead to a large environmental disaster. At a single disaster in Bhopal, the poisonous gas leak led to the loss of life of 3787 people.

Nuclear accidents and nuclear holocaust are not regular occurrences but when they do occur, they are catastrophies of unimaginable proportions. Nuclear energy was researched and developed as an alternate source of clean and cheap energy compared to fossil fuels. Along with the benefits of nuclear energy came its destructive effects due to accidents. In the short history of nuclear energy, there have been a number of accidents that have surpassed any natural calamity in their impact. A single nuclear accident causes loss of life, long-term illnesses and destruction of property on a large scale which continues over a long period of time. Radioactivity and its fallout leads to cancer, genetic disorders and death in the affected area for decades after the accident, thus affecting all forms of life for several generations.

The use of nuclear energy in war has had devastating effects on humans and on the earth. In the only use of nuclear power in war, the United States dropped two atomic bombs over the Japanese towns of Hiroshima and Nagasaki in 1945. These killed thousands, left many thousands injured and destroyed everything for kilometres around the bombed sites. The effects of the radiation can still be seen today in the form of cancer and genetic mutations in the children of the survivors of the incident.

Steps towards mitigation of disasters

The speed of delivery of active measures and relief through well-coordinated actions from various government departments is the key to reducing loss of life. This begins with timely information to the appropriate authorities to provide urgent assistance. Geoinformatic platforms provide the key to this type of information so that disaster relief can be initiated even before the peak of the disaster is reached.

All these disasters have severe economic, social and environmental costs. This can only be mitigated if we are prepared for these disasters 24×7 so that when they do occur, the machinery, manpower, expertise and governance mechanisms can move swiftly into action. Many of the natural disasters are augmented and exaggerated by human misuse of land. By avoiding building towns and cities close to the coast, we can minimise loss of life and resources when cyclones occur. Disruption by cyclones, floods and avalanches are due to mismanagement of landuse. There are known areas that are frequently hit by these natural conditions. Thus, planning of development must mark these critical areas as 'no development zones'. The Coastal Regulation Zone and the rules for a setback in flood prone areas mitigate the effects of these terrible events. Forest cover tends to hold water in the soil. Thus, forests should be preserved.

The GOI enacted a comprehensive 'Disaster Management Act' in 2005. A 'National Policy on Disaster Management' was passed in 2009. This changed a relief centric approach to a more proactive risk reduction strategy and action plan. A National Disaster Management Authority headed by the PM was set up in 2005. The National Disaster Management Plan (NDMP) was developed further in 2016. The frequency of extreme weather events has led to incorporating disaster risk reduction in Sustainable Development Goals (SDGs) and it was discussed in CoP 21-Paris Agreement, as this phenomenon is linked to climate change. It draws upon the Sendai Framework for Disaster Risk Reduction planned at the global level for 2015–2030.

The Sendai approach envisages the following.

⬥ Understanding disaster risks.
⬥ Strengthening disaster risk governance.
⬥ Investing in disaster risk reduction.
⬥ Enhancing preparedness for an effective response.

This strategy includes recovery, rehabilitation and reconstruction. Its aim is to reduce risk, loss of life, livelihoods and health risks. The framework addresses economic, physical, social, cultural and environmental aspects that affect people, business, communities and nations. India has a National Institute of Disaster Management and a National Disaster Response Force.

Vision of the NDMP

The primary aim is to make India disaster resilient, reduce risk, decrease loss of life, livelihoods and assets—attend to economic, physical, social, cultural and environmental aspects. India, due to its geography, topography and monsoonal climate variability, is one of the world's most disaster prone areas—NDMP 2016. The role of the central agencies is to support the disaster affected states.

Response measures: Immediate action on receiving an early warning and anticipating an impending disaster, or post disaster, if there was no warning when the disaster was too sudden.

Focus: Search, rescue and evacuate victims, or those likely to be affected by a secondary disaster.

⬥ Identify roles and responsibilities of congruent ministries, agencies and voluntary organisations.
⬥ Provide emergency support.
⬥ Coordination.

Recovery and 'building back better':

⬥ Rehabilitation approach has been altered to better reconstruction.
⬥ Incorporating better risk reduction into redevelopment.

Capacity development: Strengthening institutions for,
◇ prevention or mitigation of risks,
◇ response preparedness,
◇ effective responses, and
◇ recovery based on 'build back better'.

Financial arrangements: Disaster relief is a state responsibility supplemented by the GOI through a National Disaster Response Fund provided to each district which has been planned and organised for mitigative measures.

Structure of NDMP

The Plan incorporates several key issues and suggests ways in which the country can be better prepared for managing different types of disasters. The components include:

(i)   introduction to disaster management plan,
(ii) hazard risk and vulnerability profile of India,
(iii) reducing risk, enhancing resilience,
(iv) planning needs for preparedness and response,
(v)   strengthening governance,
(vi) planning effective recovery—'building back better',
(vii) capacity development,
(viii) financial arrangement,
(ix) international cooperation, and
(x)   maintaining and updating the plan.

The disaster management cycle

The organisation, planning and application of measures preparing for, responding to and initial recovery from disasters is termed disaster management.

Aspects of disaster management

◇ Disaster risk reduction
◇ Disaster risk management
◇ Disaster preparedness
◇ Disaster responses
◇ Post disaster recovery

Disaster Management Act 2005

Pre-disaster → Disaster → Rapid response → Post disaster activities → Mitigation → Recovery through 'build back better' → Prevention and risk reduction

(Adapted from NDMP 2016)

The repeated disasters that led to the loss of life and property led to the need for an Act that would lead to a more comprehensive approach to disasters. It has become evident that though most disasters were from natural untoward events, it was the mismanagement of the environment by several human-induced mistakes in environmental planning, implementation and actions that had led to enhancing the ill effects of natural disasters, loss of life and property. The Act thus addressed, types of disaster, triggers, natural hazards, climate change + enhanced weather events, human induced hazards, combined factors + natural hazards aggravated by human-induced hazards, and vulnerability of human life to chemical/biological/radiological/nuclear disasters.

Natural disasters are of five categories—(i) Geographical, geological processes, (ii) hydrological events, (iii) meteorological—small atmospheric processes, (iv) climatologically—long lived effects, and (v) biological—organic origin or caused by pathogenic vectors, bioactive substances.

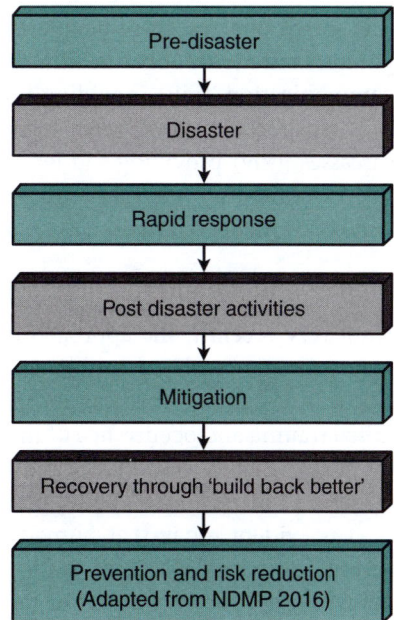

Natural hazards typologies can be cyclones and winds, tropical cyclones, storm surges, floods—urban, earthquakes, tsunamis, landslides and avalanches, droughts, and cold waves and frost.

Human induced disasters include chemical (industrial) disasters, nuclear and radiological emergencies (NRE), and fire risks.

Preparedness and response

Preparedness: United Nations International Strategy for disaster reduction (UNISDR), defines preparedness as 'the knowledge and capacities developed by governments, professional response and recovery, organisation, communities and individuals to effectively anticipate, respond to and recover from the impacts of likely, imminent or current hazardous events or conditions'.

Response measures: The actions taken immediately after receiving an early warning from a relevant authority includes, 'The provision of emergency services and public assistance during or immediately after a disaster in order to save lives, reduce health impacts, ensure public safety and meet the basic subsistence needs of the people affected' — United Nations International Strategy for Disaster Reduction (UNISDR).

Disaster management tasks: The tasks include, (i) early warning with the use of maps/satellite images and information dissemination, (ii) evacuation of people and animals (domestic and wild), (iii) search and rescue of people and animals, (iv) medical care, (v) drinking water/dewatering pumps/sanitation/public health facilities, (vi) providing food and essential supplies, (vii) communication, (viii) providing housing and temporary structures, (ix) power supply, (x) fuel, (xi) transportation, (xii) relief logistics and supply chain management, (xiii) disposal of animal carcasses, (xiv) providing fodder for livestock during scarcity, (xv) rehabilitation and ensuring safety of animals—veterinary care, (xvi) data collection and management, (xvii) relief employment, and (xviii) media relations.

From management to mitigation of disasters

Until very recently, the approach towards dealing with natural disasters has been post-disaster management, involving problems such as warnings, evacuation, communications, search and rescue, fire-fighting, medical and psychiatric assistance, provision of relief and shelter. After the initial trauma and occurrence of the natural disaster, the reconstruction and rehabilitation is done by people themselves, NGOs and the Government. The memories are considered history and irrelevant for learning on better prevention of future disasters.

It is evident today that human activities are responsible for accelerating the frequency and severity of natural disasters. Natural occurrences such as floods, earthquakes and cyclones will always occur. They are a part of the environment that we live in. However, the destruction from natural hazards can be minimised by the presence of well-functioning warning systems combined with preparedness on the part of the community that is the potential target. Although traditionally disaster management consisted primarily of reactive mechanisms, the past few years have witnessed a gradual shift towards a more proactive, mitigation-based approach.

Disaster management is a multidisciplinary area in which a wide range of issues including forecasting, warning, evacuation, search and rescue, relief, reconstruction and rehabilitation are included. It is multi-sectoral as it involves administrators, scientists, planners, volunteers and communities. These roles and activities span the pre-disaster, disaster and post-disaster periods. Since their activities are complementary as well as supplementary to each other, there is a critical need for coordinating these activities.

In order to transfer the benefits of scientific research and newer developments to the risk-prone communities, links must be developed between scientific organisations and field agencies. Coordination between government agencies and NGOs needs to be built so that any unnecessary overlap of activities may be avoided and ongoing links between the Government and communities are established.

There are a number of early warning systems for anticipating a range of natural hazards. Although they are more accurate than before and can help in prediction, it is not enough to ensure that communities are safe from disasters. This is where disaster mitigation can play an important role. *Mitigation* includes lessening the negative impact of the natural hazards. It is defined as sustained action taken to reduce long term vulnerability of human life and property from natural hazards. While the preparatory response and the recovery phases of emergency management relate to specific events, mitigation activities have the potential to produce repetitive benefits over time. Certain guidelines, if followed correctly, can result in an effective mitigation programme.

⬥ Pre-disaster mitigation can help in ensuring faster recovery from the impact of disasters.
⬥ Mitigation measures must ensure protection of the natural and cultural assets of the community.
⬥ Hazard reduction methods must take into account the various hazards faced by the affected community and their desires and priorities.
⬥ Any mitigation programme must also ensure an effective partnership between the Government, scientific community, private sector, NGOs and the community.

Risk assessment and vulnerability analysis: This involves the identification of hotspot areas of prime concern, collection of data on past natural hazards, information of the natural ecosystems and information on population and infrastructure. Once this information is collected, a risk assessment should be done to determine the frequency, intensity, impact and time taken to return to normalcy after the disaster. The assessment of risk and vulnerabilities will need to be revised periodically. A regular mechanism will therefore have to be established for this. The use of Geographical Information Systems (GIS), a computer-based programme, can be a valuable tool in this process as the primary data can be easily updated and the corresponding assessments can be made.

Applied research and technology transfer: There is a need to establish or upgrade observation equipment and networks, monitor the hazards, improve the quality of forecasting and warning, disseminate information quickly through the warning systems and undertake disaster simulation exercises.

Space technologies such as remote sensing, satellite communications and Global Positioning Systems (GPSs) have a very important role to play. Government organisations such as ISRO (Indian Space Research Organisation) play a vital role. Similarly, other government organisations like the National Building Research Organisation, the Meteorological Department and Irrigation Department can undertake applied research for devising locale-specific mitigation strategies in collaboration with educational institutions or universities.

Public awareness and training: One of the most critical components of a mitigation strategy is the training to be imparted to the officials and staff of the various departments involved at the state and district levels. This enables the sharing of information and methodology. The success of a mitigation strategy will depend to a large extent on inter-sectional, inter-departmental coordination and efficient team work. Thus, a well-designed training programme taking into account the gaps in knowledge, skills and attitude with respect to the various tasks that need to be undertaken, is a vital component.

Institutional mechanisms: The most important need at the national level is to plan, strengthen and develop the capacity to undertake disaster mitigation strategies. There is a need to emphasise proactive and pre-disaster measures rather than focus only on post-disaster responses. It is thus essential to have a permanent administrative structure that can monitor the developmental activities across congruent departments and provide suggestions for necessary mitigation measures. The National Disaster Management Centre (NDMC) has to perform these wide-ranging tasks. Professionals such as architects, structural engineers, doctors and chemical engineers (who are involved with management of hazardous chemicals), must be asked to form groups that can design specific mitigation measures.

Incentives and resources for mitigation: To a very large extent, the success of mitigation programmes will depend upon the availability of continued funding. Thus, there is a need to develop mechanisms to provide stable sources of funding for all mitigation programmes. This includes incentives for the relocation of commercial and residential activities outside the disaster-prone areas. Housing finance companies should make it mandatory for structures in such hazard-prone areas to follow special building specifications. The introduction of disaster-linked insurance should be explored and should cover not only life but also household goods, cattle, structures and crops.

Land-use planning and regulations: Long term disaster reduction efforts should aim at promoting appropriate landuse in the disaster prone areas. These are in industrial, urban residential areas and rural sectors. Maintenance of wetlands as buffer zones for floods, creation of public awareness of proper landuse practices and formation of landuse policies for long term sustainable development are all imperative.

Hazard-resistant design and construction: In areas that are prone to disasters, protection can be enhanced by the careful selection of sites and building technologies. Thus, it is essential to promote the knowledge of disaster-resistant construction materials, techniques and practices among engineers, architects and technical personnel.

Structural and constructional reinforcement of existing buildings: It is possible to reduce the vulnerability of existing buildings through minor adaptations or alterations, thereby ensuring their safety. This can be done by inserting walls on the outside of the building, buttresses, walls in the interior of the building, portico fill-in-walls, specially anchored frames covering columns and beams, constructing new frame systems, placing residential electrical equipment above the flood level and designing water storage tanks to be able to withstand cyclonic winds, earthquakes and floods.

7.4.1   Floods

Monsoonal rainfall patterns and irregularity combined with topographic situations through which our rivers flow from their origins in the mountains, through flat plains and into the sea, are the root causes of floods. However, human induced activities that are imprudent and short-sighted add to the impact of floods. As water is a driver of agricultural societies, urban development and industry, it is evident that some of India's most populous areas are on river banks. This makes such areas risk prone. While the frequency and magnitude of floods appears to be a part of climate change due to extreme weather events, the actual number of people affected is related to population expansion near river courses.

The Government of India has initiated several strategies to minimise the effects of floods. It is the unpredictable nature that leads to floods. Damage to life and property occurs as society continues to occupy flood prone areas.

There are several complex physical, geographical and hydrological interlinkages that are a part of how rivers behave after torrential rains hit an area. Thus, recognising the importance of minimising the damage, if and when rivers expand over lands that include residential, agricultural and industrial areas, can limit damage to human life by appropriate planning. In natural landscapes, rivers have flood plains and the high flood line beyond the normal level of the river. This permits the river to expand and flood adjacent areas without affecting people. Once such natural flood plains are altered to other forms of landuse, the ill effects are felt on society. Thus, unsustainable development for economic, social gains, or lack of caring for the environment adds enormously to the mortality and loss of livelihood for people who have spread into these flood prone areas.

As our rivers originate in the Himalayas, Western Ghats and Central Highlands, it is the people of the low lands that bear the brunt of this topographic situation. However, the people on the banks of rivers in the Himalayas, especially in areas where deforestation makes the riverbanks highly erosion prone, are also affected by landslides during heavy downpours.

The Central Water Commission has developed maps of flood prone areas. These show that the flood prone rivers are in the Indo-Gangetic plains, the Brahmaputra valley and river deltas. The central parts of India including Gujarat, Madhya Pradesh, Maharashtra, Odisha and Andhra Pradesh are considered 'rain storm' areas.

Every year between six and eight floods occur somewhere in India, usually between August and September. The country also sees intense short-lived weather events and storms causing catastrophic local flooding. This affects several cities paralysing the lives of urban people over several days. Flooding is often related to inadequately managed drainage systems. This is referred to as the *flash flood magnitude index* of a river.

Types of floods

(i) *Rainfall floods:* These occur annually during the monsoon around many flood prone rivers for a number of days. The Brahmaputra, due to its wide flood plain overflows frequently. In Assam, the Brahmaputra destroys crops, but also enriches paddy soil with fresh silt. It also damages the surrounding high grassland Terai vegetation. This is the habitat of several endangered species in the Kaziranga National Park such as elephants, rhinos and swamp deer. Many animals and their young have to be rescued from flooded areas every year by the Forest Department and cared for in the rescue centre set up by Wildlife Trust of India.

(ii) *Rainstorm floods:* The frequency of thunder claps leading to exceptional torrential rains over a couple of days are the cause of rainstorm floods which are associated with cyclonic storms that begin in the Bay of Bengal. These affect the east coast repeatedly.

(iii) *Coastal floods:* These affect the eastern coast disrupting both agricultural people and fisherfolk who are frequently lost at sea.

(iv) *Dam burst floods:* These floods have devastated villages and even towns and cities. They occur if small earthen dams give way causing local flooding. Larger old dams have also given way and caused serious flooding. Thus, all dams over a hundred years old are strengthened to prevent untoward effects. Although uncommon, whenever this has occurred over the last 100 years, the effects have been catastrophic. In some Himalayan rivers, excessive erosion of deforested mountain slopes has formed dams of silt and rocks downstream. These then give way after further heavy rains and create floods downstream. The root cause of such events is deforestation.

(v) *Extreme weather events:* These are meteorological causes of floods. The monsoon is unpredictable due to the large *El Nino* weather patterns in the oceans. Monsoon depressions, cyclones and mountain barriers affect these weather events leading to storms and floods. This occurs in the Ganga basin and the Punjab plain but sporadically affects other areas in Central India and the Northern half of Peninsular India.

(vi) **Landuse alterations:** These are anthropogenic (human) effects of landuse and landcover changes that add to the ferocity of thunder storms. Embankments built to contain rivers and prevent expansion into natural flood plains give way during heavy rains, leading to an increase in the speed and expansion of water into agricultural and urban landscapes. Breaching of such embankments is a frequent cause of serious local flooding.

Short-term, mid-term and long-term effects of floods

Apart from the immediate catastrophic death toll from rivers that breach their banks into human landscapes, there are several side events that require to be dealt with.

Short-term effects: In the immediate aftermath, the lives of people are seriously compromised. Health care and transport (both rail and road) are disrupted. Thus, diseases that spread in the aftermath of the disaster do not receive immediate attention and the impact of the disaster increases manifold. Drinking water and food are in short supply and what is available is frequently contaminated and spreads gastrointestinal diseases. Rodents and pests take over grain stores. The loss of transport due to roads and bridges that are damaged prevents access to health care. The armed forces are often deployed to clear debris and save lives. This requires timely rapid action.

Mid-term effects: In the mid-term, hospitals, water supply lines, sewage plants, schools and other infrastructural life support are denied for weeks or months. This causes severe hardships for flood victims.

Long-term effects: In the longer term, recovery from both physical and psychological trauma suffered by people is not quantifiable. People who lose their family and friends, their homes, their agricultural land and its produce or their livestock, rarely get adequate compensation. Some victims put it down to ill fate or their gods' vengeance, but others see it as the government's mismanagement. This requires counselling over long periods of interaction with the affected community for results to be achieved. The lives of children who have lost their parents or those who have lost their education facilities are among the saddest late impacts on those people living in the flooded areas.

Flood management in India

The severe hazards occurring due to floods in the country, was first studied in the 1950s. This led to the National Flood Control Programme in 1954 which looked at engineering and structural measures with little hard data to go by. In more recent years, satellite data has been in use for mapping floods and damage assessment. The inability to manage floods more effectively is that 24×7 multi-focal data over long periods is still not available despite the fact that a large proportion of India's population is concentrated in riverine and coastal habitats that are inherently flood prone. This requires an integrated multidisciplinary flood forecasting, warning system, management policy, strategies for prediction and mitigation through several actionable programmes which must be dealt with by different national and state level government departments. This has to be a part of sound environment management. If this is not stressed, India will continue to lose lives, suffer serious environmental disasters and have severe long-term environmental degradation.

Disaster prevention: Current strategies use measures that involve the structural change in river flows, such as dams, barrages and embankments. While this does reduce the ferocity of floods to some extent, it has long-term ill-effects on riverine and riverside ecology. Siltation is altered and bank erosion is exacerbated. Breaching of such flood barriers adds to the severity of floods.

Mitigation methods

The lower plain regions of India, in particular Bihar, Uttar Pradesh and West Bengal with respect to the river Ganga, and Assam with respect to the Brahmaputra, suffer from adverse effects of floods every year. The Ganga–Brahmaputra basin receives the maximum run-off in the three monsoon months. Based on hydrological studies, it is estimated that only 18% of the rain water can be stored in dams or reservoirs, while 82% of the rain water flows through rivers and ultimately into the sea. Floods will therefore be a recurring phenomenon in our country.

Floods can be caused by natural, ecological or anthropogenic factors, either individually or as a combined result. Anthropogenic activities, such as deforestation and shifting cultivation also contribute to floods. Forests on the hill-slopes normally exert a 'sponge effect', soaking up the abundant rainfall and storing it before releasing it gradually over several months. However, when forests are cleared, the rivers turn muddy and swollen during the wet monsoon season and run dry later on in the year during the drier periods. A large proportion of the rainfall is therefore released immediately after precipitation in the form of floods. The mitigation measures for floods include both structural and non-structural measures.

The structural measures include:

- reservoirs for impounding monsoon flows to be released in a regulated manner after the peak flood flow passes,
- prevention of over-bank spilling by the construction of embankments and flood walls,
- improvement of flow conditions in the channel and anti-erosion measures, and
- improved drainage.

The non-structural measures include:

- flood-plain management such as flood plain zoning and flood proofing, including disaster preparedness,
- maintaining wetlands,
- flood forecasting and warning services,
- disaster relief, flood fighting and public health measures, and
- flood insurance.

7.4.2   Earthquakes

After the devastating earthquake struck Gujarat on 26 January 2001, rehabilitation has been given greater attention. Gujarat's experience has shown that building shelters that are less vulnerable to earthquakes should also take into consideration the specific needs of the victims, instead of merely being a top-down approach. The role of NGOs in this is very important. Their strength lies in their resources and informality in operations. Their ability to reach out to the community and their sensitivity to local traditions is a key asset in such situations.

7.4.3   Cyclones

In India, tropical cyclone warning was initiated in 1875. The monitoring process has clearly arrived since the advent of space programmes, remote sensing and satellite tracking.

Tropical cyclones result from the development of low pressure areas in the atmosphere over warm oceans. The cyclones are associated with strong winds, lashing rain that mainly hits the coastal belts when the cyclone reaches land. Cyclones are most common in November but also occur in May. Studies on cyclones require data from the surface of the sea, from the upper atmosphere which uses aircraft reconnaissance, information from shipping, radar observations and now from sophisticated satellite data.

Structure of the tropical cyclone

The cyclone core is the centre of the cyclone and is surrounded by a spiral of precipitation. The inner winds can reach 90 m/s. There are different methods to try and predict the track that cyclones are likely to take. Most cyclones tend to follow the same track of those in the past.

Mitigating damage by cyclones

An *'Early Warning System'* is the key to preparedness in coastal and marine environments. Different agencies can thus take timely actions. Communication and early response are an essential component of reducing damage to lives. This must reach ships at sea, trawlers and small fishing vessels, through the coast guard, navy, commercial shipping, port authorities, airlines and airports, coastal governing bodies and the press.

Communication during and after disasters has several components. It must be delivered in an understandable language, must be easily received by potential victims and be acted on immediately. This can be done by radio and Very Small Aperture Terminal (VSAT) technology, telefax and websites. The country has a pre-cyclone watch and a 'four stage warning system'. This reaches the chief secretaries of high risk states so that highly coordinated action can be taken in time. A 'Cyclone Alert' is usually provided to the authorities 48 hours prior to commencement of the storm.

Response time

It is known that warnings often go unheeded at the recipient level. There is an absence of stimulus–response reaction where the receiver assesses his/her own personal risk. Administrative responses are also unpredictable and often driven by political concerns. Crisis management teams must not only be informed but should be able to mobilise a response in the fastest possible time. Kolkata, Chennai and Mumbai have 'area cyclone warning centres'.

Mitigation measures

Tropical cyclones are the worst natural hazards in the tropics. They are large revolving vortices in the atmosphere extending 150–1000 km horizontally and 12–14 km vertically from the surface of the earth. These are intense low pressure areas. Strong winds spiralling anti-clockwise in the northern hemisphere blow around the centre of the cyclone at the lower level. At the higher levels, the rotation is the opposite. They generally move 300–5000 km per day over the ocean. While moving over the ocean, they pick up energy from the warm water of the ocean and some of them grow to a devastating intensity. On an average, about 5–6 tropical cyclones form in the Bay of Bengal and the Arabian Sea every year, out of which 2–3 may be severe. More cyclones form in the Bay of Bengal than in the Arabian Sea. The main dangers from cyclones are very strong winds, torrential rains and high storm tides. Most of the causalities are caused by coastal inundation by storm tides. This is often followed by heavy rainfall and floods. Storm surges cause the greatest destruction.

Although one cannot control cyclones, the effects of cyclones can be mitigated through effective and efficient mitigation policies and strategies.

Installation of early warning systems: Warning systems installed along the coastlines can greatly assist forecasting techniques, thus helping in early evacuation of people in the storm surge areas.

Developing communication infrastructure: Communication plays a vital role in cyclone disaster mitigation. However, this is one of the first services that gets disrupted during cyclones. Amateur radio has today emerged as a second-line, unconventional communication system and is an important tool for disaster mitigation.

Developing shelter belts: Shelter belts with plantations of trees can act as effective wind and tide breakers. Apart from acting as effective wind breakers and protecting crops from being damaged, they also prevent soil erosion. A major component is in vegetating coastal areas with mangrove plantations.

Developing community cyclone shelters: Cyclone shelters at strategic locations can help in minimising the loss of human life. In the normal course of life, these shelters can be used as public utility buildings.

Construction of permanent houses: There is a need to build appropriately designed concrete houses that can withstand high winds and tidal waves.

Training and education: Public awareness programmes that inform people about their response to cyclone warnings and preparedness can go a long way towards reducing causalities.

Landuse control and settlement planning: Ideally, no residential and industrial units should be permitted in the coastal belt (5 km from the sea) as it is the most vulnerable belt. No further growth of settlements in this region should be permitted. Major settlements and other important establishments should be located beyond 10 km from these risk-prone areas.

7.4.4 Landslides

Landslides are recurring phenomena in the Himalayan region. In recent years, however, intensive construction activities and the destabilising of hill slopes due to deforestation have aggravated the problem. Landslides occur as a result of sudden or gradual changes on a slope, either in its composition, structure, hydrology or vegetation. The changes can be due to geology, climate, weathering, changing landuse or earthquakes.

A significant reduction of the hazards caused by landslides can be achieved by preventing the exposure of people and facilities to landslide-prone areas, and by physically controlling the landslides. Developmental programmes that involve modification of the topography, exploitation of natural resources and change in the balance load on the ground should not be permitted. Some critical measures that could be undertaken to prevent future landslides are drainage measures, erosion control measures such as check dams, terracing, jute and coir netting, and rock fall control measures such as grass plantation, vegetated dry masonry walls, retaining walls and, most importantly, preventing deforestation and improving afforestation.

Disasters cannot be totally prevented. However, early warning systems, careful planning and preparedness on the part of the vulnerable community helps in minimising the loss of life and property due to disasters.

CASE STUDY 7.3   Bhopal gas tragedy

The Bhopal gas tragedy in 1984 was a catastrophe that had no parallel in the world's industrial history. On 3 December 1984, a poisonous grey cloud from the Union Carbide Plant in Bhopal, Madhya Pradesh (India) covered large areas around the plant. Forty tons of toxic gas (methy isocyanate, MIC) was accidentally released from the Union Carbide's Bhopal plant, leading to severe respiratory problems due to the suffocating gas. An estimated 10,000 people died and 500,000 people suffered agonising ill effects with disastrous life-threatening conditions.

CASE STUDY 7.4 Bhuj earthquake

The 2001 Gujarat earthquake occurred on 26 January. The epicentre was 9 km south-southwest of the village of Chobari in Bhachau Taluka, Kutch district, Gujarat, India. The earthquake killed between 13,805 and 20,023 people (including 18 in south-eastern Pakistan), injured another 167,000 and destroyed nearly 400,000 homes.

CASE STUDY 7.5 Forest fire in Uttarakhand

In 2016, several forest fires occurred across Uttarakhand. These fires occurred mainly in pine forests. It led to the generation of large clouds of smoke. As a course of action, the Government deployed the National Disaster Response Force and made use of Indian Air Force Mi-17 helicopters fitted with Bambi buckets to control fires with water. According to the Forest Department, 3,500 hectares (8,600 acres) of forest was severely burnt. Nearly 1,600 incidents of fires were detected across the mountain slopes.

CASE STUDY 7.6 Landslides in Maharashtra

A landslide in the village Malin of Pune district occurred on 30 July 2014. The landslide was believed to have been caused by a large cloud burst. 200 people were believed to have been buried in the landslide in 44 separate houses. The reason has been attributed to deforestation of the upper slopes of the hills of Bhimashankar Wildlife Sanctuary.

CASE STUDY 7.7 Oil spill in Mumbai

The 2010 Mumbai oil spill occurred after the Panama-flagged MV MSC Chitra and MV Khalijia-III collided off the coast of India near Mumbai. MSC Chitra, which was outbound from South Mumbai's Nava Sheva port, collided with the inbound Khalijia-III, which caused about 200 cargo containers from MSC Chitra to be thrown overbroad into the Arabian Sea. Khalijia-III was apparently involved in another mishap on 18 July 2010. A spreading oil slick devastated the biodiversity of marine life and covered the beaches of the coast.

CASE STUDY 7.8 Tsunami at Andaman and Nicobar Islands

This disaster affected the coastal and forest ecosystems and disrupted the life of the locals who were engulfed by a giant and sudden wave from the oceans. According to official estimates in India, 10,136 people were killed and hundreds of thousands made homeless when the tsunami was triggered in the Indian Ocean due to an earthquake in 2004. The earthquake registered 9.1–9.3 on the Richter scale and was the largest in five decades.

CASE STUDY 7.9 Cyclone at Odisha

The 1999 Odisha cyclone was the strongest recorded tropical cyclone in the North Indian Ocean and among the most destructive in the region to date. The highly favourable conditions intensified the storm and it attained super cyclonic storm intensity on 28 October. The next day the increasing intensity of the storm led to a raging disaster with 260 km/h (160 mph) wind and a record-low pressure of 912 mbar (hPa; 26.93 mmHg). The storm maintained this intensity as it reached the coast of Odisha on 29 October. The cyclone steadily weakened due to persistent land interaction and dry air, remaining more or less stationary for two days, before slowly drifting offshore as a much weaker system. The storm finally dissipated on 4 November over the Bay of Bengal damaging homes, farms and plantations of hundreds of poor coastal inhabitants.

CASE STUDY 7.10   Brahmaputra floods

The 2012 Brahmaputra floods were an unprecedented event along the Brahmaputra river. It killed 124 people by the extensive and sudden flooding accompanied by landslides, and about six million people were displaced. Flooding significantly affected the Kaziranga National Park, where 540 animals died including 13 rhinos. Helicopters were deployed to drop food supplies for nearly 10,000 affected people in six villages where highway access was cut off by the flooding, 550 km west of Guwahati.

7.5 ENVIRONMENTAL MOVEMENTS

Environmental movements have led to measured changes in the governance of environment management in India over the last several decades. These have been led by NGOs (also known as community-based organisations). These movements have led to support for local peoples' needs and equity bioresource distribution. These movements have frequently had local leaders who have supported the rights of the poverty stricken people.

7.5.1 The Chipko Movement

About 300 years ago, a ruler in Rajasthan decided to fell the khejri trees in his state to create lime. The local women, led by a Bishnoi woman, Amrita Devi, clung to the trees to prevent them from being felled by soldiers sent by the ruler. They were dependent on the scarce resources from these trees. The women were ruthlessly massacred. It is said that the ruler later realised the importance of these trees for his own people. The story, however, was remembered and revived in the 1970s when massive felling of trees for timber in the Himalayas prompted local women, supported by people such as Sunderlal Bahuguna and Chandi Prasad Bhat, to lead a people's movement to prevent deforestation by timber contractors in the region. They called their movement the *Chipko movement* in memory of the event during which women had clung to their trees and given up their lives. The movement followed the path the 300 Bishnoi women had taken three centuries ago in Rajasthan.

Chipko is a movement primarily begun and supported by the local women in the hills of Uttarakhand and Garhwal, where the women who are the traditional fuelwood collectors have had to bear the brunt of deforestation. They realised that their fuelwood and fodder resources have receded from their 'resource use areas' around their settlements due to commercial timber extraction. It is also appreciated that this has led to serious floods and the loss of precious soil.

Chipko activists have undertaken long *padayatras* across the Himalayas, protesting against deforestation. The movement has been very successful and has been primarily supported by empowering local women's groups, who are the most seriously affected segment of society. The movement has proved to the world that the forests of the hills are the life-support systems of local communities and are of immense value in terms of local produce, and that the forest provides less quantifiable but even more important ecological services such as soil conservation and the maintenance of the natural water regime of the whole region.

The ability of the local women to band together in the foothills of the Himalayas goes back to the pre-independence days when women such as Miraben, a disciple of Gandhiji, moved to this region and understood that it was deforestation that led to floods and devastation of villages in the valleys and in the Gangetic plains below. They also appreciated that substitution of oak and other broad-leaved forests of the Himalayas with the planting of fast-growing pine for timber and resin was an ecological and social disaster which reduced the forest resources used by traditional hill communities.

7.5.2 Silent Valley

The save Silent Valley movement was triggered in 1973 when the Government of Kerala decided to develop a hydroelectric dam on the course of the Kunthipuzha river in the evergreen forest of Silent Valley. When ecologists and BNHS scientists pointed this out to Dr Salim Ali, he intervened and explained its ecological value and the presence of endangered and endemic plants to then Prime Minister, Mrs Indira Gandhi, who prevented the construction of a dam and had the site notified as a National Park. This has led to a major conservation gain of national and global significance.

7.5.3 The Bishnois of Rajasthan

The Bishnois are a group of traditional farmers in the Rajasthan desert area who follow their religious leader, Jambheshwar. He has given them several tenets to follow. This includes asking his followers to protect plants and animals. The Bishnois see that water and fodder is provided for wildlife around their villages even today. They have prevented hunting of wildlife even though their croplands are damaged. The Bishnois apprehended a famous celebrity film star while shooting blackbuck in their area. The star has been severely punished by the SC as a violator of the Wildlife Protection Act in 2018.

7.6 ENVIRONMENTAL ETHICS: ROLE OF INDIAN AND OTHER RELIGIONS AND CULTURES IN ENVIRONMENTAL CONSERVATION

Environmental ethics deals with issues related to the rights of individuals that are fundamental to life and well-being. These concern not only the needs of each person today, but also those who will come after us. It also deals with the rights of other living creatures that inhabit the earth.

The world is such a beautiful place. We as a species, endowed with so many evolutionary gifts, have a duty towards keeping our world a beautiful, liveable and happy place. In fact we don't know if life as we experience it occurs anywhere else in the universe. If we are the only ones, then we have a unique responsibility to the universe itself. So, we have global, national and local environmental responsibilities. However, the earth's resources have limits. We cannot overuse or misuse them. In many situations what we want may not be good for our environment. Our ethical behaviour towards the environment must be in consonance with environmental integrity. In fact, we need to better our environment by promoting values for managing nature sustainably, supporting and maintaining naturalness, and preserving our natural resources and biological diversity. This is what environmental ethics is about. It is also about the equal rights of all people to environmental resources such as clean air, drinking water, secure food, good environmental health management and education.

Mahatma Gandhi—The philosophy behind need / greed

Mahatma Gandhi's philosophy was based on what we refer to today as sustainability of the earth's resources. His ideas were far ahead of his times as he had brought home to people that we should use only what we need that the earth can provide equitably and sustainably for all of us. But the earth can never support people's greed for more and more products made from nature's resources. Gandhiji's own life was simplistic. He used minimal resources and did not waste any.

When you first read about the environment you will think it concerns governance and government and official functionaries who play a role in its management. Dig deeper and you realise it is closely linked to your own life and lifestyle.

Five principles of environmental ethics

The five principles (1985) are as follows.

(i) The diversity of species and communities must be preserved—because people like nature through a sentiment called biophilia (E O Wilson, 1984).

(ii) The ultimate decimation of populations and species must be prevented—because humankind has accelerated the loss of species. This was normally a very slow process which took hundreds of thousands of years. Human activities have in the last few decades accelerated the processes that can lead to extinction of threatened species of flora and fauna.

(iii) Ecological complexity must be maintained—because the processes within ecosystems are closely interlinked with many dependent species and their habitats, as well as their specific genetic characteristics.

(iv) Evolution should continue—because this process which created the earth's biological diversity is an inherent aspect of nature which creates new and better adapted species.

(v) Biological diversity has intrinsic value—because human beings value its inherent characteristics. It has ethical and existence value—people appreciate biodiversity just for its presence. We appreciate evergreen forests rich in species and colourful coral reefs even if we don't see them. They are all part of the virtual world that we see in our minds eye.

Human rights

Several environmental issues are closely linked to human rights. These include the equitable distribution of environmental resources, the utilisation threshold of resources and Intellectual Property Rights (IPRs), conflicts between people and wildlife especially around PAs, and resettlement issues around development projects such as dams and mines. Access to health care to prevent environment-related diseases are aspects that have implications on the rights of every human being.

Value education

Value education in the context of the environment is expected to bring about a new sustainable way of life. Education, both through formal and non-formal processes, must thus address the understanding of environmental, natural and cultural values, social justice, preservation of human heritage, equitable use of resources, managing common property resources and understanding the risks of ecological degradation.

Essentially, environmental values cannot be taught. They have to be inculcated through a complex process of appreciating our environmental assets and experiencing the problems caused due to the destruction of our environment. The problems that are created by technology and economic growth are the result of improper thinking on what 'development' means. Since we still put a high value only on economic growth, we have no concern for aspects such as sustainability or equitable use of resources. This mind set must change before concepts like sustainable development can be acted upon.

Environment education must bring in several new values. Why and how can we use less resources and energy? Why do we need to keep our surroundings clean? Why should we use less fertilisers and pesticides in farms? Why is it important for us to save water and keep our water sources clean; why should each of us segregate our garbage into degradable and non-degradable types before disposal?

All these issues are linked to the quality of human life and go beyond simple economic growth. They deal with a love and respect for nature. These are the values that will bring about a better humanity, one in which we can live healthy, productive and happy lives in harmony with nature.

Environmental values

Every human being has a variety of feelings for different aspects of his or her surroundings. The western or modern approach to life values the resources of nature for their utilitarian importance alone. However, true environmental values go beyond valuing a river for its water, a forest for its timber and Non-Timber Forest Produce (NTFPs), or the sea for its fish. Environmental values are inherent in feelings that bring about sensitivity for preserving nature as a whole. This is a more Eastern, traditional value. There are several writings and sayings in India's ancient cultures that support the concept of the oneness of all creation, of respecting and valuing all the different components of nature which includes flora, fauna, air, water and earth. Our environmental values must also translate into pro-conservation actions in all our day-to-day activities. Most of our actions have adverse environmental impacts unless we consciously avoid them. The sentiment that attempts to reverse these trends is enshrined in our cultural values and our constitution.

Values lead to a process of decision making which creates pro-environmental action. For value education in relation to the environment, this process is learned through an understanding and appreciation of nature's oneness and the importance of its conservation. Humans have an inborn desire to explore nature and to unravel its mysteries. However, modern society and educational processes have suppressed these innate sentiments. Once exposed to the wonders of the wilderness, people tend to bond closely with nature. They begin to appreciate its complexity and fragility and this awakens a new connectedness and the desire to protect our natural heritage. This feeling for nature is a part of our constitution which strongly emphasises this value.

The concepts of what constitutes 'right' and 'wrong' behaviour changes with time; values are not constant. It was once considered 'sport' to shoot animals. It was considered a royal, brave and desirable activity to kill a tiger. In today's context, with wildlife reduced to a tiny fraction of what there was in the past, it is now looked down upon as a crime. Thus, the value system has altered with time and circumstances. Similarly, with the large tracts of forest that existed in the past, cutting a few trees for fuelwood or housing was not a significant criminal act. Today, this constitutes a major concern. With the small human numbers in the past, throwing away a little household degradable garbage into the street could not have been considered wrong. But with the enormous numbers of people throwing away large quantities of non-degradable waste, it is indeed extremely damaging to the environment. We must prevent this through strong environment-related values through our education system.

Appreciating the negative effects of our actions on the environment must become a part of our day-to-day thinking. Our current value system is based on economic and technical progress as being what we need in our country with its emerging economy as a world power. While we need economic development, our value system must change to one that makes people everywhere support a sustainable form of development. This is essential so that we do not have to bear the cost of environmental degradation in the future.

Environmental problems created by development are not necessarily due to the need for economic development, nor to the technology that produces pollution, but rather to a lack of awareness of the consequences of unlimited and unrestrained anti-environmental behaviour. Looked at in this way, it deals with concepts of what is appropriate behaviour in relation to our surroundings and our duty towards other species on earth. This is what environmental values are about.

Each action by an individual must be linked to its environmental consequences in his/her mind, so that a value is created that strengthens pro-environmental behaviour and prevents environmentally damaging actions. This cannot happen unless new educational processes are created that provide a conservation orientation to what is taught at school and college level. Every child asks questions like 'What does this mean?' They want an explanation for things happening around them that can help them make decisions and they develop values through this process. It is this innate curiosity that leads to a personalised set of values in later life. Providing appropriate 'meanings' for such questions related to our own environment develops a set of values that most people in society begin to accept as the norm. Thus, pro-environmental actions begin to move from the domain of individuals to that of a community, a nation and the world. At the community level, this occurs only when a critical number of people become environmentally conscious and constitute a pro-environment lobbying force that makes governments and other people accept good environmental behaviour as an important part of development.

What are the various professions that have to make value judgments that greatly influence our environment? Nearly every profession can and does influence our environment, but some do so more directly than others. Policy-makers, administrators, landuse planners, media, architects, medical personnel, healthcare workers, agriculturalists, agricultural experts, irrigation planners, mining experts, foresters, forest planners, industrialists and most importantly, teachers at the school and college levels, are all closely related to pro-environmental outcomes.

Environmental values are linked to varied environmental concerns. While we value resources that we use as food, water and other products, there are also environmental services that we must appreciate. These include nature's mechanisms in cleaning air by the removal of carbon dioxide and addition of oxygen by plant life, recycling water through the hydrological cycle and maintaining climate regimes.

However, there are other aesthetic, ethical values that are equally important aspects of our environment that we do not consciously appreciate. While every species is important in the web of life, there are some which man has come to admire for their beauty alone. The tiger's magnificence, the whale and elephant's giant size, the intelligence of the primates, the graceful flight of a flock of cranes, are all parts of nature that we cannot help but admire. The lush splendour of an evergreen forest, the great power of the ocean's waves, and the tranquillity of the Himalayan mountains are things that each of us value even if we do not experience it ourselves. We value nature itself for its presence. This is nature's 'existence value'.

Urban gardens and open spaces are also valuable as aesthetic and life enriching places for recreation. This must be a prime concern to urban planners. These green spaces act as not only the 'lungs' of cities, they also provide the much-needed psychological support. The mental peace and relaxation provided by such areas needs to be valued, although it is difficult to put a price tag on these environmental assets. Nevertheless, these centres of peace and tranquillity give urban dwellers an opportunity to balance their highly modified environments with the splash of green of a garden space. Dr Ernest Wilson believed, as many do today, that these green spaces are vital to our mental and physical well-being. He coined the term *biophilia* to describe this phenomenon.

Environmental values must also stress the importance of preserving our ancient heritage structures. The characteristic architecture, sculpture, art and craft of ancient cultures are invaluable environmental assets. They tell us where we have come from, where we are now, and perhaps (if we are willing to learn from them) where we should go. Architectural heritage goes beyond preserving old buildings, to conserving whole traditional landscapes in rural areas and streetscapes in urban settings. Unless we learn to value these landscapes, they will disappear and our heritage will be irrevocably lost.

As environmentally conscious individuals, we need to develop a set of values that are linked with a better and more sustainable way of life for all people. There are several positive as well as negative aspects of behaviour that are linked to our environment. The positive feelings that support the environment include a value for nature, traditional culture, heritage and equity. We also need to become more sensitive towards preventing aspects that have a negative impact on the environment. These include our attitude towards the degradation of the environment, loss of species, pollution, poverty, corruption in environmental management, the rights of future generations and animal rights.

Valuing nature

The most fundamental environmental sentiment is to value nature herself. The oneness of our lives with the rest of nature and a feeling that we are only a miniscule part of nature's complex web of life becomes apparent, when we begin to appreciate the wonders of nature's diversity. We must appreciate that we belong to a global community that includes 1.9 million known living species of plants and animals. We know that the earth's life-forms are unique. We have a great responsibility to protect life in all its glorious forms and must therefore respect the wilderness and the natural habitats of all its living creatures. We need to develop a sense of value that makes us protect what is left of the wilderness. On the one hand, we need to protect natural ecosystems, while on the other we must protect the rights of local people. Apart from valuing the diversity of life itself, we must also learn to value and respect diverse human cultures. Many of the tribal cultures of our country are vanishing because people with more dominant and economically advanced ways of life do not respect the lifestyles of tribal folk that are, in fact, closer to nature and frequently are more sustainable. We believe that our modern technology-based lifestyles are the only way for society to progress. However, this is only a single dimension of life that is based on economic growth.

While currently the environmental movement focuses on issues that are concerned with the management of the natural environment for the 'benefit' of humans, 'deep ecology' promotes an approach that is expected to bring about a more appropriate ecological balance on earth.

Valuing cultures

Every culture has a right to exist. Tribal people are frequently most closely linked with nature and we have no right to foist on them our own modern way of life. The dilemma is how to provide them with modern health care and education that gives them an opportunity to achieve a better economic status without disrupting their culture, beliefs, myths, art and way of life. This will happen only if we value their culture and respect their way of life.

Social justice

As the gulf widens between the haves and the have-nots, it is the duty of the former to protect the rights of the latter. If this is not respected, the poor will eventually rebel, anarchy and terrorism will spread and the people who are impoverished will eventually form a desperate seething revolution to better their own lot. The developing world will face a crisis earlier than the developed countries, unless we protect the rights of the poor people.

Modern civilisation is a blend of homogenous cultures, based until recently on a belief that modern science holds the answers to everything. We are now beginning to appreciate that many ancient and even present day sequestrated cultures have a wisdom and knowledge of their own environments that is based on a deep sense of respect for nature. Tribal cultures have, over many generations, used indigenous medicines which are proving to be effective against diseases. They have produced unique art forms, such as paintings, sculptures and crafts, which are beautiful. They have their own poetry, songs, dance and drama. All these art forms are unfortunately being rapidly lost as we introduce a different set of modern values to them through television and other mass media. The world will be culturally impoverished if we lose this traditional knowledge. They will soon lose the beauty within their homes that is based on the things made from nature. The art of the potter will be lost forever to the indestructible plastic pot. The bamboo basket-weaver who makes a thing of beauty that is so user-friendly and aesthetically appealing, will give place to yet another plastic box. Much that is beautiful and hand-crafted will disappear if we do not value these diverse aspects of human cultures.

Human heritage

The earth itself is a heritage left to us by our ancestors, not only for our own use but also for the generations to come. There is much that is beautiful on our earth—the undisturbed wilderness, a traditional rural landscape, the architecture of a traditional village or town and the value of a historical monument or place of worship. These are all part of our human heritage.

Heritage preservation is now a growing environmental concern, because we have undervalued much of this heritage during the last several decades and it is vanishing at an astonishing pace. While we admire and value the Ajanta and Ellora Caves, the temples of the 10th–15th centuries that led to different and diverse styles of architecture and sculpture, the Mughal styles that led to structures such as the Taj Mahal, or the unique environmentally-friendly colonial buildings, we have done little to actively preserve these assets. As environmentally conscious individuals, we need to lobby for the protection of the wilderness and our glorious architectural heritage.

Equitable use of resources

An unfair distribution of wealth and resources, based on a world that is essentially only for the rich, will bring about a disaster of unprecedented proportions. The equitable use of resources is now seen as an essential aspect of human well-being and must become a shared point of view among all socially and environmentally conscious individuals. In spite of the great number of people in the more populous developing countries, the people in the developed countries (though smaller in number) use more resources and energy. This is equally true of the small number of rich people in poor countries whose per capita use of energy and resources, and the generation of waste based on the one-time use of disposable products, leads to great pressures on the environment. As we begin to appreciate that we need more sustainable lifestyles, we also begin to realise that this cannot be brought about without a more equitable use of resources.

Common property resources

A major portion of the environment does not belong to any individual. There are several commonly owned resources that all of us use as a community. The water that nature recycles, the air that we all breathe, the forests and grasslands that maintain our climate and soil, are common property resources. When the government took over the control of the community forests during the British rule, the local people who until then had controlled their use through a set of norms that were based on community use, began to overexploit these resources in which they now had no personal stake. Bringing back such traditional management systems is extremely difficult. However, in the recent past managing local forests through village-level Forest Protection Committees (FPCs) has shown that if people know that they can benefit from the forests, they will begin to protect them. This essentially means sharing the power to control forests between the Forest Department and the local people, referred to as co-management or participatory management.

Preserving resources for future generations

Can we utilise all the resources of the world, leaving nothing for future generations? This ethical issue must be considered when we use resources unsustainably. If we overuse and misuse resources and energy from fossil fuels, what will happen to our future generations? A critical concern is to preserve species and natural undisturbed ecosystems that are linked with bioresources, which must be protected for the use of future generations. Just as our ancestors have left resources for us, it is our duty to leave them behind for our future generations. They have a right to these resources. We only hold the world as trustees for the future generations.

The rights of animals

Can humans, a single species, use and severely exploit the earth's resources, which we share with millions of other plant and animal species? The variety of plants and animals that share the earth with us also have a right to live and share the earth's resources and living space. We have no right to push a species that has taken millions of years to evolve, towards extinction. Not only do wild and domesticated animals have a right to life, they have the right to a dignified existence. Cruelty to an animal is no different ethically from cruelty to another human being.

Human beings are just one small cog in the wheel of life on earth. We frequently do not realise that we have exploited nature and other species well beyond what we should have used justifiably. Every plant and animal has a right to life as a part of the earth's community of living things. Nature if left to itself has natural prey–predator relationships and maintains a balance in each ecosystem. While evolution has developed a system whereby species become extinct and new ones evolve to fill the world's ecosystems with new plant and animal species, it is human beings alone that have been responsible for the recent rapid decline in the number of species on earth. More importantly, we are now reducing the population of so many species that in the near future we will in all probability create a major extinction spasm that will seriously endanger the existence of mankind. Thus, endangering the existence of wild plants and animals and bringing them close to the brink of extinction is not only unfair to those species but also to future generations of people.

7.7　ENVIRONMENTAL COMMUNICATION AND PUBLIC AWARENESS

It is of utmost importance to create an ethos that will support a sustainable lifestyle in society. This brings us to the need for environmental education. The Supreme Court has ordered that every young individual at school and college level should be exposed to a course on the environment. It is not only to create an awareness of environmental issues, but also to bring about pro-environmental action. Among the variety of tools that can bring home the ethical issues of the environment, nothing is as powerful as real-life experiences. Creating a love for nature brings about strong pro-environmental action. Our current educational processes at school and college level are being reoriented to bring about this change.

Environmental education and education for sustainable development include both formal and non-formal learning. In 1992 M C Mehta, the famous environmental lawyer filed a PIL in the Supreme Court that brought out the fact that the Government had done nothing to create public awareness on environmental issues. The Supreme Court addressed the Ministry of HRD and the Ministry of Environment to initiate urgent measures to correct this.

In today's world where many of us are far removed from nature, Communication Education and Public Awareness (CEPA) forms a strategy to appreciate that everything we use, if tracked back to its source, has come from nature. We depend on an intact unpolluted world that is based on nature's goods and services. No life is possible without these measures. Nature's resources, that we all use and depend on, can only be optimised if they are equitably shared by all of us. If the disparity is too great it can only result in anarchy.

Bringing back an ethic for nature conservation through educational processes and media requires a focus on environment education and conservation awareness. The best way to do so is to make our young people aware of our dependence on natural resources from the wilderness and also expose them to the beauty and wondrous aspects of nature, which form a sharp contrast to the sad plight of degraded areas and polluted sites in which most of humanity now lives in the developed and developing world. In view of the Supreme Court directive, environment education has been introduced in all school textbooks. The Supreme Court has ordered the UGC to implement a Core Module Course [now referred to as Ability Enhancement Compulsory Courses, (AECC) – Environmental studies, 2017] which is the purpose of this textbook for all courses at the undergraduate level. India is perhaps the only country in the world to implement such an interdisciplinary compulsory course for all college students at the UG level.

Public awareness is the responsibility of the media in using both print and electronic outreach on the environment to the public at large. The messages must include issues related to natural resource management and biodiversity conservation, pollution, energy conservation, public health and ethical considerations which can move the world towards a sustainable future.

Environmental sensitivity in our country can only grow through major public awareness campaigns. This is referred to in environmental terms as Communication Education and Public Awareness (CEPA) strategies. This has several tools—electronic media, press, school and college education, adult education, which are all essentially complementary to each other. Green movements can grow out of small local initiatives to become major players in advocating methods for environmental protection to assist the Government. Policy makers will only work towards environmental preservation if there is a sufficiently large bank of voters that insist on protecting the environment. Orienting the media to report pro-environmental issues is an important aspect as several advertising campaigns frequently have messages that are negative to environmental preservation. Programmes such as 'Smart cities' and '100 resilient cities' further the cause of public awareness on urban environmental issues.

Using an environmental calendar of activities

There are several days of special environmental significance, which can be celebrated in the community and can be used for creating environmental awareness in your vicinity, city or village (Table 7.1).

Appreciating the beauty of nature and treasuring the magnificence of the wilderness

We often take nature for granted. We rarely take the opportunity to gaze at a scenic sunset, spend time to sit in the incredible silence of the forest, listen to the songs of birds and the sound of the wind rustling through the leaves, take the trouble to watch the magic of a seed germinating from the ground and gradually growing into a seedling over several days, observe a tree through a round of seasons as it gets new leaves, flowers, fruit and seeds, or reflect on the incredibly large number of links between all the different animals and birds that depend on the seasonal changes

Table 7.1   Days of environmental significance

Month	Dates	What you can do
Feb 2	*World Wetland Day*: To raise awareness about wetlands and their importance. This day marks the anniversary of the 1971 Ramsar Convention on Wetlands of International Importance at Ramsar in Iran.	You can initiate a campaign for the proper use and maintenance of wetlands in the vicinity of your city or village.
Mar 21	*World Forestry Day*: To raise awareness about the rapid disappearance of forests.	Initiate an action-oriented programme including activities such as tree plantation in your area.
26	8.30 pm: *Earth Hour*: To take a stand against climate change by turning off lights for 1 hour in your home and/or business.	Pledge your support for the planet by switching off all lights during Earth Hour.
Apr 7	*World Health Day*: To raise awareness about issues of public and occupational health.	Organise a campaign for personal sanitation and hygiene. Topics that deal with environment-related diseases and their spread can be discussed and preventive measures suggested.
18	*World Heritage Day*: To protect and preserve our cultural heritage.	Organising a visit to a local fort or museum can create awareness among the local people about their very valuable heritage sites.
22	*Earth Day*: To draw attention to increasing environmental problems caused by humans on earth.	This day is now celebrated all over the world with rallies, festivals, clean-ups, special shows and lectures.
Jun 5	*World Environment Day*: A day to stimulate worldwide awareness of environmental issues and encourage political action. This day marks the anniversary of the 1972 Stockholm Conference on Human Environment in Sweden.	This day can be used to project the various environmental activities that a college has undertaken during the year. New pledges must be made to strengthen an environmental movement at the college level.
11	*World Population Day*: To raise awareness on global population issues.	The vital link between population and environment could be discussed in seminars held at colleges and other NGOs.
Aug 6	*Hiroshima Day*: To remember the dropping of the first atomic bomb on the Japanese city of Hiroshima in 1945, that killed thousands.	Could be a day to discuss the impacts and consequences of nuclear weapons. The Bhopal gas tragedy and the Chernobyl disaster should also be discussed as examples.
Sep 16	*World Ozone Day*: International day for the preservation of the ozone layer. This day marks the Montreal Protocol signed in 1987 to control the production and consumption of ozone depleting substances.	A day for students to find out more about the threats to this atmospheric layer and initiate discussions on how the Montreal Protocol has been a success in mitigating this global threat and how its model could be applied in other international agreements such as the Kyoto Protocol.
28	*Green Consumer Day*: To create awareness (at a household level) amongst consumers, about various environmentally friendly products available in the market.	Students could talk to shop keepers and consumers about excess packaging and a campaign to use articles which are not heavily packaged could be carried out. Prevent the use of plastics in consumer goods through a campaign.
Oct 1–7	*Wildlife Week*: To protect and conserve our species and threatened ecosystems.	State Forest Departments usually organise various activities in which every student should take part. Poster displays or a street play to highlight India's rich biodiversity can be planned.

in their habitat. It is the beauty of nature, the green of the forest, the blue of the sea, and the gold of the sunrise that has intrinsic value, and that we tend to ignore. These are not mundane day-to-day events, they are magical aspects of nature's clock that is ticking silently all around us. If we experience these wondrous aspects of nature, our lives will be enriched immeasurably.

Once we realise that wilderness has a value of its own, it puts humans in their rightful role as the custodians of nature rather than as exploiters. If one visits a wilderness area, a forest, lakeside, waterfall, or seashore, one begins to value its beauty. Without the wilderness, the earth would be a bleak, human-dominated landscape. The problem is how much of the wilderness can we preserve in the presence of an ever-growing hunger for land and resources for its utilitarian value? Unless we begin to see the ecological value of the wilderness, an ethic for its conservation cannot become part of our daily lives. And without wilderness, the earth will eventually become unliveable.

Ahimsa or non-violence towards life, which includes all plants and animals, provides India with its basic philosophy which early Hindu philosophers and later, Buddha, Mahavira and Mahatma Gandhi spoke of. Buddhist and Jain philosophy is intrinsically woven around non-violence and the great value of all forms of life. This philosophy ascertains that animals are not to be viewed purely for their utility value but are a part of earth's oneness which is linked with our own lives. In Indian philosophy, the earth itself is respected and venerated. In contrast, in Western thought, nature is to be subjugated and used. These are basic differences in thinking processes. Several modern philosophers in the West have now begun to appreciate these Eastern patterns of thought as a new basis for human development. This shift from a purely utilitarian or scientific exploitation of nature to one of harmony with nature, can however only occur, if each of us understands and respects nature's great 'oneness'.

The conservation ethics and traditional value systems of India

In ancient Indian tradition, people have always valued mountains, rivers, forests, trees and animals. Thus, much of nature was venerated and protected. Forests have been associated with the names of forest gods and goddesses, both in the Hindu religion as well as in tribal cultures. 'Tree' goddesses have been associated with specific plant species. *Ficus religiosa*, the peepal tree, is venerated and is not to be cut down. The banyan tree in some regions, such as Maharashtra, is venerated once a year by tying a thread around it as a symbol of respect. The *tulsi* plant is grown in every home. Patches of forests, now called 'sacred groves', have been dedicated to local forest gods and goddesses in many Indian cultures, especially in tribal areas. These traditionally protected forest patches depict the true nature of undisturbed vegetation and have a large number of indigenous plant species. Their exploitation has been controlled through local sentiments.

Certain species of trees in India have been protected as they are valued for their fruit or flowers. In most farms, the mango tree is protected for its fruit even when wood becomes scarce. The Mahua tree (*Madhuca indica*) is protected by tribal people as it provides edible flowers, oil from its seeds and is used to make a potent alcohol. Many plants, shrubs and herbs which were once available in the wild in plenty are used in Indian medicines. These are now rapidly vanishing. Many species of animals are venerated as being the *vahana* or vehicle of various gods who are said to travel through the cosmos on them!

In Indian mythology, the elephant is associated with Ganesha. The elephant-headed Ganesha is also linked to the rat. Vishnu is associated with the eagle. Rama is linked to monkeys. In mythology, Hanuman rendered invaluable help to Rama during his travels to Lanka. The sun god, Surya, rides a chariot drawn by several horses on which he moves through the sky. The lion is linked to Durga and the blackbuck to the moon goddess. The cow is associated with Krishna, the snake with Shiva and Vishnu, the swan with Saraswati. Vishnu's incarnations have been represented as taking various animal forms which serially include a fish, a tortoise, a boar, a dwarf and a half-man–half-lion form.

The associations with various plants that have been given a religious significance include the tulsi, which is linked to Lakshmi and Krishna. The tulsi plant is also linked to the worship of one's own ancestors. The peepal tree is said to be the tree under which Buddha attained enlightenment. It is also associated with Vishnu and Krishna. Several trees are associated with the goddess Lakshmi, including amalaki, mango and tulsi.

Traditionally these species, which were considered important aspects of nature, were the basis of local life-support systems and were integral to a harmonious life. In societies of the past, these examples were all a part of the ethical values that protected nature. As modern science based on the exploitation of nature spread into India, many of these traditions began to lose their effectiveness in conserving nature. Concepts that support nature's integrity must thus become a part of our modern educational systems. This constitutes a key solution to bring about a new ethic of conserving nature and living sustainable lifestyles.

What can I do? Learning by action

Most of us are always complaining about the deteriorating environmental situation in our country. We also blame the government for inaction. However, how many of us actually do anything about our own environment? Think about the things you can do that support the environment in your daily life in your profession and in your community. You can make others follow your environment-friendly actions. A famous dictum is to 'think globally and act locally' to improve your own environment. 'You' can make a difference to our world.

CASE STUDY 7.11 Air pollution in Delhi

In 1993, during the England cricket tour of India, when they lost a match, they attributed part of their loss to the air pollution in Delhi. During that time, vehicular emissions, which accounted for 70% of the air pollution, would morph into deadly smog during the foggy winters. This resulted in an increase in respiratory illnesses, with children and senior citizens being the worst affected. The increasing concentration of pollutants is also because of the winters experienced by Delhi characterised by cold, dry air and ground-based inversion with low wind conditions. Pollutants are trapped by the warm air layer. The crop residues and stubble burning have been reaching Delhi due to the prevailing wind pattern in autumn and winter months (Anfossi D, et al., 1990; CPCB-NAAQM, 2000).

Due to the rapid increase in vehicles and the poor condition of the vehicles, the load of automobile pollutants in ambient air was extremely high. The Central Pollution Control Board estimated that city traffic added as much as 2000 tons of pollutants a day in the year 2000. Due to this, the Supreme Court ordered the government to take action. The order directed the government to improve the air quality, initiating the use of CNG, especially in the public transport support system. A time frame was provided for solving the problem. Various policy measures were taken over the next decade.

- Providing CNG for motor vehicles.
- Public transport was made to switch to CNG.
- Vehicles older than 15 years were scrapped.
- Highly polluting industries were shifted out of Delhi. A total of 1328 units of H-category industries were shut down in 1996.
- Metro rail was introduced in the year 2000.
- The Central Pollution Control Board found that the total pollution load in Delhi was due to vehicular pollution, domestic pollution, industrial emission, garbage burning, road dust and construction activities (ENVIS CPCB, 2017).
- The National Clean Air Program was launched in January 2019 by the Government of India.

These measures have led to controlling air pollution in Delhi.

SUMMARY

- With the increasing world population, natural resources are under increasing pressure, threatening sustainable development. Environmental challenges of water shortages, loss of biodiversity, air and water pollution and climate change, greatly impact human health.

- A sustainable way of life is linked to a better health status of the community. This is further associated with stabilising the human population and sustainable use of natural resources. This is a balancing act between economic growth, equity in use of resources and long-term environment management. These concerns are closely linked with enhancing public awareness and creating a sense of environmental responsibility.

- A better health status of the society will result in a better way of life only if it is coupled with stabilising the population growth.

- Various development projects often require relocating people to an alternative site. Displacing people is a serious issue. It reduces their ability to subsist on their traditional natural resource base and also creates great psychological pressures. In such situations, alternative land and finances for developing a new village are required for resettlement.

- India is very vulnerable to droughts, floods, cyclones, earthquakes, landslides and forest fires. The National Disaster Management Plan (NDMP) aims to make India disaster resilient, reduce risk and decrease loss of lives and property. NDMP incorporates several key issues and suggests ways in which the country can be better prepared for managing different types of disasters. The National Disaster Response Force (NDRF) is constituted for the purpose of special response to build a safer and disaster-resilient India. Training and technological development are the key to preventing loss of life and property through advanced technology and training of various stakeholders. This is achieved through a culture of prevention, mitigation and preparedness to generate a prompt and efficient response at the time of disasters.

- Environmental ethics is about creating a better environment by promoting values for managing nature sustainably, supporting and maintaining naturalness, and preserving our natural resources and biological diversity. Environmental ethics looks at a philosophical way of valuing our natural environment. It looks at ownership of resources and the way they are equitably distributed. Environmental ethics expresses a need for equity and respect for all other species that we share the planet with. Environmental ethics deals with the fact that our generation holds nature's assets and services which are to be preserved for the future generations and their wellbeing.

QUESTIONS

1. Why is stabilising our population important?
2. What are the problems and concerns involved in the resettlement and rehabilitation of people in various projects?
3. How do you reduce your environmental footprint and enhance your handprint? Give at least five actions for decreasing footprint and five actions for increasing handprint.
4. Discuss with an example, how environmental degradation affects human health.
5. What are the main elements of a disaster mitigation strategy? Mention at least three.
6. Describe mitigation measures that can be taken to manage floods.
7. What are the immediate and delayed mitigation measures for cyclones?
8. What is the importance of public awareness in environmental management?
9. Write a note on environmental ethics. Mention at least three issues/challenges of environmental ethics.
10. Explain with examples role of religion and cultures in environmental protection.

Field Work

Learning Objectives _____

In this chapter you will learn,

◆ The procedure to visit a local area to document environmental assets—river, forest, grassland, hill or mountain
◆ The procedure to visit a local polluted site
◆ How to study common plants, insects and birds
◆ How to study a simple ecosystem

8.1 VISIT TO A LOCAL AREA TO DOCUMENT ENVIRONMENTAL ASSETS: RIVER / FOREST / GRASSLANDS / HILL / MOUNTAIN

Background

Documenting the nature of an ecosystem gives us a deeper appreciation of its value to mankind. Each ecosystem has something different to offer. It may contain natural resources that local people depend on; provide important ecological functions; have tourist or recreational potential; or simply have a strong aesthetic appeal that is difficult to quantify in economic terms. In fact, it can have multiple benefits for mankind at the global, national and local levels. An ecosystem is not only used in various ways by people belonging to different cultures and socio-economic groups, but has a different significance for different individuals, depending on their way of life. A tribal from a wilderness setting, an agriculturalist from farmlands, a pastoralist from grasslands or a fisherman, looks at his or her environment very differently from an urban resident who is mainly focussed on the management of the quality of air and water, and the disposal of garbage. In many cultures, men and women have different views and relationships with nature. In rural India, for example, it is mostly women who collect resources and see the degradation of their ecosystem as a serious threat to the existence of their family. They are thus more involved and will fight against processes that lead to the loss of their resource base.

Tribal people who live by hunting and gathering have a deep understanding of nature and what it provides for them to survive. Farmers know how to utilise their land and water resources, and also appreciate what droughts and floods can do to their lives. A shepherd or livestock owner knows the grasslands intimately. In contrast, urban dwellers are far removed from the sites from where they get their natural resources. As these have originated from a remote area and have been collected by rural people, urban dwellers cannot relate so easily to the value of protecting the ecosystems from which resources are obtained.

In assessing an ecosystem's values, it is not enough to look at its structure and functions, but also at who uses it and how the resources reach the users. One also needs to appreciate what it means to oneself. The wilderness provides a sense of wonderment for all of us, if we experience it in person. This helps to bring about a desire to conserve natural resources.

Guidelines for the study of environmental assets

There are two parts to this study:
(i) documenting what you see, and
(ii) documenting the findings of what you ask local user groups.
There are several key questions that one should attempt to answer in the study of the natural resources of any ecosystem.
(i) What are the ecosystem's natural resource assets?
(ii) Who uses these and how?
(iii) Is the ecosystem degraded? If so, how?
(iv) How can it be conserved?

One could go into enormous detail in answering these four basic questions. You will need to refer to the relevant units in this textbook, the guidelines provided in this Unit, as well as field guides to plants, insects, birds and others. You should begin your field study by observing the abiotic and biotic aspects of the ecosystem and documenting what you see. Ask questions to local user groups about their environment. Is their utilisation sustainable or unsustainable? Look for and document signs of degradation. Finally, study aspects that can lead to its conservation.

- Describe the ecosystem as you see it. Its structural nature, its quality and the differences one can perceive in its geographical features, and its plant and animal life. This takes time and patience. The more time one spends in careful scrutiny, the more one begins to appreciate its intricacies.
- How does the ecosystem function? What are the links between different species with each other and with their habitat? Observe its food chains. Look at it as if it is an intricate machine at work.
- By interacting with local residents and multiple user groups, decide if this is sustainable or unsustainable utilisation. If it is undisturbed, why has it remained so? If it is sustainably used, how is its use controlled? If it is degraded, how did it get to this state and when? If it is seriously degraded, what measures would you suggest to restore it and to what extent could it be used so that the utilisation would be sustainable?

You may not be able to observe and find out the answers to all these questions during a single visit. You will thus have to ask questions of local people who have a stake in the area to get answers. You may need the help of an ecologist, botanist, zoologist, geologist, hydrologist or a forester to get deeper insights. A historical background frequently helps to clarify many of these questions, as landscapes are not static and always change over time.

Documenting the environmental assets of each ecosystem

Documenting general features during the field survey: Describe the site and its features as provided in the proforma for fieldwork under the following headings—Aims and Objectives, Methodology, Observations on the Site, Findings of Interviews with Local People, Results and Conclusions.

Documenting the special resource features of individual ecosystems

Once the general features are documented, observations pertaining to the specific features of the ecosystem must be documented. The checklist on resource use of each ecosystem can help in creating an environmental profile of an area and will help in your appreciation of the ecosystem's goods and services, which include its important assets. However, this is to be used only as a guideline and a note needs to be prepared on each finding once you have made your observations and asked local people relevant questions about the ecosystem's resources in detail. Unless one does this for several different areas, one cannot really appreciate the assets of an ecosystem in clear terms, as these are often qualitative judgments that one makes by comparing the resources available in the study area with many others.

> **Proforma for field work on documenting environmental assets of each ecosystem**
>
> Use the format below as a general guideline for your field analysis. The points provided in the guidelines can be used to fill in the answers to the various issues for each ecosystem. The field work should be recorded in your journal as:
>
> (i) Aims and objectives: To identify and document:
>   - What are the ecosystems goods and services (checklist of resources)?
>   - Who uses them and how?
>   - Is the utilisation sustainable or unsustainable (signs of degradation)?
>   - How can the ecosystem be used sustainably?
> (ii) Methodology:
>   - Observation of the ecosystem.
>   - Questioning local people on the use of resources and sustainability.
>   - Discussion: Observations on levels of resource use found during the field work.
>   - Findings: Specific concerns relevant to the study site's sustainable utilisation as discussed with local people.
> (iii) Results and conclusions

River ecosystem

Guidelines on what to look for on river resource use:

Observe what local people use from the river, wetland or lake: They collect drinking water and use it for other domestic needs. They catch fish and crabs; graze their cattle and buffaloes in or near the water. They use the water from the lake by installing pumps to irrigate their fields.

Mapping land use in terms of its water resources: Document the pattern of land use around the aquatic ecosystem – river, tank or lake – and assess the importance of the water resources in the ecosystem. Observe that all the animals, both wild and domestic, must come to the water source, or have its water brought to them.

Field observations on a river-front:
- Observe a clean stretch of river in a wilderness area. The water is clear and full of life. In its many pools fish dart about, tadpoles swim around and crabs crawl along the bottom of the water.
- In a rural area observe all the different ways in which people use the water from the river.
- Observe a river in an urban area, the water cannot be used for drinking as it is dirty.
- Observe the water in a glass – it is coloured – can we drink it? Who has polluted it and how? This is a sign of unsustainable use of water.

Possible observations:
- Along a river in a forest, observe all the different animal tracks at the edge of the water. All wildlife depends on this resource for their day-to-day survival.
- Identify the different fish that local fishermen have caught. Ask if the fish catch has decreased, remained the same, or has increased over the last decade or two.
- Resource use: Observe and document the different types of fish and other resources used by local people. Are these for consumptive or productive purposes?
- Observe how the ecosystem is utilised and document these assets—water distribution, fish, crustaceans, reeds, plants used as food, any other resources.
- In your report, compare and contrast an unpolluted and a polluted body of water. Only the more robust species remain in polluted water while the more sensitive ones disappear.

Water—the greatest of all resources
- What do you use water for during the course of one day? How much do you use?
- Can you stop wasting water by using it carefully?

- How can you reduce the water you use for bathing and other tasks? Discuss how wastewater can be used in the garden.
- How can water be recycled?

Observations on the site that should be recorded

- *Type:* Permanent flow/seasonal flow; slow moving/rapid flow; deep/shallow.
- *Qualitative aspects:* Describe its abiotic and biotic aspects; is the flow natural or disturbed by a dam upstream?
- Describe its aquatic plant and animal life.
- What are the characteristic features of its components—banks, shallow areas, deep areas, midstream areas, islands? How is the land used?

The findings on the site that should be recorded through interviews

- What is the water used for – domestic use/agriculture/industry – and in what proportion?
- What other resources are used—fish, crustaceans, reeds, sand, others? What impact does the level of use have on the ecosystem?
- Is the water potable? If not, what are the sources of pollution—domestic sewage/agricultural runoff/industrial effluents? Which of these affect it most seriously?
- What is the extent of pollution—severe/high/moderate/low/nil? Explain why.
- Test the quality of water. What are the results of your water-quality tests?
- What efforts are made to keep the river clean, or to clean it up?
- Is its utilisation sustainable or unsustainable?
- Provide a historical profile of, and changes in, its environmental status by asking local people.
- Does it flood? If so, how frequently? How does this affect people? What preventive steps can be taken to prevent the adverse effects of floods?
- How can you enhance public awareness about the need for keeping the river clean?
- How are you dependent on the river ecosystem? How is it linked to your own life?
- Results of the water analysis.

Forest

Guidelines on what to look for in forest resource use:

Assessing forest use: Ask local people, especially the women, what products they collect from the forest. Document what is for household use, what is sold in the local marketplace and what is taken and sold to other areas. Fruits, leaves, roots, nuts, fuelwood, timber, grass, honey, fibre, cane, gum, resins and medicinal products are all forest products of great value.

Looking for signs of forest use: Several signs tell us how the forest is used by people. Look for human footprints and hoof marks of domestic animals, which demonstrates the dependence of man and his animals on forest vegetation.

Observe the number of cattle tracks and cow dung piles, which tell us where the local people graze their domestic animals. Look especially for cattle tracks near watering places. The zigzag paths on a hill-slope that have very little vegetation cover are a sign of overgrazing.

People cut the branches of the trees and shrubs for fuelwood. The number of cut stumps of branches can be used to assess the level of utilisation. If the forest is seriously lopped all around, it is clearly degraded. Most of the energy required to cook meals and heat their homes in winter is forest-dependent. Ask the local women how far they must travel for fuelwood. Larger tree-stumps show the number of trees used for building houses, or that have been felled and sold as timber.

Observe the environment in a neighbouring village. Look for the various forest products used by the people, or marketed by them.

Where do local people get their water? The presence of water in the streams is dependant on the existence of the forest.

Document the level of forest loss: Observe areas around villages where the forest is overused and contrast this to the intact vegetation of sanctuaries and national parks. Are there signs of degradation of the canopy, formation of wasteland, or signs of soil erosion?

◈ What are the products that you use in daily life that originate in forests?

Examples: water, paper, wood, medicines. The oxygen we breathe is produced by vegetation. Draw up a list of articles you use that could have originated from a forest ecosystem.

Observations on the site that should be recorded

◈ Identify the forest type—evergreen/semi-evergreen/deciduous/dry-deciduous/thorn forest.

◈ Is it a natural forest or a plantation?

◈ Observe its qualitative aspects—undisturbed/partially disturbed/mildly degraded/severely degraded.

Findings on the site that should be recorded through interviews

◈ List its natural resources, goods and services.
   Goods—food, fuelwood, fodder, NTFPs, water and others.
   Services—water regime, climate control, oxygen, removal of carbon dioxide, nitrogen cycle.

◈ Who uses the ecosystem's natural resources and to what extent? List the level of use of each of its natural resources (sustainable/unsustainable). Are these for personal use, for marketing, or for both? What proportion of the income of local people comes from the sale of fruit, fodder, wood, NTFPs?

◈ Make a map of the study area showing the different land uses and where resources are collected from.

◈ Provide a historical profile of its utilisation and changes in its environmental status by asking local people about their resource dependency.

◈ Is the ecosystem overused due to the number of people that depend on it, or the greed of a few, or both?

◈ Is it protected? If so, how?

◈ If it is to be restored, how can one make this possible?

◈ What forest produce do you use in your day-to-day life?

Grassland

Guidelines on what to look for on grassland resource use:

Utilisation pattern of the grassland: Discuss with local people how they use grasslands, graze cattle, cut fodder and collect fuelwood.

Grassland carrying capacity: Observe the enormous quantity of grass needed for the number of domestic herbivores dependent on it. This is an indication of the 'carrying capacity' of the grassland; that is, how many animals it can support.

Mapping landuse in grassland areas: Near a village, make a landuse map showing where the cattle are sent for grazing and for water, where people collect fuelwood and other products.

Documenting grassland degradation: Document if there has been a change in landuse patterns during the last few decades by asking local people. Observe differences in protected and degraded areas.

◈ What are the products you use that come from grasslands? Examples: milk, meat.

Observations on the site that should be recorded
◇ Identify the type of grassland—Himalayan/terai/semi-arid/shola/area developed for grass collection/common grazing land/forest clearing.
◇ Qualitative aspects—Describe its abiotic and biotic features. Document the nature of its soil, plant and animal species (wild and domestic). How do they use their habitat?
◇ What changes occur seasonally?

Findings on the site that should be recorded through interviews
◇ Who uses it and to what extent?
◇ Estimate the extent of free grazing by cattle, sheep, goats and their proportion.
◇ Estimate the extent of fodder collection.
◇ What is the productivity of the grassland? Ask the local people whether the fodder is—not enough for their own livestock/just enough for their own livestock/enough for their own livestock and to sell to other fodder-short areas.
◇ Provide a historical profile of its utilisation and changes in its environmental status by asking local people.
◇ Is this utilisation level sustainable or unsustainable?
◇ Is the grassland burned too frequently? Document why local people burn the grass.
◇ Can they do rotation grazing of their common grasslands and thus manage it better?
◇ What products do you use from grassland ecosystems in your daily life?

Hill-slope

Guidelines on what to look for on hill-resource use:

Hills are fragile ecosystems that are easily degraded. They are utilised by a variety of user groups. Understand the level of pressure and its utilisation patterns by asking the local people.

Observations on the site that should be recorded
◇ Identify the type of hill—steep/gradual slope; summit/peak/plateau top.
◇ Qualitative aspects—Describe its abiotic (soil characteristics) and biotic (vegetation) characteristics.
◇ Describe its contour and make a map marking its features such as nala courses, rocky outcrops, precipices, springs.
◇ Describe if its soil cover is intact/degraded/partially or severely eroded.
◇ Is it covered with cattle tracks? Do hoof marks of domestic animals and their dung piles indicate excessive grazing?
◇ Describe its vegetation profile and map different vegetation patterns (tree cover, scrub, grass cover, bare rock).
◇ Identify the plants that grow on it (trees, shrubs, herbs, grasses) and wild and domestic animals present.

Findings on the site that should be recorded through interviews
◇ What is it used for—urban housing/slum development/tourism/fuelwood collection/ grazing livestock/collecting water from its watercourses/greening. Observe the proportion or extent used for each purpose. Is it sustainable?
◇ Provide a historical profile of its utilisation and changes in its environmental status by asking local people.
◇ If it is eroded, what measures can be suggested to reverse the trend?
◇ How is this linked to your own life?

Mountain

Guidelines on what to look for on mountain resource use:

Mountains have very specialised ecosystems with clearly defined altitudinal variations. They are used by several different stakeholders. The effects of human interference on the mountains affect people in the valleys below.

Observations on the site that should be recorded

- Identify the type—Himalayan range/foothills/ghats.
- Qualitative aspects—describe its abiotic and biotic features.
- Describe its topography and soil characteristics. Make a map marking its features—snowcaps/rocky precipice/grassy slopes/tree-line. Discuss the proportion of each type.
- Describe its plant and animal species. How do they use their habitat?
- Identify the forest type with its dominant (common) tree species.
- Describe its soil cover/degree of erosion.

Findings on the site that should be recorded through interviews

- Describe the utilisation pattern of any forest cover and its grassy slopes.
- Who uses it?
- Do local people get as much natural resources from it today as they did in the past?
- If yes, how is this managed?
- If no, why not and what measures can be taken to remedy these trends?
- Provide a historical profile of its utilisation and changes in its environmental status by asking local people. Have there been landslides or floods in the valley below?
- How is our own life linked to this ecosystem?

8.2 VISIT TO A LOCAL POLLUTED SITE

Pollution occurs from a variety of sources and affects different aspects of our environment and thus our lives and health. Polluted sites include urban, rural, agricultural and industrial areas. Identify the site type and describe the sources of pollution.

Pollution can affect:

- Air (smoke, gases),
- Water (urban sewage, industrial chemical effluents, agricultural pesticides and fertilisers),
- Soil (chemicals, solid waste from industry and urban areas), and
- Biodiversity: effects on plant and animal life.

(Observations on pollution must include all the above aspects.)

Proforma for fieldwork

(i) Aims and objectives: To study the cause and effects of pollution at the site.

(ii) Methodology: Certain key questions related to the polluted site are given below. Explore the site to answer the questions about the area you have visited.
- What is the site?
  Rural—agricultural area, polluted water body, polluted industrial area; Urban—solid waste management site, polluted industrial area
- What do you observe at the polluted site?
  Solid waste—garbage dump, polluted water at a river or lake, gaseous effluents or smoke coming out of an industrial area or other observations.
- Explore the reasons for pollution. Observe and document the components in the garbage/the polluted water body/industrial chimneys.

- Observe the area and list the waste that is seen in the garbage dumping site. Categorise the waste into the three types:
- Degradable wastes are those which are easily decomposed by microorganisms. These include food wastes, plant material, animal carcasses and others.
- Non-degradable wastes are those which are not easily decomposed such as plastic and glass.
- Toxic wastes are those that are poisonous and cause long term effects such as paints, sprays and several other chemicals.

(iii) Findings:
- What are the effects of the pollutant?
- What actions can you take to get the pollution reduced?

General observations

The following aspects need to be observed and documented:

- The type of land or water use in the polluted area, its geographical characteristics, who uses the area, who owns it.
- Map the area to be studied.
- Identifying what is being polluted—air, water, soil; the cause(s) of pollution and the polluting agent(s).
- Assess the extent of pollution—severe/moderate/slight/nil, to: the air, water, soil, biodiversity.
- Assess from literature, the health aspects associated with the pollutant.
- Ask local residents about its effect on their lives.
- Make a report of the above findings.

8.2.1   Solid Waste Study Site

Guidelines for the study of solid waste polluted sites

Pollution caused due to solid waste can be seen at various places:

Garbage dumps: One of the urban or rural environmental problem sites that can be easily studied is a garbage-dumping area. This problem is basically due to increase in population, the over-utilisation of non-biodegradable disposable consumer goods, and the lack of awareness about the management of waste at the household level. How much garbage is produced everyday is not given much thought. No one really thinks about where the garbage goes or what happens to all the things we throw away.

Interview

Interview some rag-pickers at a roadside disposal area or at a dump and understand their problems. Prepare a survey sheet and ask them:

- What is the area covered in a day?
- How many hours are spent in collecting the waste?
- What are the types of waste collected?
- What are the problems faced while collecting waste?
- What do they prefer to collect and why?
- What is done with the waste collected?
- If it is sold, where?
- Would it be better to collect waste from homes rather than from roadside bins?
- Do they feel that segregation of waste would help them?

It is essential to understand that rag-pickers perform an environmentally important activity for all of us. While we throw away our own waste insensitively, it is they who segregate various types of waste for recycling and reuse. They are thus performing a great pro-environmental function for most of us.

Garbage is a source of various diseases. The improper handling of organic waste leads to a large population of flies, cockroaches and rats that are responsible for the spread of diseases. Products like plastics are not degraded in nature and hence remain for a long time in the environment, thus adding to the need for more dumps. For many years, waste has also been dumped into oceans, rivers or on land. These methods of disposing off waste contribute to the contamination of soil and groundwater under the dumping site, foul up the air and aid in the spread of diseases.

Households: The garbage generated in our homes is termed as *domestic waste*, while a community's waste is referred to as 'municipal waste'. Domestic waste is further classified as kitchen waste which is degradable, wet waste and non-biodegradable recyclable home waste, which consists of plastic, glass and metal. Observe and document what happens in homes of different economic groups. What happens to your own household waste? Could it be managed better?

> ### Study of a dump site
> - What is the location of the site and where does the garbage come from?
> - How is it collected, by whom, and at what frequency?
> - How is the waste managed?
> - What are the different types of garbage?
> - What proportion of it is non-degradable and what is degradable?
> - What can we do to reduce the quantity of garbage?

Agriculture: Agricultural waste consists of biomass, including farm residues such as rice husk, straw and bagasse. This biomass could be effectively used for generating power or producing paper. Waste material from fields includes fertilisers and pesticides that are a serious health hazard.

Industries: Industrial solid waste includes material from various industries or mines. Industries produce solid wastes during the manufacturing processes. Some of these are chemicals that have serious environmental ill-effects, since they are toxic. Visit an industry and ask what the waste products are and how they are disposed of.

The waste generated during mining is non-biodegradable; it remains in the environment nearly indefinitely. Solid waste is also generated as a result of excavation and construction work.

Hospitals: The waste generated from hospitals contains cotton dressings and bandages with blood or other tissue fluids and pus, all of which can contain pathogens. It can spread bacteria, fungi and viruses. Used needles, syringes, bottles, plastic bags, operation theatre waste, such as tissues, blood and plastic disposable equipment all need very careful disposal. The hospital should have a waste separation system at source to separate the waste into biomedical waste, glass and plastics. The biomedical waste can then be autoclaved or incinerated so that microorganisms are killed.

8.2.2   Water Pollution Site

Guidelines for study of polluted water sites

Observe if the river/lake/tank can be considered unpolluted/slightly polluted/moderately polluted or severely polluted by looking at the water and doing simple tests using a water- monitoring kit.

- Document the name of the river and the nearby urban or industrial site from where the pollution is generated.
- Is there urban garbage dumped on the bank?
- Are there industrial units near the site?
- Do the industries discharge their wastewater into the site? Is this treated or untreated?

- What is its colour and odour?
- Are there any sources of water contamination from the surface run-off from adjacent agricultural land on which fertilisers and pesticides are used?
- Ask the fishermen if this has affected their income.
- Identify the plants, birds and insects found on the banks.

8.2.3 Air Pollution Site

Guidelines for the study of polluted air sites
Air pollution sites include cities due to traffic congestion in urban centers, and industrial areas due to gaseous products released during manufacturing processes.
- Ask people from the area what the effects on their lives are.
- How can these be reduced?
- How can you make more people aware of this issue and the effects on their health?

8.3   STUDY OF COMMON PLANTS, INSECTS, BIRDS

Guidelines for the Study

Common plants, insects and birds have been selected as they occur nearly everywhere. While one may have to visit a national park or sanctuary to see mammals or reptiles, several plants, insects and birds can be seen around an urban or rural setting where there is some vegetation. If you have an opportunity to visit a national park or sanctuary, you can add other animals.

One needs a little equipment—a journal to take notes, preferably a pair of binoculars, and field guides to identify plants, insects, birds, reptiles and mammals. These are available from Bombay Natural History Society (BNHS).

Field reference books:

(i) S H Prater, *The Book of Indian Animals*, BNHS
(ii) Salim Ali, *The Book of Indian Birds*, BNHS
(iii) J C Daniel, *The Book of Indian Reptiles*, BNHS
(iv) P V Bole and Y Vaghani, *Field Guide to the Common Trees of India*, BNHS
(v) R J Daniels, *Amphibians of Peninsular India*, Universities Press (I) Ltd
(vi) Abdul Jamil Urfi, *Birds: Beyond Watching*, Universities Press (I) Ltd
(vii) Krushnamegh Kunte, *Butterflies of Peninsular India*, Universities Press (I) Ltd
(viii) R J Daniels, *Freshwater Fishes of Peninsular India*, Universities Press (I) Ltd

Plants

(i) Identify and list common plant species at the study site (at least 20; 10 trees, 5 shrubs, 5 herbs).
(ii) Identify if there are rare species by using a field guide or asking a botanist.
(iii) Identify and list the types of plants—trees/shrubs/climbers/ground cover: herbs, grasses. Observe their abundance levels.
(iv) Describe five plant species. Document the characteristic features that help in the identification of the selected species:
- Describe the specific characteristics of the leaves/flowers/fruit/seeds.
- Describe the plant's role in the ecosystem.

- How is it used and by whom?
- Is it being collected sustainably or over-harvested?
- Is it common or rare? If it is rare, why?
- Is it a keystone species? If so, why?

Animals

- Identify and list all the species you see in the study site.
- What are the major field identification features of the common animals and birds that you observe?
- Look for and document at least 10 species for each group—insects, birds (and mammals, if possible).
- Document the characteristic features for each of the ten species and record the following:
  - The role of the species in the ecosystem: What role does the species play in nature— producer, herbivore, carnivore, decomposer, pollinator, seed dispersal agent or pest.
  - The level of abundance at the site—classify as abundant/common/rare/very rare.
- Watch and document the area unobtrusively to observe all the links between the different species and between a species and its habitat. What role does each species play in the food chain and energy pyramid?
- Observe the habits of each of the selected species, such as feeding behaviour, nesting (for birds), breeding and territorial behaviour.
- Refer to a relevant field guide and document the following:
  - The distribution of each of the selected common species in the country.
  - The current status from a field guide—abundant, common, uncommon, rare, endangered.
  - If it is rare, is it on the endangered list?
  - Is it used by people? For what purpose?
  - How can it be protected?

8.4   STUDY OF SIMPLE ECOSYSTEMS

Field studies to be documented

- Describe any two ecosystems in the same way—specific forest type, marine, coastal, mangrove delta, lake or cave that you have visited for documentation of an ecosystem.
- Discuss its abiotic and biotic aspects.
- Describe its common species and their habitats.
- Describe its food chains, food web, and food pyramid.
- Discuss its biogeochemical cycles.
- Describe all its habitat characteristics.
- Describe its utilisation.
- Discuss its conservation potential.

Remember to take the time also to just enjoy the feeling of being with nature. Learn to appreciate the beauty of natural vegetation. It can become a thrill to watch wild species of animals and birds without disturbing them in their habitat.

General guidelines on aspects that can be observed and documented during ecosystem field studies:

The major questions that must be addressed during a field visit to any ecosystem – such as a forest, grassland, semi-arid, desert, hills, mountain ranges, lake, river or seacoast – include:

- What is the ecosystem called on the basis of its typical features? What are its abiotic andbiotic characteristics?
- Are its goods and services being misused or overused? What are the signs of degradation of the ecosystem that can be observed, that have occurred in the area? Deforestation, pollution of a water body and soil erosion are signs of degraded ecosystems.
- How can this degradation process be prevented by sustainable use of the ecosystem's goods and services by changing ones own habits, such as by saving water, electricity and paper?
- How can we all care for mother earth in our own way? Many small actions together reduce the adverse impacts of human activities on the ecosystems.

Observing the water cycle: During a monsoon field trip observe the effect of the rain.
- The type of vegetation is an indicator of the amount of rainfall. Classify the ecosystem – forest grassland semi-arid/desert type – on the basis of rainfall.
- Observe how the rain percolates into the subsoil. This recharges groundwater, which charges wells, streams and rivers.
- Document if the rain is eroding the soil. This can be judged by observing if the water is brown in colour. The colour is an indicator of the extent of soil erosion and is darker wherever the plant cover has been destroyed. It takes thousands of years for new soil to form. Excessive silt eventually changes the course of the river and leads to flooding of the surrounding land.

Observing the carbon cycle: Since plants absorb carbon dioxide that we exhale, and split it into carbon and oxygen that we breathe, we are dependent on the plant life on earth. Eventually large-scale deforestation could make life on earth impossible. Document this as an ecosystem service.

Carbon is a component of the food we eat in the form of carbohydrates, which come from plant material. Thus, we need plants to give us oxygen and food, without which we cannot survive.

Observing the oxygen cycle: While on the field trip, focus attention on the amount of green material that plants contain. Without this, there would not be enough oxygen for animals to breathe. Sunlight is essential for plant photosynthesis which produces grass, new leaves, branches and induces the growth of the trunks of trees. Sunlight is essential for plant growth in the water, including microscopic algae and underwater vegetation, which is the source of food for all aquatic forms of animal life.

Observing the nitrogen cycle: Observe the quantity of dried leaves on the ground in a forest, or the dried leaves of plants planted in the area that have collected as detritus. This material can be seen to be decaying. Ants, beetles and worms that feed on this dead material are breaking it up into small fragments. Microscopic bacteria and fungi are acting on this material to convert it into nutrients for plants to grow.

Observing the energy flow: Look for the different types of insects and birds in the trees. Frugivorous birds feed on fruit, insectivorous birds and spiders feed on insects; these form food chains. There are thousands of such food chains in an ecosystem. These interlinked chains can be depicted in the form of a 'web of life'. Observe that in our surroundings there is a great amount of plant material. There is much less animal life, more herbivores than carnivores. Estimate and document the differences in the number of plants, herbivores and carnivores in an area. This can be depicted as a food pyramid.

Objectives of a field visit to an ecosystem

- Identify the local landscape pattern in the forest, grassland, desert, river or hills. These are unmodified 'natural' ecosystems.

◆ Identify the modified ecosystems such as farmland, grazing land, industrial land and urban land.
◆ Compare and contrast natural and intensively-used areas.

The study site may have a mosaic of landscapes and aquatic ecosystems. Use the observations to create a map of the area and its ecosystems. Also, document the following.

(i) Common plants you see (trees, shrubs, grasses).
(ii) Animals observed (mammals, birds, fish, insects).

The natural landscape is beautiful. Describe how you feel about it. The plants and animals have several exciting features that can be 'discovered'. Observe and document their abundance or rarity, their habitat, their behaviour and their links to other species.

8.4.1 Field Visit to a Forest

Visit the nearest or most convenient reserve forest, national park or wildlife sanctuary. Meet the concerned forest official to explain your study. Check if there is an Interpretation Center where there may be local information and ask for brochures or other material. The officials may agree to address a group of students. Observe the forest type and make notes on the ecosystem.

Classify the forest type: During the field visit to the forest identify which type of forest is found in the area. Is there only one type or are there several types? If so, why?

Coniferous, deciduous, evergreen, thorn forest and mangroves are some examples.

◆ Interpret the connection between the abiotic and biotic aspects of the ecosystem.
◆ Observe differences in types of vegetation during the field visit and relate this to abiotic features such as temperature, rainfall, soil and topographic patterns, wherever possible.
◆ Understand food chains and food pyramids.
   ▪ Observe the abundance of different species in the ecosystem.
   ▪ Observe which plants are found commonly in the forest. Only a few species are very abundant but there are a large number of less common species of trees, shrubs and climbers and small ground plants that add to the diversity of plant life in any forest.
   ▪ Observe and document the names of animals seen. Classify them as mammals, birds, reptiles, amphibians or insects. Classify these into herbivores and carnivores. If these are counted, you will appreciate that there is a relative abundance of herbivores over carnivores.

Identify the structural levels in a forest: Identify the layers of the forest. Draw profiles of the structure and label the levels—ground, undergrowth, trunk, branches, canopy.

Document the micro-habitat for species in different levels of the forest: Observe which animal uses which different parts of a forest habitat. Some live on the ground among the fallen leaves (worms and insects such as ants, termites and beetles), others live in the middle layer on branches and tree trunks (lizards and woodpeckers), many others live in the canopy of the tree tops, (such as fruit- and nectar-dependent birds like sunbirds, parakeets and mynas). There are insectivorous birds, (flycatchers, drongos and bee eaters) in the canopy. Several insects live under the ground. If one turns over the dead leaves on the forest floor there are a large number of animals (millipedes, ants, beetles). Document what you have seen and estimate their abundance at different levels.

◆ Observe the food chains and interpret the food web.

Field observation—examples of food chains that are easily seen:

Flower → butterflies → spiders
Flower → sunbirds → birds of prey
Fruit → parakeet → birds of prey

Seeds → rodents → birds of prey
Flowers → bees → bee eaters
Seeds → munias → small carnivorous mammals and birds of prey
Leaves → monkey → leopard
Grass → chital → tiger

Observe what all the animals are feeding on and reconstruct as many food chains as possible.

Observe that a single species can play a role in several food chains. Thus, the chains form a food web.

Write about what you have seen about the food chains and food web in the area.

Interpreting the food pyramid and biomass distribution: Observe that in the forest the number of trees, shrubs and ground cover of plants constitutes an enormous amount of living material (biomass). Compared to this plant life, the number and biomass of herbivores is lower while the number and biomass of carnivores is even smaller. Write an explanation for this phenomenon using the examples you have observed in the study site.

Though ants are very small, together the thousands of ants form a large amount of living material. Thus, they have a great influence on the ecosystems functions. Observe and document how an ant colony works together.

Explaining the detritus cycle: Observe the large number of ants and beetles in the dead leaves fallen on the forest floor. Taken together, they constitute a very large mass of living animals. They break down an enormous amount of dead plant and animal waste material. Without this process the forest ecosystem would lose its integrity.

See what earthworms, millipedes, ants and beetles do without disturbing them. Insects and earthworms moving on the forest floor are breaking down the detritus so that microscopic fungi and bacteria can recycle this material into nutrients for the forest plants to grow. Look for the larger fungi such as mushrooms and bracket fungi, that also do this work. This shows how the cycle works.

Interpret the temperature and moisture control functions of the forest: Observe temperature differences under a closed forest canopy and outside in the open. Feel the moisture in the detritus and superficial layers of soil, which can be compared with the dry soil elsewhere.

Appreciate species diversity: Make a rough estimate of the number of species of trees, shrubs and different plants that form the ground cover. One need not name them all. Appreciate the wide variety of plants in a forest in comparison to a monoculture plantation.

Appreciate abundance of different plant species: Look around at the trees in the forest. Only a few are very abundant, while a much larger number of species are uncommon. Identify the most commonly-observed trees. Appreciate that there are some rare species of plants. These can become extinct if the forest is cut down.

Appreciate abundance of different animal species: On the forest floor, the most commonly-observed animals are ants; there are more ants than any other animal species on earth. Look for beetles; there are more species of beetles in the world than any other group of species. Observe how many different types of beetles there are, even if you cannot name them all.

The integrity of the ecosystem is based on these small but very important species that are a major part of the web of life of the forest. They are the prey of insectivorous birds, amphibians and rodents. Insects break down the detritus of the forest, which is the nutrient material on which the forest plants grow. Mankind thus cannot survive without them as they produce the nutrient material on which plants depend. Man is dependent on plant life, and is thus indirectly dependent on the insects in the forest detritus.

How do forests influence the water cycle?

The forest acts as a sponge: Feel the moisture and coolness of the forest air and compare this with the drier and warmer temperature outside the forest cover. The difference is obvious; the shade of the trees reduces the local temperature. Feel the level of moisture on the forest floor and compare this to the dryness of the ground outside the forest. A considerable amount of moisture is retained in the dead leaves (humus) of the forest floor. Dig a small hole in the ground. It is moist and cool under the forest floor. This demonstrates how the forest acts as a sponge and releases water gradually into streams after the monsoon is over. This can continue for the rest of the year and provide water for people outside the forest.

Understanding prey and predator relationships; food chains: There are spiders on the ground, which form tunnel webs to catch the crawling insects. In the trees, wood spiders make giant webs, three feet in diameter, to catch flying insects. Look for insect life in the canopy of the trees, on trunks, on the ground and especially under dead fallen leaves. Identify which species are predators and what is their prey.

Searching for examples of food chains, food webs and food pyramids: Different species of lizards are found on the tree-trunks and on the forest floor. There are chameleons in the trees and skinks on the ground. They feed on insect life, which in turn feed on the plants. This is a simple food chain that can be easily observed.

A spider catching an insect in its web is another demonstration of a simple food chain. The same insects are used by spiders and lizards as prey. Thus, multiple food chains are linked to each other. This forms a small part of the food web of the forest. There are several insectivorous birds, such as bee eaters, flycatchers of many species, babblers and many others, that form many different food chains.

There has to be a very large amount of plant life to provide enough food for the herbivores which are prey species for the very few carnivores in the forest. This demonstrates how a food pyramid works and how energy moves from one level to the other. The energy is used for the day-to-day functions of animals such as hunting for food, respiration, metabolising food and breeding. Observe that there is a very large amount of plant life, a smaller number of herbivores and very few carnivores. This observation explains the concept of the food pyramid.

Document the links between food chains and processes such as pollination of plants in the forest: Animals such as monkeys, squirrels and birds feed on leaves, fruits and seeds. Insects (ants, butterflies) and birds (sunbirds and mynas) use flower nectar for food. These flowers have bright colours to attract them. During this process, the insects and birds pollinate the plants. At night when most animals sleep, the bats and moths pollinate flowers. These flowers are usually white in colour so that they can be seen at night. Thus, many plants depend on animals to pollinate them so that the seeds can develop. The regeneration of forests thus depends on these animals. These links are important aspects that maintain the forest's web of life.

Look for the pollinators—butterflies, moths, beetles, ants and nectar-feeding birds are easy to observe. Look for birds that eat berries and fruit and disperse seeds; these include bulbuls, parakeets. Finally, look for birds of prey (eagles, vultures) that complete the food chain.

Seed dispersal: Observe that monkeys, squirrels and birds (parakeets, mynas and hornbills) are agents of seed dispersal as they feed on fruit and spread the seeds far and wide. Birds such as bulbuls eat small berries along with their seeds. After passing through the birds' intestines they are excreted and can germinate. They germinate more effectively as their covering is removed in the bird's intestine. The birds thus help in the dispersal and regeneration of plants. Though a plant is rooted to the same spot, evolution has linked plants with animals that help the plant species to spread by dispersing its seeds.

Other seeds are light and have wings or hair. These are dispersed by the wind. Observe how these float through the air for long distances.

Regeneration: Look for seeds and seedlings growing on the forest floor. Observe that while there are plenty of small seedlings there are fewer saplings, as a majority of the seedlings die. Only a few of the surviving young saplings will finally grow into large trees. Plants thus need to generate a very large number of seeds. Only those seeds that find a spot that provides all the conditions needed for their germination and growth can end up as large trees. Many seedlings die due to forest fires, grazing or trampling by domestic animals.

Forest animal communities and interrelationships: Forest birds form feeding parties of many different species. Together, they feed on different parts of plants such as flower nectar, fruit or on insects. Observe that when fruit-eating birds search for their food in the foliage they disturb the hiding insects, which are then caught by the insect-dependent birds. Others, while looking for berries in the bush layer, disturb insects that are caught by other insectivorous birds. Thus, birds of different species help each other in finding their food and stay together in large mixed feeding parties that move from one tree to another. Identify what each bird species feeds on.

When the langurs feed on fruit, a part of the fruit is dropped uneaten on the forest floor. The *chital* and *sambar* following the langurs are able to eat this fallen fruit. From the top of the tree, the monkeys can easily spot an approaching tiger or leopard more easily than the deer. The monkey gives an alarm call at the approach of a predator alerting the deer to its approach; this is a mutual interdependency.

Major prey–predator behaviour: Carnivorous animals are very wary of human beings, as man has killed them for thousands of years. These animals move very stealthily so that they can catch their prey. They become almost invisible, as their colour and pattern camouflage them in the undergrowth to be able to approach the prey unnoticed. The predators have to make several unsuccessful attempts before they catch their prey, as the prey is extremely sensitive to any movement or sound. For all these unsuccessful attempts they have to spend a large amount of energy. The deer have sharp eyesight and a good sense of smell to avoid being caught. Predators like tigers or leopards counter this by moving very cautiously in the forest. Other predators such as wild dogs hunt in packs. Omnivorous birds such as hawks and eagles swoop down from the sky on their quarry at great speed. Even though they are superb hunters, their prey is frequently able to escape.

Searching for camouflaged species: The beautiful stripes of the tiger and the rosettes of the leopard match the light and shade in the forest and thus camouflage these animals so that they cannot be easily seen. Predators such as tigers and leopards frequently see us before we see them and disappear stealthily into the deeper forest. However, we can observe their pug marks on the forest floor. It is exciting to see a fresh pugmark in the forest. It tells a tale. The track, if fresh, can end up at a sleeping tiger or a leopard!

The colour of moths is similar to that of the brown tree bark on which they rest during the day. Chameleons change their colour to suit their surroundings. The green colour of grasshoppers matches the foliage they live in. Stick insects look like twigs. Look for other signs of camouflaged species in the forest. Each has a reason and is linked with evolutionary processes.

Study of wildlife signs: The wildlife of the forest leaves behind several signs even if we cannot see the animals themselves. Each animal has its own footprint, which can be identified. Animals also leave their characteristic droppings that can be easily identified. Thus, these signs in the forest can tell us which species live there and indicate their day-to-day activities.

Listen to the birdcalls: There are many different kinds of beautiful calls. This indicates that there are many more birds than we can see in the forest.

Observe feeding patterns of animals: Observe the feeding patterns of forest animals unobtrusively. Learn about those that we cannot observe easily. Some of these are given below:

- Tigers feed on sambar, chital and monkeys
- Leopards feed on barking deer, hare and occasionally village cattle
- Jackals feed on hare, mice and birds
- Jungle cats feed on hare and mice
- Pangolins feed on ants
- Mongooses feed on snakes and mice
- Chital feed on grass
- Sambar feed on grass, young leaves and fruit of shrubs and trees
- Barking deer feed on fruit and leaf buds
- Elephants feed on grass in the monsoon and tree leaves in the winter and summer
- Squirrels feed on nuts and fruit
- Porcupines feed on fleshy roots of plants and bark of trees
- Birds of prey feed on rodents, reptiles, frogs and small birds
- Birds like flycatchers, feed on insects and worms
- Bulbuls feed on fruit. When they have young ones in the nest, the parents feed the chicks with worms and caterpillars
- Sunbirds feed on flower nectar
- Spiders catch insects
- Insects like beetles and bugs feed on plant material
- Insects like the praying mantis and dragonfly feed on other smaller insects
- Ants and termites feed on dead plant material
- Beetles and bugs feed on leaves and sap of plants
- Butterflies feed on flower nectar in the day
- Moths feed on flower nectar at night
- Worms feed on forest detritus

Habitat use by different species: Observe how different species of animals use various layers, starting from the forest floor, upwards along the trunk and branches, to the canopy of the trees. These species show that the forest habitat consists of various layers, each forming a microhabitat within the forest. Among the more common insects, the termites build their homes out of mud, which is present on the forest floor. The large red fire-ants build homes out of leaves in the tree canopy.

Langurs and macaques use the tree canopy for leaves and fruit as well as the forest floor, where they look for fallen flowers, buds and fruit. The leopard hunts on the forest floor for prey such as the barking deer, chital and hare. Eagles hunt for their prey in trees and on the ground. Their prey consists of small birds, rodents, snakes and frogs. The giant squirrel is rarely ever seen on the ground. It looks for fruit and nuts by crossing from one tree branch to another. It requires forests with an unbroken canopy. The rat snake is usually on the ground while the vine snake twines onto branches of shrubs and trees; both are non-poisonous. Birds like bee eaters and drongos catch insects while they fly through the canopy by swooping through the branches. Babblers most often look for insects and worms by disturbing the dead leaves on the forest floor. The hornbill looks for fruit in trees and makes a nest in a large hole after carefully selecting a very tall old tree. The crow pheasant hunts for grasshoppers and worms on the forest floor and in trees. The shrews look for insect life underground. Millipedes, centipedes and scorpions move around on the forest floor. Many of these animals live in holes under the ground, under rocks or among the dry leaves.

8.4.2   Field Visit to a Grassland

Observe the variety of plant and animal life in the grassland: Document the food used by each animal that is identified.

Describe the seasonal changes in the grassland: Describe how the grassland would look in different seasons. Describe the anticipated changes in colour and the condition of the grasses—growing phase, flowering phase, dying phase, dry phase.

Abundance of grassland species: Try to count the number of grasshoppers that jump out of a disturbed 1 sq. m quadrant on the ground. Try and count at least 20 such quadrants. You may find this hard! Repeat the count for ants, beetles and other insects. This will be nearly impossible, as there are just too many of these insects. This will demonstrate the great abundance of insects in the grassland. Compare this to the much smaller number of first-order consumers, birds and mammals, that can be counted in the grassland. The predators, mammals and birds of prey (raptors) are the least abundant.

Birdwatching in grasslands: Make a checklist of common grassland birds by identifying them from the *Book of Indian Birds* by Salim Ali. Read what each species feeds on.

Observing the insect world: Observe how the ants live and collect food. Observe how beetles behave in the grassland. Observe the abundance of grasshoppers, beetles and ants. Compare this with the number of their predators.

Observe a spider catch its prey: See the different types of webs. Tunnel-web spiders make a tunnel and sit inside waiting for prey, which are pulled in and eaten. Other spiders in the grassland make small orb-webs that have radial and spiral threads. Some spiders build a colony which is like a mass of web material.

Document animal behaviour: Make a general list of behavioral patterns of all the animals, birds and insects you see. What are they doing? How and where did you find them in the grass? What is their relationship to the grassland as a habitat?

Understanding grassland food chains: Identify as many plant and animal species as you can. Use the list to form as many food chains as possible.

8.4.3   Field Visit to a Desert or Semi-Arid Area

Observe desert and semi-arid landscapes: Observe the sparse but specialised nature of vegetation in a desert or semi-arid landscape. Document the number of animal species that are seen in the vicinity. There are very few in comparison to other types of ecosystems.

Observe the fauna of a semi-arid landscape: Identify the birds and insects which are most easily seen. Document how each species is using its habitat. What do these species feed on in this harsh environment?

◆ Observe typical species, such as dung beetles that roll dung into a ball, in which they lay their eggs so that their young get food.

- Observe birds of prey that use this ecosystem.
- There are rare birds in a few areas, such as the Great Indian Bustard.
- There are rare mammals, such as the wolf.

8.4.4   Field Visit to an Aquatic Ecosystem

Document the nature of aquatic ecosystems: Visit an aquatic ecosystem such as a pond, lake, river or seacoast. Observe if the water is clean or polluted. A simple kit can be used to study the water quality.

Guidelines for a write-up on a pond ecosystem

A pond is a highly dynamic mini ecosystem. It changes rapidly during the year. To study a pond one should, as far as possible, cover all its phases. The monsoon phase is when the pond changes from a dry to a wet (aquatic) state. It is the growing phase, when it is colonised by microflora and fauna and then other forms of macroscopic life. At the height of the monsoon, it is in a mature aquatic phase, which is full of life. Once the rain stops, the pond begins to shrink. Its periphery becomes dry and is colonised by terrestrial plants like grasses and herbs. As it shrinks, its aquatic flora and fauna die, giving way to land flora and fauna. Eventually, it may only remain in the form of a ditch or depression containing terrestrial forms and dormant aquatic invertebrates such as insects that must await the next monsoon. This process, when repeated year after year, leads to the silting up of the pond, which eventually gets shallower and shallower and in the course of time, gives way to a grassland, scrubland and after many decades to a forest. This is the process of succession.

- Observe the pond. What seasonal stage is it in?
- What do you expect will occur over the next 3 months, 6 months, 9 months and 12 months?
- What are the vegetation zones in the pond?
- What species use the pond as a habitat?
- Take some water from the pond and examine it under a microscope. What do you see?
- Describe the pond's periphery—its soil, vegetation and fauna.
- Describe the pond's floor—its soil, vegetation and fauna.
- Write a note on the food chains you observe.

Studies on the ecology of a pond

Make observations on a seasonally active pond if possible on several occasions before, during and after the monsoon. Document the seasonal changes in the plant and animal life.

Observe the vegetation zones at the water's edge: Observe the different zones of vegetation— grasses on the periphery, emergent reeds, floating vegetation and underwater plants in the pond.

Seasonal field observations on a pond: Early stage, soon after the pond fills with water in the monsoon—Observe the algae and microscopic animals. These can be observed under the microscope.

Fully active phase—Observe the submerged and emergent vegetation as well as the fish, frogs, snails, worms and aquatic insects.

Shrinking phase—Note the drying aquatic plant life, with dead and dying plant material and terrestrial plants growing on the exposed mud of the pond.

Dry phase—See it overgrown with grasses and shrubs with hidden dormant animals in the mud, which cannot be seen.

Laboratory exploration: Observe the water collected from the pond in a glass. Document its colour and what it contains. Observe the water collected from a pond under the microscope. There are a large number of algae and zooplankton that form the basic food chains of the aquatic ecosystem.

Observations on a lake ecosystem

Document the way in which different water birds use the various habitats both on the shore and in the water. Each of the different species of aquatic birds shares its habitat with only a few other species. Each species specialises in certain types of food and feeds at different depths. The length of the legs of different wading birds is an indicator of the depth at which they feed. The length of their beaks indicates the depth of mud or sand into which they can probe.

Diversity and abundance of life: Make a checklist of all the visible aquatic flora and fauna. Identify those that are most abundant. Observe and document the food chains. Estimate or count the population (abundance) of different species observed in the aquatic ecosystem.

Observations at a wetland

Visit a wetland and observe the varied vegetation zones within the ecosystem. Document and map its vegetation patterns—underwater/emergent/floating/none. Describe if the water is clean or turbid. Describe the level of algal growth and weeds. What is the nature of its bed— rocky/sandy/silt/muddy/mixed (in what proportion?). Develop a map of the vegetation of the aquatic ecosystem and its relationship to species of aquatic birds. Ask local fishermen to show you their fish catch. Observe the ducks, waders and other birds. These are most abundant in the winter, as most of them are migrants from across the Himalayas.

Guidelines for a write up on a river ecosystem

A river is an aquatic ecosystem that is influenced by the monsoon. It may be a perennial river or a seasonal one. The river ecosystem has abiotic and biotic components. While many of its species are aquatic, there are also terrestrial species that use its banks; both these need water. Aquatic species live in the water, while the terrestrial species live on the banks but are highly dependent on the proximity of water. Many species, such as amphibians and aquatic insects, use both aquatic and terrestrial habitats.

- Describe the aquatic ecosystem in the river water and the terrestrial ecosystem on the riverbank.
- Describe the characteristics of the bed of the river, the depth of the water and the flow rate in different sectors—rapid/slow/stagnant.
- Describe the various habitats of different species of flora and fauna in and around the river.
- Document what you see in the water under a microscope.
- Document how different species use the water and the banks of the river.
- Describe how each of the habitat parameters is linked with the species that live there.
- Observe the food chains. Document the aquatic food chains, the terrestrial food chains on the bank, and those in which both aquatic and terrestrial species occur.

Observation on a field visit to a beach

Beaches can be sandy, rocky, shell-covered or muddy. On each of these different types, there are several specific species, which have evolved to occupy a separate niche. Observe all the different crustaceans such as crabs that make holes in the sand. Observe how the various shore birds feed on their prey by probing into the sand.

Observations at a river

Depending on the location of the river, the study can demonstrate its ecological status. The river is a dynamic system with seasonal fluctuations in flow rates that affect its plant and animal aquatic life as well as the flora and fauna along its banks. Observe and document how life is dependent on the river's integrity.

8.4.5 Field Visit to a Hill/Mountain

The ecosystem of the hill you are observing is linked to its altitude, slope, soil characteristics, vegetation and animal life. It has different vegetation patterns that create specific microhabitats for a variety of fauna. The habitat changes seasonally. What do you expect will occur in 3 months, 6 months and 9 months from the present scenario?

Guidelines for a write up on a hill/mountain ecosystem:

◈ Describe the hill-slope, soil, watercourses and other characteristics.
◈ Describe its various plants and animals.
◈ Observe and document its food chains.
◈ Describe the water cycle, the nitrogen cycle, energy flow and the detritus cycle, with specific reference to the hill/mountain ecosystem.
◈ What would happen if all domestic animals were to be prevented from grazing?
◈ What would happen if ants were to be completely eliminated from this ecosystem?
◈ What would happen if all the vegetation were removed from the slopes?

8.4.6 Field Visit to Delhi Ridge

Make a visit to Delhi ridge or a similar wilderness area in your city or locality and do following observations.

◈ Describe the hill-slope, soil, watercourses and other characteristics.
◈ Describe the various plants and animals that are seen.
◈ Document the various services provided by Delhi ridge for the people of Delhi.
◈ If you see people visiting the area, ask them the purpose for their visit (exercise, clean air, mental peace, doctors advise, bird watching, cultural value, and so on).
◈ Make a report of the various services provided by such areas in an urban environment.

Index